Second Edition

Environmental Chemistry in Society

Second Edition

Environmental Chemistry in Society

James M. Beard

CRC Press is an imprint of the
Taylor & Francis Group, an **informa** business

CRC Press
Taylor & Francis Group
6000 Broken Sound Parkway NW, Suite 300
Boca Raton, FL 33487-2742

Printed on acid-free paper
Version Date: 20130503

International Standard Book Number-13: 978-1-4398-9267-1 (Paperback)

Library of Congress Cataloging-in-Publication Data

Beard, James M.
Environmental chemistry in society / author, James M. Beard. -- Second edition.
pages cm
Includes bibliographical references and index.
ISBN 978-1-4398-9267-1 (alk. paper)
1. Environmental chemistry. 2. Pollutants. 3. Environmental management. I. Title.

TD193.B43 2013
628--dc23 2013003676

Visit the Taylor & Francis Web site at
http://www.taylorandfrancis.com

and the CRC Press Web site at
http://www.crcpress.com

This edition is dedicated to Susan, the love of my life;
to my children: Kristin, Amy, Brian, Kelly, David,
and Landon; and to all my grandchildren.

Contents

Preface

I wrote this book because of my belief that it is important for non-science-oriented students to understand the environment. It is, after all, these very individuals who will influence the course of public policy on the environment in any democracy. Many books present environmental science to the non-science student, but few look specifically into environmental chemistry. Among the numerous chemical issues that are important to any understanding of the environment around us, all of us need to have some understanding of global warming, ozone depletion, energy sources, air pollution, acid rain, water pollution, waste disposal, and hazardous waste.

To the students who will read this book, I would point out that, although this is a college text, there is no assumption of any background in chemistry. Within the text you will find all of the background information necessary to understand it. To the faculty who will use this book, I would like to note that this is a self-contained environmental chemistry text, in which students can find all of the background they need. The book is structured in such a way as to give students a background in science, chemistry, and toxicology before delving into such areas as energy in society, air quality, global atmospheric concerns, water quality, and solid waste management. The basic structure is given below:

Introduction	Chapter 1
Foundational material	Chapters 2, 3, 4, 5, 6, 7
Specific background material	Chapters 8 and 12
Environmental content chapters	Chapters 7, 9, 10, 11, 13, 14, 15

The intention is that all students cover Chapters 1 to 6, with Chapter 7 strongly encouraged but not essential. All other chapters could be covered in any order, provided Chapter 8 precedes Chapters 9 and 11, and Chapter 12 precedes Chapter 13. This arrangement of material allows instructors the freedom to cover the material in a manner that can be customized to the needs of their courses.

In the second edition the environmental data have been updated and material has been added concerning fracking, the Fukushima Daiichi power plant disaster, and the Deepwater Horizon oil rig blowout. Most of the homework questions have been rewritten to provide questions that require more critical thinking skills.

I have enjoyed writing this book and have tried to make it very readable for any college student. Any text, of course, can be made better. I would welcome any suggestions for improvement in or corrections to the text. I can be reached by email at jbeard@catawba.edu.

Acknowledgments

I would like to thank many people for their understanding and support during the writing of this book. A special thanks to my wife, Susan, for being tolerant while I wrote and wrote and wrote. Her continued support made this book possible. I would like to express my appreciation to Catawba College for their flexibility during this effort and for the sabbatical semester which accounted for a considerable amount of the writing time. My thanks to Delores Imblum, the biology–chemistry administrative assistant, who provided valuable help with the manuscript. I also thank Maegen Worley and Tracy Ratliff of Catawba College for providing the original artwork and layout for the book's cover. I would like to acknowledge my daughter Amy's contribution of some of the illustrations in the book; my colleague Dr. Carol Ann Miderski for helping me with the ins and outs of thermodynamics; the late Dr. Bruce Griffith, professor of history, for much of the background information used in Chapter 1; and Dr. Wayne E. Steinmetz, professor emeritus of chemistry at Pomona College, for numerous suggestions to improve the book. Finally, I would like to express my appreciation to the late Dr. Richard H. Eastman of Stanford University for important advice and encouragement when I started the book.

Author

James M. Beard, PhD, earned a BA in chemistry from Manchester College (now Manchester University) in 1965 and a PhD in organic chemistry from Stanford University in 1969. He then spent a year at Iowa State University as a postdoctoral research associate in organic photochemistry. After two years of teaching chemistry at Manchester College, he pursued a career as a clinical chemist. During the next 12 years, Dr. Beard worked as a product developer, quality control consultant, and administrator in both hospital and industry settings. In 1985, he returned to chemical education as an associate professor of chemistry at Pikeville College. Since 1988, Dr. Beard has been a member of the faculty at Catawba College in Salisbury, North Carolina, where he is a professor of chemistry. Dr. Beard's major interest at Catawba has been in environmental chemical education.

CHAPTER 1

Background to the Environmental Problem

As we move into the 21st century, we humans are trying to come to terms with what we can and cannot do regarding the world around us—our environment. In the middle of the past century, there was a widespread feeling that the world was our apple. We could do with it as we pleased, and the Earth would continue to sustain us as it always has. We felt that resources were limitless and that pollution, although disagreeable, would be diluted somehow. Gradually, the world around us began to send messages that all was not well. Nature was trying to indicate that there were limits, and if we ignored them there would be consequences. The purpose of this book is to explore some of these limits and consequences.

We look at our environment from the point of view of the chemistry that takes place in it. In environmental studies, it is almost impossible to separate the biological issues from the chemical ones; therefore, our chemistry emphasis will be a matter of focus. This book focuses on environmental issues that are more chemical in nature but inevitably covers some aspects of biology also.

Discussions on the environment tend to be very "human centered." The environment includes our total surroundings. As this is a science text, the emphasis is on the physical surroundings. The environment and the laws and principles that regulate the function of the environment are discussed. From environmental discussions, it is often easy to get the impression that there is a great natural world around us from which we are separate, yet this was not always the case. Let us explore how we came to consider ourselves as separate from the natural world around us.

PREAGRICULTURAL DEVELOPMENT

Humans began to emerge as a species about 1.7 to 2 million years ago. The species that evolved first was not the same as humans today, but was our early ancestor. The species was known as *Homo habilis* and probably emerged in southern Africa. It is believed that these early relatives of ours were scavengers who survived by eating wild plants and the meat of dead animals. This very early human species was close to nature and survived the same way as other animal species. The only thing that may have separated this very early human species from the other animals of the time was the humans' ability to make and use simple tools. This activity was something our ancestors shared with only a few other species, in which this activity remained fairly rudimentary. Tools set humans ever so slightly off from the nature around them. In other ways, these very early humans blended with nature. They took what they required directly from the natural world around them and returned their biological wastes directly to it.

The next human species was *Homo erectus*, which evolved about 1.5 million years ago. These early humans engaged in a new activity—hunting. Hunting became a method of survival that humans used up to about 10,000 years ago. *H. erectus* was a species of **hunter–gatherers**, as was the next species, *Homo sapiens*, which emerged about 100,000 to 300,000 years ago and was followed by human beings as we know them, *H. sapiens sapiens*. For most of our time on Earth, *H. sapiens sapiens* have been hunter–gatherers. We appeared on the planet about 40,000 years ago and were hunter–gatherers for about 30,000 years. In this early period, humans became more and more sophisticated in their methods. As time passed, the tools used for hunting got better. Our ancestors learned to hunt in groups and to use fire to drive away wild animals from the forest. Over this period of time, the impact of humans on the environment increased dramatically. Humans could kill large numbers of animals and may have contributed to the extinction of some. The use of fire allowed these early ancestors to convert forests into grasslands. Despite all of these changes, the impact of early humans was minor when compared to today's standards. There were only a few of them, they were nomadic, and they had only their own muscle power to use. The environmental impact of these early humans was local and transient.

HORTICULTURE AND AGRICULTURE

Sometime about 10,000 years ago (8000 BCE), humankind began to take more direct control over the environment. Humans began to domesticate animals such as cattle and pigs and cultivate specific plants that were useful for food or other purposes. This early system of production of food and useful materials was known as **horticulture**. Horticulture had certain hallmarks that distinguished it from agriculture. A modern analogy to many of the elements of horticulture would be the family garden. The plots were small and contained a large variety of plants. In growing these plants, depletion of the soil was rarely a problem because most of the people were still semi-nomadic. After 2 to 5 years, they would simply move on and find a new plot of ground. Such a system could go on indefinitely.

The fact that the cultivated plots were small was probably more a matter of necessity than of choice. Early humans had very limited means of breaking up the soil and had great difficulty if the ground was too hard. In very early times, cultivation was done with little more than a pointed stick. (Over a long period of time the stick evolved into the hoe.)

In about 5000 BCE, the plow was invented. The plow changed things radically. By using domesticated animals to pull the plow, much larger fields could be cultivated. **Agriculture** developed when people figured out how to grow crops and raise livestock on what was for them a relatively large scale. When cultivation was done on a large scale, it was more logical to specialize and grow larger plots (fields) of only one crop. Because the farmer would grow more of the one crop than he needed, he could trade the excess to get other things of value. These exchanges of crops and other needed items became the basis of early trade and commerce.

Crops and livestock became the earliest forms of wealth, and land was needed to produce both. Farmers needed to acquire land, settle on it, and protect it. Farming led to the end of the nomadic lifestyle as the principal way of life. Nomadic hunter–gatherers gradually became sedentary farmers. The increased food supply and the sedentary nature of farm life led to an increase in population, as more children could be supported and the children could work on the land. Nomadic families tended to be smaller because food was less plentiful, and moving children from one place to another was difficult. Agriculture also brought with it the first indications of environmental problems. Some fields, after they were used continuously over a period of time, produced less and less as the nutrients in the soil were used up. In some cases, farmers had to pull up stakes and move on.

As farming spread to arid areas, water became just as important as land. Fairly early on in the development of agriculture, water was diverted from rivers by means of canals or ditches for the purpose of **irrigation**. Early agriculture in the Nile Valley depended on water from the Nile River as it does even today.

DEVELOPMENT OF TOWNS AND CITIES

The success of agriculture in producing more food than that required by a family allowed some people to leave food production altogether. By making other products or providing services that the farmer needed, these craftsmen could trade these goods and services for food. Such individuals began to form small groupings called villages or towns. As agriculture improved and people became more sophisticated, some of the towns grew larger. Eventually, these towns became **cities**.

Cities are not merely large towns. Towns and cities differ in that cities have some type of administrative structure and organization. Often the very early cities were organized around a central temple, and the temple servants often became the city administrators. It was not uncommon for the city government to be headed by a priest king or in some cases a god king.

Human history is very much entwined with the history of cities. In fact, the words "city" and "civilization" are both derived from the same Latin root, *civitas*. Based on the best archaeological data (Macionis and Parrillo, 2007), the oldest city in the world is Jericho, which was founded sometime between 8000 and 7000 BCE; however, large areas with related cities began to appear about 4000 BCE in Mesopotamia and Egypt.

Early cities were generally built on the banks of rivers or by the sea. The rivers were a source of water and a means of removing waste. The rivers and seas also served as routes for trade and commerce. Early cities had many restrictions. The cities were limited by the primitive transportation available for bringing food into the city. Cities also had to be small in area, as the citizens traveled on foot wherever they went. Shops, markets, jobs, and dwellings had to be close to one another. Even the increased use of horses and wagons did not change this dynamic very much. Some large cities such as Rome and Constantinople did develop, but they were exceptions rather than the rule. With the advent of the Renaissance, cities began to grow and prosper. Technological innovations improved life in the cities, which allowed them to attract more and more people. Finally, all of these innovations came together into what came to be known as the **Industrial Revolution**.

INDUSTRIAL REVOLUTION, PHASE I (APPROXIMATELY 1760–1860)

Despite tremendous changes in the cultural, political, and religious nature of the world since ancient times, other elements of life were still relatively the same until 1760. Cities, although very important, were small compared to today's standard because they required a fairly large agricultural base to support them. In the previous 50 years, there had been some innovations in the field of agriculture, but most of the world's agriculture was being carried out as it had long been, with most of the energy provided by domesticated animals.

In many areas of life things remained unchanged as well. The forms of energy used, for example, were those that had been known since antiquity. These energy sources were animal power, wood burning, water power, and wind power. There were some occasional local exceptions to these traditional power sources, such as the early and extensive use of coal as a fuel in England, as mentioned in Chapters 7 and 9, but generally, the sources of energy remained unchanged. Of the 17 elements known in 1760, 9 of them had been known since antiquity. Everyday life changed very slowly, if at all.

The Industrial Revolution changed nearly everything about everyday life. Things such as hourly wages, standard workweeks, vacations, and many other related concepts did not exist before the Industrial Revolution. So much of how we organize our lives both individually and corporately is directly a result of the Industrial Revolution. It is impossible to present a complete account of the historical and cultural ramifications of the Industrial Revolution in a few pages; however, an attempt has been made to convey some of the significance of these revolutionary changes. One of the problems in discussing the Industrial Revolution is that it is difficult to trace the origins of the revolution to a single primary event from which everything else occurred; rather, the revolution was the result of many interconnected events occurring over a period of several decades or more.

The early Industrial Revolution occurred for the most part in England. The most obvious changes occurred in the textile, coal, and iron industries. The textile industry had been slowly mechanizing for at least 200 years before 1760. By the middle of the 18th century, factories were needed to house the large amount of machinery necessary to produce cloth. One of the essential elements needed to operate machinery was energy, and in 1750 the best source of energy was water power. To get easy access to water power, most textile factories were located in the hills of the English countryside where there were streams. Because there were fewer inhabitants in the hills, it was necessary to import labor from the cities.

With the invention of an efficient **steam engine** by **James Watt** in 1769, the location of textile factories suddenly became independent of water power. Textile factories could be located in large cities where the workforce resided, as steam could be used as the power source. These changes led to rapid development of the modern city, where large numbers of people were involved principally in factory work. In addition to steam power, which brought the jobs to the city, advances in agriculture and transportation also increased the amount of food available in the city.

The invention of James Watt's steam engine in turn depended on advances in the iron industry. One of the early giants of the iron industry was **John Wilkinson**, who had been involved with the British Navy in the production of cannons. To get the maximum power from a cannon, a precise bore of the barrel was necessary, and Wilkinson had learned how to produce such a precise bore. When Watt and his partner, Matthew Boulton, needed precisely manufactured steam cylinders and pistons, they turned to Wilkinson.

It was Wilkinson who had pushed the iron industry harder and further than anyone else. He realized that when more things were made of iron, more iron would be needed, and it would become less expensive to produce each ton of iron. Wilkinson was one of the first to understand the economy of scale. With plentiful supplies of cheap iron, factories became easier to outfit because the machines in them were made out of iron. As more factories developed, one thing all of them needed was power.

Once the steam engine was widely established, coal became the principal energy source. Coal was needed not only to power the steam engines but also to provide the heat needed to extract iron from its ore, to soften the metal in order to remove impurities, and to melt the metal to cast it into shapes. The steam engine also made coal mining more efficient. Underground coal mines had a tendency to fill up with water from seepage, and a method was needed to remove the water. Once the steam engine was perfected, water was removed from coal mines by steam power rather than by people, horses, or water power. The steam engine made the coal industry more efficient, which in turn made the coal industry better able to supply coal to produce iron and to operate the steam engines of the factories in England.

INDUSTRIAL REVOLUTION, PHASE II (APPROXIMATELY 1860–1950)

Some profound changes were seen between 1830 and 1860. Gradually, inventors depended on scientific information to chart their course in the development of new ideas. This change is evident in the evolution of the electric industry. Until the beginning of the 19th century, electricity was an interesting novelty but not of any practical value; however, many scientists carefully studied electrical phenomena. The well-known experiment carried out by Benjamin Franklin using a kite in a thunderstorm is one example out of many. Yet, electricity continued to be impractical as long as there was no means of producing a continuous electric current.

In 1800, **Alessandro Volta** invented the forerunner to what is known as the battery today. Despite their limited electricity-producing capacity, these batteries allowed people to use an electric current. Once electricity was available, exploration of its use began to move ahead rapidly. The idea of the mass use of electricity for lighting emerged in the last half of the 19th century. It was clear, however, that batteries could not generate the amount of electricity that was necessary. During this period the first electric **generators** were invented. Generators could convert mechanical action into electricity. Mechanical energy could be provided by devices such as a waterwheel or a steam engine. Hence, coal, by fueling a steam engine, could provide the energy to produce electricity. Electricity began to spread to all areas of the industrial society. By the end of the 19th century, there were electric streetlights and electric streetcars in many cities, electric lights in many homes, and electric motors in many factories.

None of this would have been possible were it not for the steel industry. **Steel** is harder and stronger than most of the iron available for use. For many applications, including large electric generators, iron is either too soft or too brittle. The properties of the iron depend on the carbon content of the metal. Iron is extracted from its ore and processed using charcoal or **coke** (relatively pure carbon produced from coal). The iron picks up some of the carbon. Most iron as it is initially produced after being extracted from its ore is 2 to 4% carbon and contains many impurities. This type of iron is known as **cast iron** and is extremely brittle. The iron can be worked to remove nearly all of the carbon and some of the impurities to produce **wrought iron**. Wrought iron is not brittle but is relatively soft and as a result does not hold an edge well. With the proper percentage of carbon, one can get a product with just the right combination of strength and hardness. Such a product is steel, which is an **alloy** made up of iron and 0.5 to 1.5% carbon. An alloy is a metal containing more than one element; in the case of steel, those elements are iron and carbon. Steel must also have most of the impurities removed, as such materials will degrade the quality of the product.

Steel (as distinct from iron) has been known since the 12th century BCE. Steel has been in use in various cultures. The quality and carbon content of the steel vary from situation to situation. The Japanese Samurai sword has been made from steel of very exacting quality for centuries. The properties of steel for various applications are different because of the variability in carbon content. Before the middle of the 19th century, steel could be made only in small quantities because it could only be made from fairly pure batches of wrought iron. This purified iron was difficult to obtain and had to be heated with charcoal or coke for several days to produce the steel.

To produce steel on a larger scale, it was necessary to make it from cast iron, which is available in large quantities. Any process had to remove nearly all of the impurities and adjust the final carbon percentage to that desired in the final product. About 1847, William Kelly, a businessman in the United States, started work on a device that would make steel from cast iron; however, because of financial problems he was not able to complete his project. A similar device was patented in 1856 by **Henry Bessemer**, who was working independently in England. The **Bessemer process** revolutionized steel making, but ran into competition from another process rather quickly. A German named **William Siemens**, who was living in England, developed the **open-hearth process**, and steel was being produced in France by 1864. The Bessemer process was an important development, as it opened the way to the advent of the age of steel, but it was gradually replaced by the open-hearth process, which was the major steel production method until the middle of the 20th century.

With the practical problems of steel production solved, the output of the steel industry soared. England produced 60,000 tn of steel in 1850, before the Bessemer converter was invented. By 1898, the annual English steel output was 5,000,000 tn. Worldwide production of steel went up from 560,000 tn annually in 1870 to 12,000,000 tn in 1890. The United States alone produced 10,000,000 tn in 1901.

An industry that was intertwined in many ways with the steel industry and the electric industry was the chemical industry. More than any other industry, the chemical industry typified how the second phase of the Industrial Revolution operated. In 1856, a young student from the Royal College of Chemistry in London acted on the advice of his professor and attempted the synthesis of quinine. Quinine was used to combat the effects of malaria, and the British had an empire scattered all over the world, including the tropics. To attempt this synthesis, the student, **William Henry Perkin**, worked with compounds from coal tar. Coal tar is a byproduct of the steel industry produced when coal is converted to coke. The synthesis did not work, but Perkin produced a good yield of a purple dye. Perkin immediately realized that he had a great opportunity because at that time purple dye was very expensive. Purple was the color of royalty because its only source was a Mediterranean mollusk, and producing the dye took a very large number of mollusks. Perkin left school and, with the support of his father, went into business for himself. Thus, the chemical dye industry was born. In the dye industry, science was used to systematically search for more and more dyes. The more dyes available, the greater the profits; therefore, chemical science and the chemical industry became linked forever. This partnership spilled over into fertilizers, explosives, drugs, and plastics, to name a few areas.

The influence of the steel industry on the chemical industry can be illustrated by some events around the beginning of the 20th century. In 1898, William Crookes observed that the vast deposit of nitrates in Chile was being rapidly depleted. Nitrates were one of the important components of fertilizer, and many of the advances in agriculture depended on it. In 1904, a German chemist named **Fritz Haber** (see Figure 1.1) began work on this problem and developed a process for the production of ammonia. Ammonia could be converted to nitrates and used for manufacturing fertilizer. By 1913, with the help of Carl Bosch of the Badische Anilin- & Soda-Fabrik (BASF), an industrial plant was in operation based on this process (now generally referred to as the **Haber process** but also known as the Haber–Bosch process). This process requires considerable heat and pressure. The reactants must be heated to 1000°F (550°C) and put under 150 to 200 times normal atmospheric pressure. The equipment necessary to carry out such a reaction can only be made of steel. This reaction is one of the most important industrial chemical processes ever devised, and it would not have been possible without steel.

Figure 1.1 Fritz Haber.

The life of Fritz Haber also illustrates how the advances of science and technology can be used for good as well as bad. As it turns out, nitrates are a key ingredient of not only fertilizers but also conventional explosives, as well. During World War I, the Germans were cut off from their usual sources of nitrates, and the Haber process was important in supplying the explosives for their munitions during the conflict. Haber was a very patriotic German who did not stop with explosives. A key proponent of the use of poisonous gases on the battlefield during World War I, Haber went to the front and personally supervised the use of these lethal chemicals. After the war, Haber was involved in several projects to support Germany during what was a difficult time for the country. With the rise of the Nazis, however, Haber, who was Jewish by birth, was forced into exile in 1933.

The connection between chemicals and electricity can be seen in the aluminum industry. Before 1886, aluminum was considered a rare and expensive novelty, despite the fact that aluminum was a plentiful component in the Earth's crust. A bar of aluminum was displayed in 1855 at the Paris Exposition next to the Crown jewels. The price of aluminum in 1859 was about $17 per pound, which was about the same price as silver. In 1886, a chemistry student at Oberlin College in Ohio, **Charles Martin Hall**, experimented with methods of extracting aluminum from its ore. The method that this young man developed involved melting the ore and then passing electricity through it. About the same time, a Frenchman by the name of **Paul Héroult** independently developed basically the same process. This basic method, known as the **Hall–Héroult process**, is still used today. Such a procedure is not possible without the availability of large quantities of electricity; therefore, commercial production of aluminum and development of the electric industry go hand in hand. Charles Hall founded the Aluminum Corporation of America (ALCOA) and became a multimillionaire. Héroult became quite wealthy as well and was involved in many ventures in addition to the production of aluminum. Both men died in 1914, within eight days of each other.

Another hallmark of the second phase of the Industrial Revolution was the tremendous increase in energy consumption. Before the Industrial Revolution, energy consumption was modest and for the most part involved animal, human, or water power. Coal and wood were used mainly for heating. The first phase brought the need to provide fuel for steam engines and iron works. Energy consumption increased even more dramatically in the second phase of the revolution, as everything was done on a massive scale. Energy-intensive steel production began to increase rapidly. The chemical industry churned out all kinds of products that had not been produced before, and most of them required energy for their production. Electricity supplied power easily to all types of functions, some of which either had been done by hand in the past or had not been done at all.

The key locations of the second phase of the Industrial Revolution were also different from the earlier phase. The Industrial Revolution began in England, but the second phase occurred in Germany and the United States, as well as England. These countries became the great powers during the second phase. Before 1860, many of the things that are taken for granted now did not even exist. Some of these items include aspirin (1900), mechanical refrigeration (1861), transportation of frozen meat (1879), the tractor (1898), the airplane (1903), the internal combustion engine (1882), and the diesel engine (1892). The effect of these developments on our lives has been overwhelming.

We now assume that certain conveniences will always be available to us. We expect to have cars, jet planes, dishwashers, clothes washers, air conditioners, packaged food, microwave ovens, and various electric appliances. All of these items consume natural resources, require energy, or produce wastes. Some of them do all the three, whereas all of them do at least one of the three. Natural resource consumption, energy consumption, and waste production lead us to the following key questions: "Is our lifestyle sustainable? Can we live this way for a long time?"

SCIENCE AND THE SCIENTIFIC METHOD

Much of what has been described so far could be described as the history of technology. Contrary to popular belief, science and technology are not the same, and in fact have not always been closely related. First, science will be dealt with and then technology. **Science** as we know it is a fairly recent development, which only began to evolve in the 17th century. Originally known as natural philosophy, science took form as scholars began to suggest that theories about nature should be subjected to experimental testing. As will be seen in Chapter 2, when this approach began to be used the resulting insights very quickly led to the reexamination of many long-held ideas.

Science is built around what has come to be known as the **scientific method**. The scientific method comes from the concept that ideas about how the universe functions should be able to be tested to see if the universe actually operates in this way. The scientific method is, in fact, a continuous cycle, and as a result the method has no beginning and no end. Step 1 is usually given as *defining the problem* for which one wishes to find an understanding. The problem must be of the type that one would expect to better understand it as a result of well-designed experiments. Step 2 is to make a guess about how things function in this problem. This guess is referred to as a **hypothesis** and may be described in some cases as a *model*. Step 3 involves designing and carrying out an *experiment* to see if the hypothesis is correct. The results of this experiment can lead to one of the three outcomes: The experiment might suggest that the hypothesis is correct, that the hypothesis is totally wrong, or that the hypothesis is partly wrong and requires modification. These observations will take us back to step 2, where we keep our hypothesis, abandon it, or modify it. Once the status of our hypothesis is decided, it will again have to be tested by experiment. This is a never-ending process because testing and revision of the hypothesis continue. Even if the experiment agrees with the hypothesis, the process does not end because the next experiment might show that it is incorrect. If a hypothesis is confirmed for the most part repeatedly, then at some point it may come to be known as a **theory**.

To illustrate how the scientific method works, consider the issue of how smog develops in southern California, a problem discussed further in Chapter 9. Historically, there were several ideas about how the smog was formed. If all of the materials that are found in the polluted air over Los Angeles are put into a box, a toxic mix is obtained, but this mix does not have the same properties as the smog. The problem, then, is, "Why do the gases being put into southern California air not produce the smog usually found in the air when they are put in a box?" Someone eventually surmised that perhaps light is needed to convert the gases to smog. This statement would become the hypothesis. The experiment would be to shine light on a transparent container of the gases. If smog is produced as a result of adding light, then the hypothesis is verified. If smog is not produced, then the hypothesis would be refuted. If the gases change in the presence of light but do not produce something identical to the smog, then the hypothesis may have to be modified. Maybe light and something else are required. A new or revised hypothesis requires more experimentation. Ultimately, this process of questioning, experimenting, answering, and questioning can go on indefinitely.

There are many situations in which one cannot run an experiment in the classical sense but can still apply the logic of the scientific method. Generally, these are fields of study concerning reconstructing the past, experimenting on systems that are so complex as to not be able to be directly tested, or testing systems in which direct experimentation would be considered unethical. In these situations, one tries to state the hypothesis in such a way that data can be collected from current or past situations that occur naturally. Some examples are archaeological findings, ancient gases trapped in polar ice cores, natural occurrences of disease rates under various existing circumstances, and political polling data.

The scientific method cannot be applied everywhere, nor should it be. Some fields of study are based on an approach that is very different from the scientific method. Mathematics, for example, is based on deductive logic. In mathematics, one makes a series of assumptions and then logically deduces whatever follows from those assumptions. Mathematics is not based on the input of empirical data of any kind. Many other fields of study may use empirical data from time to time but not in a systematic manner; that is, they are not data driven. Other issues such as style, reflection on the great thinkers in the field, statements of faith, analysis of the works of others, and argument based on opinion may be somewhat more important than in the study of science. These areas of study are not less valid than science but rather are different ways of knowing. Some fields of study where application of the scientific method is not generally valuable include religion, history, literature, and language studies.

Fields that are data driven and rely on the scientific method are known as the sciences. Based on earlier discussion, science can be defined as the pursuit of knowledge by observation and testing. Fields of study often included in the sciences are physics, chemistry, biology, psychology, sociology, and political science. This list is not exhaustive but mentions some of the major fields.

Usually when we think of science, it is often about **natural science**. Natural science deals with the natural world around us, including biological species; the water, air, and all of the materials of the Earth; the planets of the solar system and their moons; the sun and stars; and the far reaches of space. Most commonly this group of studies includes biology, chemistry, and physics.

SCIENCE AND TECHNOLOGY

Science and **technology** are closely related in today's society. This chapter has so far discussed science, but now the emphasis will be on technology. Technology dates back to antiquity. Early tools such as the garden hoe are examples of technology. Other examples include cooking, making pottery, extracting metals from their ores, making beer and wine, and extracting drugs from plants. All of these things were carried out before the artisans or practitioners were able to comprehend the scientific principles behind these activities. One excellent definition of technology was given by Hill (1992): "Technology is the sum total of processes by which humans modify the materials of nature to better satisfy their needs and wants."

Although science began to emerge in the 17th century, it was not until the 19th century that humans began seriously to apply science to technology. Since then, inventors and businesspersons have understood the value of applying scientific discoveries to the development of new and better products for humanity (and perhaps their profit margins). Such activities have come to be known as **applied science**. Technology using applied science has developed new drugs, lighter aircraft, hybrid plants, and many other products. Not all of these advances have been in a positive direction, however, as can be seen from the story of Fritz Haber earlier in this chapter. Also, science and technology can be applied, for example, to making chemical warfare weapons and atomic bombs. The problem is that by the middle of the 20th century, the close connection between science and technology had become so commonplace that many people easily confused the two. The distinction between the two fields began to blur. This book focuses on the study of the science of chemistry as it pertains to the environment, but chemical technology is also an important part of this picture.

SCIENCE AND THE ENVIRONMENT

The environment, broadly, can be construed to include nearly everything on the Earth, both living and nonliving. Biology (which is the study of living things) has always been a focal point of environmental studies because it is the living things and their interactions that define much of the nature of the environment. However, living organisms are affected by other substances in the environment, and all substances are chemicals of some type or the other. Additionally, all organisms are themselves made up of chemicals. Clearly, there is a chemical dimension to what goes on in the environment, and it is this aspect of the environment that will be studied.

As we study the chemistry of the environment, we must be aware that science can only tell us certain things. Science can inform us about relationships. Scientists can note that certain chemicals and dead fish in a stream usually occur together. Perhaps science can demonstrate that these chemicals cause the death of fish. Science may be able to suggest ways of keeping the chemicals out of the stream or keeping the fish alive in some other way. *What science cannot do is render value judgments.* The value of fish, the scenic beauty of the wilderness, a microwave oven, a car, a deer, our life, or someone else's life is a judgment made based on our ethical, moral, and religious assumptions. These assumptions are distinct from the science presented here. It is not my intention to make this book totally devoid of personal philosophy, but it is hoped that the scientific facts or theories will be clearly differentiated from personal value judgments.

ENVIRONMENT AND PUBLIC POLICY

Clearly, as a society, we constantly make value judgments about the environment. We collectively build dams, protect certain species, create national parks, and pass laws against air pollution. Such activities bring environmental studies into the area of public policy, but why does public policy have to intrude itself into environmental issues? In 1968, Garrett Hardin published an article in *Science* entitled "The Tragedy of the Commons." The title of this article is based on a parable introduced in an 1833 pamphlet written by William Lloyd. In the pamphlet, Lloyd discussed the relationship between pastureland that is held in common by everyone ("the commons") and individual herdsmen. Obviously, it was in the best interest of any individual herdsman to graze his livestock on the commons as much as possible, but if everyone did this then eventually the commons would be overgrazed and would be useless to all. Such a parable is an extremely good paradigm for our times. Many environmental issues can be viewed in this way. The commons can be saved for long-term benefit only if the individuals take their fair share at a fair rate, and it works in the same way with the common resources of the environment. There is often a conflict between the short-term welfare of a few and the long-term welfare of society as a whole. Because most of us do not graze cattle on the commons or anywhere else for that matter, then what are the commons of today? The commons are common resources. These include air, which we all breathe; water, which we all need; and living beings upon which we all depend, such as trees, algae, and fish. It is through public policy that society protects the commons from destruction by selfish individuals. The debate, of course, centers around what exactly should be included as part of the commons, and how it should be protected. Many of the topics in this book lead to such discussions.

Public policy is usually carried out by the authority of the government; however, educational factors, social pressures, economic pressures, and social protests can have an effect. The government usually uses either incentives or penalties. Incentives include encouragements such as tax cuts, payments of money, and so on. Such an approach costs the government money, which is often in short supply; therefore, governments usually resort to penalties to stop undesirable activities. These penalties include taxes and fines. There are some other restrictions under which we all operate, but these have nothing to do with the government. These restrictions are the inviolate laws of nature, which are discussed in Chapter 2.

DISCUSSION QUESTIONS

1. At what point did humans cease to be part of "nature" and become separate from their natural surroundings? Must humans have an effect on "nature" just because of their presence or can humans just remain in a natural setting and leave it unaffected?
2. Describe the lives of people who are hunter–gatherers. Also, describe the lives of those engaged in horticulture and agriculture. Compare these three groups with one another.
3. Ancient cities were different from modern cities in many ways but also had some similarities. In what ways were ancient cities similar to modern cities? In what way were they different?
4. Ancient cities were smaller and, therefore, should have had less impact on their environment. Given modern advances in sewage treatment, waste disposal, and other related issues, do modern cities have a greater or smaller impact on the environment than ancient cities? Explain.
5. Compare and contrast life in developed countries of the 21st century with life in Europe before the Industrial Revolution.
6. How were developments in iron production, textile manufacture, and the development of the steam engine related?
7. In what significant ways did the first phase of the Industrial Revolution differ from the second?
8. Is the term "Industrial Revolution" an appropriate name? Were the changes that occurred more a revolution or an evolution? Why?

9. How does the life of Fritz Haber illustrate some of the moral ambiguities associated with the use of science and technology?
10. Consider this statement: "Today's modern conveniences are detrimental to the well-being of our society." Make an argument supporting this statement. Make an argument opposing this statement.
11. Outline the scientific method and give some of its limitations. Can any subject be studied using the scientific method if one wishes to do so?
12. Explain the difference between science and technology, and suggest why it is that we tend to confuse them in the modern world.
13. What is meant by the "Tragedy of the Commons," and how does it apply to societal issues in our culture?
14. What are some of the measures that governments can take to control the environmentally harmful activities of its citizens? Which ones are used most often and why?

BIBLIOGRAPHY

Bessemer, H., *Sir Henry Bessemer, F.R.S: An Autobiography*, Offices of Engineering, London, 1905, pp. 138–151.

Binczewski, G.J., The point of a monument: a history of the aluminum cap of the Washington monument, *JOM*, 47(11), 20–25, 1995 (http://www.tms.org/pubs/journals/JOM/9511/Binczewski-9511.html).

Brain, M., *How Iron and Steel Work*, HowStuffWorks, Atlanta, GA, 2012 (http://science.howstuffworks.com/iron.htm).

Friedrich, B, *Fritz Haber (1868–1934)*, Fritz Haber Institute of the Max Planck Society, Berlin, Germany, 2005 (www.fhi-berlin.mpg.de/history/Friedrich_HaberArticle.pdf).

Garfield, S., One man and his color, *Chemical Heritage*, 24(3), 8–9, 35, 2006.

Geller, T., Aluminum: common metal, uncommon past, *Chemical Heritage*, 25(4), 32–36, 2007/08.

Hardin, G., The tragedy of the commons, *Science*, 162, 1243–1248, 1968.

Hart-Davis, A., *Henry Bessemer, Man of Steel, On-line Science and Technology*, September 1, 1995, (http://www2.exnet.com/1995/09/27/science/science.html).

Hill, J.W., *Chemistry for Changing Times*, 6th ed., Macmillan, New York, 1992, p. A37.

Hudson, J., *The History of Chemistry*, Chapman & Hall, New York, 1992.

Isenberg, A.C., Ed., *The Nature of Cities: Culture, Landscape, and Urban Space*, University of Rochester Press, Rochester, NY, 2006.

Macionis, J.H. and Parrillo, V.N., *Cities and Urban Life*, Prentice-Hall, Upper Saddle River, NJ, 2007.

NobelPrize.org, The Nobel Prize in Chemistry 1918: Fritz Haber, in *Nobel Lectures, Chemistry 1901–1921*, Elsevier, Amsterdam, 1966 (http://nobelprize.org/nobel_prizes/chemistry/laureates/1918/haber-bio.html).

CHAPTER 2

The Natural Laws

MATTER

Matter and the changes that it can undergo are a fundamental focus of chemistry. Matter can be defined as anything that has mass* and occupies space. The nature of matter has been of great interest since the days of the ancient Greek philosophers. Since 1860, it has been generally accepted that matter is made up of extremely small, almost indestructible particles known as **atoms**. These atoms have the ability to combine with one another to give rise to a rich variety of molecules, which explains the incalculable number of different substances present in the universe. A **molecule** is an atom or any combination of atoms that is capable of existing for some amount of time in the presence of other molecules. These molecules can be large, containing many atoms of different types such as the heme molecule ($C_{34}H_{32}FeN_4O_4$), or they can be small, with few atoms, such as water (H_2O), or only one atom, such as helium (He). Even multi-atom molecules may contain only one type of atom, as in normal oxygen gas (O_2). In this case, the atom is O and the molecule is O_2.

It is these atomic concepts that allow us to begin to understand the complex world of matter around us. A substance made up of molecules containing only one type of atom is known as an **element**. There are 90 of these elements that occur naturally on the Earth, including gold (Au), nitrogen (N_2), oxygen (O_2), silver (Ag), iron (Fe), helium (He), and sulfur (S_8). If the molecules contain more than one type of atom, then the substance is referred to as a **compound**; water (H_2O) and table sugar ($C_{12}H_{22}O_{11}$) are some examples of compounds.

With sufficient time and skill it is often possible to purify substances. If the effort is successful, the result is a **pure substance**, which is any matter containing only one type of molecule. Examples are pure gold, pure oxygen, pure water, or pure table sugar. Very little of the world around us is made of pure substances. Water contains other materials dissolved in it, steel is not pure iron, and ice cream has a variety of substances in it.

* We use the term *mass* here rather than *weight*. Weight refers to gravitational attraction and would have no meaning in outer space. Mass refers to the amount of material intrinsically present without regard to gravitational attraction and has universal application.

When two or more substances are mixed together, the result is called a **mixture**. Mixtures are the most common form of matter. Mixtures can be classified into two different types. The first type is a **homogeneous mixture**. Homogeneous mixtures are those in which each substance in the mixture is evenly distributed—for example, a soft drink. A soft drink is made up of a large variety of ingredients mixed together, but each drop of the soft drink contains exactly the same proportion of each ingredient as any other drop. The other type of mixture is a **heterogeneous mixture**. Heterogeneous mixtures are those in which the substances in the mixture are not evenly distributed—for example, ice cream with nuts. One bite of the ice cream might have nuts, whereas another bite might include only pure ice cream and no nuts. The composition varies depending on where one samples the mixture.

EARLY DEVELOPMENT OF CHEMISTRY

The origin of **chemistry** as a science in the modern sense probably goes back no further than the 18th century; however, the roots of chemistry can be traced back to ancient Greece and even further depending on how one defines it. The ancient Greek philosophers were among the first to have considered theories about matter and have their thoughts preserved. The most influential of these thinkers was **Aristotle**, who lived in the 4th century BCE and was such an influential philosopher that for centuries his ideas were accepted as fact. The medieval Roman Catholic Church elevated Aristotle's ideas to a level just below the Holy Scripture. Aristotle believed that all substances were composed of primary matter (*proto hyle*). This matter could express itself as one of the four elements: earth, air, fire, or water. All substances were then believed to be made up of various combinations of these four elements. The ancient Greek philosophers based their ideas about matter on observations and thought. Their weakness was that they were not inclined to test theories through experiments. If the philosopher was influential enough and the theory seemed reasonable, it was unlikely that it would be challenged. Aristotle was so well respected that his four-element theory continued to hold sway until the 16th century.

The next significant development in the area of chemistry took place within a few centuries after the time of Christ. From that time until about the 17th century, much of the activity in the area of chemistry was known as **alchemy**. The alchemists were principally interested in the conversion of common metals into gold. Although they never reached their goal, their activities were profoundly experimental in contrast to the earlier Greek philosophers. Many of the techniques of modern chemistry were originally developed by the alchemists. In addition to using their laboratory expertise, they were able to isolate many substances such as grain alcohol, various acids, and the element now known as phosphorus. In their search for a way to produce gold, the alchemists discovered few theories about the nature of matter and its interactions. During this time, the Aristotelian understanding of matter continued to dominate European thought.

In the 17th century, the synthesis of theory and experiment began to occur in a branch of philosophy known as **natural philosophy**. **Sir Francis Bacon** was an early proponent of carefully planned experiments, in which results were accurately recorded so that experiments could be repeated or modified, if necessary. He also suggested that experiments implied generalizations and theories, which could be tested by further experiments. This approach evolved into the scientific method. As noted in Chapter 1, natural philosophy was the forerunner of what we currently refer to as science.

It was not until the 17th century that a serious challenge was made to the Aristotelian view of matter. In his book *The Sceptical Chymist*, published in 1661, the famous scientist **Robert Boyle**, by using logic and experimental data, made it clear that there was no evidence to support the four-element theory.

LAVOISIER AND THE LAW OF CONSERVATION OF MATTER

Antoine Lavoisier (see Figure 2.1), a Frenchman who lived in the latter half of the 18th century, is considered the father of modern chemistry by most historians. Lavoisier developed an intense interest in the nature of burning. A key question was whether matter was destroyed during combustion. Lavoisier performed many

Figure 2.1 Portrait of Lavoisier and his wife by Jacques-Louis David (ca. 1788).

experiments related to combustion in which he very carefully weighed both the starting materials and the products. When he precisely accounted for all the products, including gases, Lavoisier demonstrated that the mass of the products was always equal to the mass of the starting materials. Eventually, Lavoisier formally summarized his observations as a general law, the **law of conservation of matter** (also known as the "law of conservation of mass"). This law states that matter can be neither created nor destroyed; it can only be converted from one form to another.

Lavoisier, for all of his creative genius, was unable to escape the wrath of the French Revolution. He had been an associate in a business that was under a contract to collect taxes for the king. Because of this association he was sent to the guillotine on May 8, 1794, and the world lost perhaps one of the greatest chemists who ever lived.

The law of conservation of matter has some powerful consequences on the environment. All of the matter the Earth had is still present and always will be. The only exceptions to this are a few meteorites that have crashed in, some gases that have escaped, and a space probe or two. In other words, everything we use has to come from somewhere and, when we are done with it, it has to go somewhere. We often throw things away. The lesson from this law is that there is no "away."

DISORDER

We are all aware of disorder around us, but most of us probably never thought of disorder as an important concept in science. An egg can be used to illustrate the idea of order. Every substance in an egg is in place for a purpose. There is a shell, enclosing a membrane, which encloses the egg white, inside which we find the yolk. Everything is in order. To illustrate disorder we only need to throw the egg on the floor. The shell would break into many pieces, the membrane would tear apart, the egg white would spill all over the place, and the yolk would probably rupture and spill as well. Now everything is a mess (i.e., disorder).

Another idea that this example may illustrate is that disorder is fairly easy to create. Consider the egg we just threw on the floor. Conversely, disorder is difficult to overcome. Can one put the egg back in order again? Recollect the nursery rhyme "Humpty Dumpty":

> Humpty Dumpty sat on a wall.
> Humpty Dumpty had a great fall.
> All the king's horses and all the king's men
> Couldn't put Humpty Dumpty together again.

One of the fundamental observations about the universe is that order is difficult and disorder is easy. Is it easier to clean one's room or to mess it up? Is it easier to put a deck of cards in a certain order or to arrange the cards randomly? Is it easier to build a house or to tear it down? However, from a scientific point of view, we must define order and disorder more carefully because they are not always that obvious; for example, which is more ordered, a sugar crystal or a glass of sugar water? To make this distinction, we require information.

To explain what is meant here, let us first consider a deck of cards and then the sugar question. A well-ordered deck of cards can be described with very little information. The suits are arranged ace through king, starting with clubs, spades, hearts, and finally diamonds. In the preceding sentence, the position of each card in the deck has been described. In a randomly arranged deck, however, 51 separate pieces of information are required to specify the position of each card in the deck. (It is 51 and not 52 cards because when 51 cards have been specified then the position of the last card is known by default.) Now consider the sugar example. The sugar crystal is considered to be in order because the molecules are in a crystal lattice. If the positions of a few molecules in the lattice are known, then the positions of all of the molecules can be calculated. In sugar water, however, the sugar molecules are randomly scattered in water. Each molecule must be specified individually. In this case, large amounts of information are required to describe the positions of the molecules; therefore, the system is disordered. Generally, *order* is viewed as matter that is highly organized or highly purified, or both—for example, a living organism or pure gold. Generally, *disorder* is viewed as a matter that is disorganized or a mixture of substances or both—for example, a crumbling building or polluted water.

Scientists have developed the term **entropy** to discuss disorder. Entropy is a measure of disorder or randomness. It is defined in such a way that the higher the entropy the greater the disorder. A fundamental concept that emerges here is that effort is required to create ordered (or low-entropy) matter. Pure salt has low entropy, whereas saltwater has high entropy. To recover pure salt from saltwater one needs to supply energy to boil off or evaporate the water. An aluminum can is a low-entropy item, and considerable amount of electricity is required to make one from aluminum ore, which is a high-entropy material.

WORK AND ENERGY

Effort, in terms of physical science, is defined as **work**, which is defined as the result of using a force to move a sample of matter over some distance. This definition may seem somewhat limited, but it can be applied to many situations. The arguments for a widespread application of the definition are beyond the scope of this chapter. Some kind of work is always required to decrease entropy or disorder.

To do work, energy is required. **Energy** is defined as the ability to do work or to raise the temperature of a sample of matter;* therefore, it logically follows that to be able to decrease disorder or entropy energy is required. The importance of energy in our society cannot be underestimated. Chapter 7 discusses many of the ways in which energy is used and produced in our culture. Doing work on matter can cause it to move. This moving matter can do work on other material. The wind, for example, represents moving air. This moving air can strike a windmill and cause it to move. Thus, the moving air has the ability to do work and therefore possesses energy. The energy that matter has because of its motion is called **kinetic energy**.

* Energy can be defined as the ability to do work; however, in this chapter, we choose an alternative definition to avoid a detailed discussion on the nature of heat energy.

Heat energy is another form of energy. Heat energy is related to the temperature of a substance. The more the heat energy, the higher will be the temperature; however, heat energy and temperature are not one and the same. The temperature is determined by the amount of heat energy and the nature and mass of the object. Metal objects require less heat energy to rise to a given temperature than does the same mass of water. The amount of heat energy required to heat the water in a teacup to 212°F (100°C) is much less than that required to heat a bathtub full of water to the same temperature. Heat energy, unfortunately, is easier to talk about than it is to define. To define it, we take advantage of the fact that heat energy flows between objects with different temperatures. Heat energy is defined as the energy that flows from one body to another when there is a temperature difference between the two.

Much of the energy in the universe is not utilized at any given moment. This energy is available to do work at a later time and is known as potential energy. **Potential energy** is defined as stored energy, which can be released to do work or produce heat. Water that is about to descend from a waterfall has positional or gravitational potential energy. When the water falls, the energy is released and can be used to turn a waterwheel. The electricity in an electric socket has electrical potential energy that can be tapped by plugging into it. A gallon of gasoline has a chemical potential energy that can be released by burning the gasoline in a car's engine.

FIRST LAW OF THERMODYNAMICS

Despite the importance of energy in our lives, most of us are unaware of it. Suppose that someone was watering a lawn and after some time turned the sprinklers off. If you were asked where the water went, most likely you would come up with an answer. You might say that the water soaked into the ground or flowed into the street, but whatever your answer it would most likely be a reasonable one. However, if you were sitting in a room and after some time the lights were turned off, what would you say if someone asked where the light went? Most people would not be able to answer because we do not tend to think of light as something that goes somewhere. In fact, most of the light is either absorbed by objects in the room and converted into heat energy or is reflected out of the window.

The realization that energy did in fact have to go somewhere gained acceptance in the 19th century. Scientists finally concluded that there was just so much energy in the universe, and that it could not be added to or subtracted from.* This idea was eventually formalized into the law of conservation of energy. It is also called the **first law of thermodynamics** because one of the early statements of the law came from the study of heat engines such as the steam engine. This field of study was known as *thermodynamics*. Currently, **thermodynamics** is the study of heat transfers and related forms of energy.

* Recently, nuclear physics has demonstrated that matter and energy can be interconverted; however, this is not taken into account in our discussion.

The first law of thermodynamics states that energy can be neither created nor destroyed; it can only be converted from one form to another. The basic thrust of this law can be seen by considering the energy involved during a trip in a car. The energy used in the car actually originated from the sun; however, this energy was not the sunlight of the last year or 10 years ago, but instead it is energy that came from the sun millions of years ago. Some of this sunlight was captured by green plants of that time and stored as biochemical energy in the actual structure of the organisms themselves. Over the long expanse of time, this biomaterial was slowly converted to petroleum containing the energy that came from the sun. This petroleum was pumped out of the ground by one of the oil companies and converted into gasoline, but the petroleum product still contained the stored solar energy as chemical potential energy. The gasoline was pumped into the car and burned in the engine, releasing the energy to heat and move the car. The heat energy used to warm the car and the kinetic energy used to move the car are the sum of all the energy released. After the trip is over, where did the kinetic energy of the car go? It was converted into heat energy as well. The moving parts of the car turned against one another, resulting in friction that heated up these parts. Also, resistance due to the car cutting through the air heated the air, and the car's brakes slowed the car through the use of friction, which heated the brake pads.

If, in fact, energy cannot be destroyed but only converted from one form to another, why do we always need more of it? The answer to this is related to some important observations about heat energy. It can be noted from the example of the car trip that the energy ends up as heat energy. This seems to be the eventual fate of all energy when one attempts to use it to produce useful work.

The problem with heat energy is that it has some definite restrictions for its use as an energy source. First, some of the heat energy on the planet is continually radiated into space; therefore, although this energy is not destroyed, it is not available for use. Second, even the portion of heat energy that remains on the Earth has some limitations on the amount that can be converted into useful work.

SECOND LAW OF THERMODYNAMICS

Sadi Carnot, one of the early researchers of thermodynamics, was a Frenchman who became intensely interested in the steam engine. He studied these engines in detail and outlined some of the principles on how heat energy can be used to do work. Based on Carnot's observations, **Rudolph Clausius**, a German chemist, formulated the **second law of thermodynamics**. This law has probably been stated in more different ways than any other scientific law. Many of the statements of this law even seem unrelated, but they can be proven to state the same thing.

The second law is fundamental to an understanding of how the universe works; yet, it can be difficult to explain. Ultimately, it is related to the idea of entropy or disorder. Let us once again consider the broken egg. If an egg falls on the floor and breaks, nothing is lost; rather, the egg has been "rearranged." Not only is all of the matter in the egg still present, but all of the energy that was converted into kinetic energy while the egg was falling is also present. This kinetic energy has been converted into heat energy that is now in the broken egg and on the floor. Based on the first law, however, there is absolutely nothing to prevent the egg from gathering up the pieces from the floor and putting itself back together. In addition, the egg could gather the thermal energy from the floor, convert that energy into kinetic energy, and jump back to the point from where it was dropped. However, we all know that such things do not happen (except when a video is reversed). What is unusual about the process of putting the egg back together? Such a process produces order from disorder (i.e., decreases entropy).

One fundamental statement of the second law of thermodynamics as given by Clausius is

> The entropy (disorder) of the universe moves toward a maximum.

This law fits with the idea that eggs break, milk spills, and ice melts. We all know that these processes occur if we allow them. You may be wondering, though, about those situations where entropy (disorder) seems to decrease. Many of us have placed water in the freezer. The next morning we find that the water has frozen to ice. This freezing represents a decrease in entropy (decrease in disorder). Does this spontaneous creation of order represent a violation of the second law? No, it does not. The flaw in the usual reasoning is that we fail to look at the universe as a whole. The freezer consists of a motor, a fan, and a compressor to pump the heat out. As the heat is blown off of the compressor coil into the external air, the air is warmed. The entropy increase of the air is greater than the entropy decrease of the water, which has now become ice.

The ideas about entropy as applied to matter in the earlier section on disorder can be applied to energy as well. Low entropy or ordered energy is the energy that can be directed to produce predictable motion that can be harnessed. Some examples include falling water, electricity, and chemical energy. Falling water can be used to turn a waterwheel, electricity can be used to turn a motor, and chemicals can be used to make a battery. What happens to the energy after the work is done? It is usually converted into heat energy. Generally, heat energy is considered to be of highest entropy; however, adding heat energy to the materials in a given situation does not always give the same entropy increase. Very hot systems are fairly insensitive to entropy increases when heat energy is added to them, as they are very chaotic already; hence, when heat energy is added to or taken away from a very hot system, the change in entropy is negligible. Conversely, adding or subtracting similar amounts of heat energy to or from a fairly cold system will have a drastic effect on the total disorder of the system. The entropy would change dramatically. As a result, we consider high-temperature systems to be less sensitive to entropy changes upon heating or cooling and more useful to do work than low-temperature systems. These high-temperature systems can do work because the increase in entropy in the transfer of heat to the low-temperature system makes up for the decrease in entropy involved in producing an ordered motion.

Heat energy in higher temperature systems can be used to produce orderly motion in, for example, a steam engine. Heat energy is used to produce steam that expands and pushes back a piston. The steam is condensed into water and the piston is pulled back, after which more steam pushes the piston back again and the cycle repeats itself. The fundamental feature of heat engines is that they operate between two temperatures. The heated material that goes into a heat engine comes out at a lower temperature and with higher entropy. The heat energy in this material is less useful for producing work.

The use of energy in most processes is intended to decrease entropy either by making low-entropy products from raw materials or by producing low-entropy energy that would be useful to do work. Products of low entropy include plastics, metals, glass, and products formed from raw materials. Low-entropy energy includes electricity, moving vehicles, and focused x-rays. Because the second law requires that processes lead to an increase in entropy, and because most useful processes lead to a decrease in entropy, processes used to do work cannot be 100% efficient. There must be high-entropy energy left over to give a net increase in entropy. When energy is harnessed to do work, there will be waste energy in the form of heat energy that will be less useful for producing work. The result of such logic is that the second law of thermodynamics can be stated in the following alternative form:

> In any process using energy to do work, some part of that energy will be converted to, or remain as, heat energy. This heat energy will be less useful for doing work than the original energy.

One can, therefore, conclude that not all energy is created equal. Energy may not be destroyed, but it can be made less useful. As a result, energy can be viewed as being of varying quality with regard to its ability to do work. High-quality energy is the energy that does work easily and efficiently. Electricity, steam, gasoline, and water moving down from high elevations are sources of high-quality energy. The best examples of low-quality energy are situations where heat energy is spread out in such a large mass that the temperature remains fairly low, such as all the heat energy in the atmosphere over Ohio in the wintertime, all the heat energy in the Atlantic Ocean, or all of the heat energy in San Francisco Bay. It is not that one could not get this heat energy to do work, but it would not be easy and even lower quality heat energy would be produced as a result of the process.

If we reexamine the car trip, we would find that not only does the kinetic energy eventually turn into heat energy but also much of the chemical potential energy that was in the gasoline was converted directly into waste heat energy. All internal combustion engines require a cooling system because the engine gets hot. None of this heat energy does any useful work. It is waste heat energy, and a system has to be designed to get rid of it. In many applications, a method can be devised to get work from the waste heat energy, but one can never convert all of it into work.

The consequence of the second law of thermodynamics is that one cannot recycle energy completely. As energy is used for work, it becomes less and less useful to do work.

MATTER, ENERGY, AND THE ENVIRONMENT

A strong unifying principle of this book is that most of our environmental problems arise when we try to violate some of the fundamental laws of nature. Our society has a voracious desire for materials and energy and a careless way of disposing the unwanted residue. In the end, these laws cannot be violated. We will all pay the price if we individually or collectively try to do things that are against the laws of nature. The following chapters will show how this occurs.

DISCUSSION QUESTIONS

1. Matter is made of atoms. Explain how these atoms are related to the existence of compounds within matter. If you had a sample of a pure element, what would need to be true of the atoms in the sample?
2. Most commonly, materials in nature are not pure substances but are mixtures. Speculate on why you feel that might be. Would it have anything to do with the second law of thermodynamics?
3. Explain the difference between a homogeneous and a heterogeneous mixture.
4. Explain why many of the techniques of modern chemistry were developed by the alchemists rather than the Greek philosophers.
5. What are some implications of the law of the conservation of matter on modern society?
6. How is entropy related to order, disorder, and the amount of information required to describe some sample of matter?
7. We have subdivided energy into kinetic energy, heat energy, and potential energy. Consider a steam locomotive from the early part of the 20th century. What kind of energy is involved with each of the following: the coal, the steam, the pistons, and the moving locomotive?
8. Ultimately, from where does the energy come that is used to propel an automobile down the road? Explain.
9. What are some of the implications of the first and second laws of thermodynamics on how we use energy in modern society?
10. Based on the two laws of thermodynamics, could you make a perpetual motion machine—that is, a machine that runs continuously without the input of energy? Why or why not?

BIBLIOGRAPHY

Hudson, J., *The History of Chemistry*, Chapman & Hall, New York, 1992.

CHAPTER 3

Underlying Principles of Chemistry

ATOMIC THEORY

The atomic theory is fundamental to any modern understanding of chemistry. The idea that matter is not continuous is old, dating back to ancient Greece. The first record of the atomic theory of matter is found in the 5th-century BCE writings of **Democritos**, who credited his teacher **Leucippos** for the idea. Democritos and Leucippos argued that if matter was divided again and again a number of times, eventually one would arrive at something that could no longer be divided. In fact, the modern word "atom" is derived from the Greek word *atomon*, meaning indivisible. This theory practically exerted no influence on science for over 2000 years because Aristotle rejected it, and therefore no one seriously considered the theory until relatively recent times.

Robert Boyle adopted ideas that seemed to include the **atomic theory**; however, it was the English Quaker schoolteacher **John Dalton** (see Figure 3.1) who first set down modern atomic theory. Although there is evidence that he formulated the theory earlier, it was first published in 1807 by Thomas Thomson, who gave Dalton full credit for the theory. Dalton's atomic theory can be outlined as follows:

1. Elements are composed of small indivisible particles called atoms.
2. The atoms of a given element are similar in all respects.
3. The atoms of one element are different from the atoms of all other elements.
4. Chemical compounds are formed when the atoms of two or more elements come together to form molecules.

Dalton's theory was a powerful and largely correct view of the microscopic nature of matter. There were some errors, such as the fact that we now know that atoms can be split under special circumstances and that not all atoms of a given element are absolutely identical, but, considering the knowledge of the time, these flaws were minor.

Figure 3.1 Engraving of John Dalton by William Henry Worthington. (Courtesy of the Chemical Heritage Foundation Collections.)

Despite the theory being powerful, it was not generally accepted by the entire scientific community in Dalton's lifetime. It was not until 1860 that opposition to the atomic theory began to disappear. Since then, the atomic theory has been universally accepted, with certain modifications.

The advent of atomic physics required a more sophisticated view of the atom. The specific event that led to this change of view was the discovery of **radioactivity** by **A.H. Becquerel** in 1896. In 1895, Wilhelm Röntgen discovered x-rays, and Becquerel was interested in whether there was any connection between fluorescent compounds and x-rays. Fluorescent compounds are compounds that, when irradiated with one type of light, will emit another type. You may have observed this phenomenon when a white shirt glows brightly when placed under a black light that is emitting ultraviolet (UV) light. Quite by chance, the fluorescent compound that Becquerel chose to study was a uranium salt. Because it was expected that strong sunlight was needed to cause the compound to fluoresce, Becquerel carefully wrapped a photographic plate in a dark material, placed the salt crystal on top of it, and set it under the sun. He reasoned that the fluorescent crystal might produce x-rays that would penetrate the wrapping and produce an image on the photographic plate. It worked. The next time he attempted to repeat the experiment it was cloudy; therefore, he put the experimental setup in a dark drawer and waited for a sunny day. Even after three days, the weather had not improved and Becquerel was tired of waiting. He decided to develop the plate to demonstrate that the crystal was not producing the image without sunlight. Much to his surprise the image was much bolder than earlier. The radiation produced was coming directly from the crystal without the sunlight. This led Becquerel to demonstrate that the radiation was emitted directly from the uranium itself. **Marie and Pierre Curie** followed up on this research and discovered other elements that emit radiation; that is, they are radioactive (Marie Curie was one of Becquerel's students). For example, they demonstrated that another known element of that time, thorium, was radioactive. The Curies discovered two new elements that are radioactive—polonium and radium—and Marie Curie received the Nobel Prize in 1911 for the discovery of these elements. Unfortunately, her husband Pierre was struck and killed in 1906 by a horse-drawn wagon when he was crossing the street. Earlier, Becquerel and the Curies received the Nobel Prize in Physics in 1903 for the discovery of radioactivity. Marie Curie is one of only four people ever to receive two Nobel Prizes.

Modification of the atomic theory was now required, but not because of the discovery of radioactivity alone, but rather due to the observation that some of the radiation was, in fact, a stream of particles. Most of this work was done between 1899 and 1906 by Becquerel, the Curies, Ernest Rutherford, and others. (For further discussion on these particles, see the section in Chapter 4 on nuclear chemistry). The inescapable conclusion was that if atoms emit particles, then they could not be absolutely indivisible.

One of the most important series of experiments concerning the nature of the atom was carried out between 1906 and 1911. Most of these experiments involved the shooting of particles from a radioactive source at very thin gold foil. Gold can be pounded so thin that it is only about 2000 atoms thick. The results of these experiments were puzzling. Most of the particles passed through the gold as if there was nothing present, although there was minor scattering of some of the particles. What was particularly strange was that a very small percentage of the particles were deflected so severely that some of them almost went back to the source. The situation is shown in Figure 3.2. The last observation was very difficult to explain based on the atomic theory of the time. One of the key investigators, **Ernest Rutherford**, remarked, "It was quite the most incredible event that has ever happened to me in

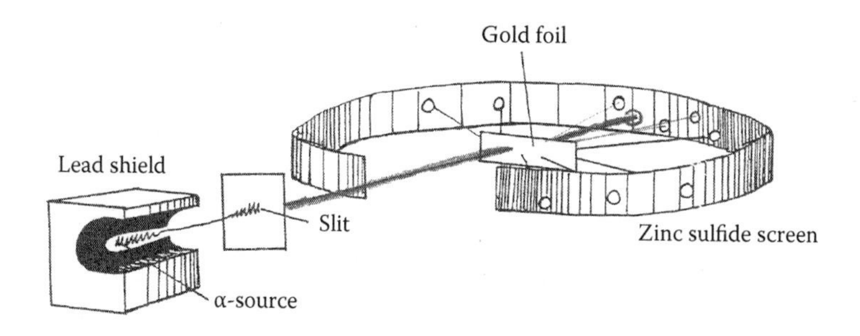

Figure 3.2 General setup of the gold foil experiment. The cylindrical screen is shown to make the diagram more understandable. A moveable screen was actually used.

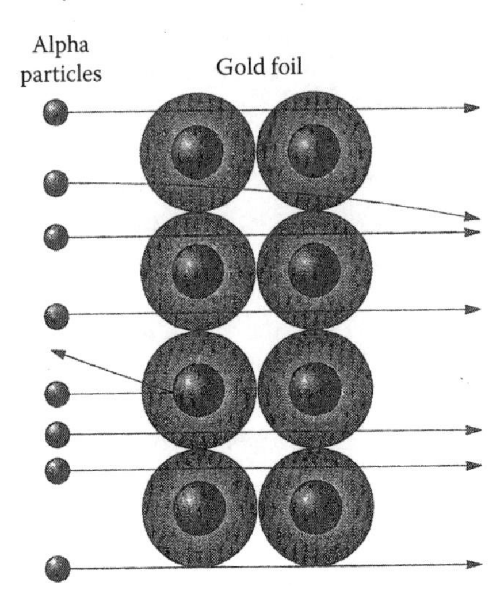

Figure 3.3 Diagram showing Rutherford's explanation of the gold foil experiment.

my life. It was almost as incredible as if you fired a 15-inch shell at a piece of tissue paper and it came back and hit you." In 1911, it was Rutherford who came up with the explanation that gave rise to the modern understanding of the atom.

Rutherford was a New Zealander working at the University in Manchester, England, when he postulated that the atom was mostly empty space with a very dense center called the **nucleus**. This nucleus contained nearly all of the mass of the atom. The rest of the atom represented most of the volume of the atom, but contained nearly zero mass. If, for example, the nucleus were the size of a baseball, the edge of the atom would be over 2 miles away. The nucleus is positively charged, and the rest of the atom contains very light negatively charged particles, now called **electrons**, in a quantity just sufficient to balance the positive charge on the nucleus. Figure 3.3 shows how the model explains the experimental results. Because the atom is mostly empty space, most of the particles directly pass through it. Some of the particles get too close to a nucleus and are deflected a little. A very few particles strike the nucleus nearly head-on and are deflected in a grand style because of the tremendous mass of the nucleus. It would be somewhat like bouncing a rock off a skyscraper.

Modern atomic theory has developed a more detailed understanding of the atom. For our purposes, we will be viewing the atom as made up of three distinct particles—the **proton**, the **neutron**, and the **electron** (see Table 3.1). The nucleus of the atom is made up of protons and neutrons. These are heavy, dense particles each having a mass of one in arbitrary **atomic mass units** (amu). The difference between these two particles is their charge. The proton has a charge of +1, whereas the neutron has no charge (0). Nuclei of the atoms of different elements have

Table 3.1 Characteristics of Subatomic Particles

Name	Mass (amu)	Charge
Proton	1	+1
Neutron	1	0
Electron	1/1837	–1

different numbers of protons and perhaps neutrons. The total number of protons and neutrons determines the weight of the nucleus and therefore the weight of the atom. (The weight of the electrons is so small that they can be ignored.) The number of protons determines the charge on the nucleus.

The charge on the nucleus is important because it determines the number of electrons that will be attracted by the nucleus. Each electron carries a charge of –1. The number of electrons will be exactly equal to the number of protons or the number of positive charges on the nucleus. As noted above, the reason why nearly all of the mass is in the nucleus is that the electron is a very light particle, weighing only about 1/1837 amu. The weight of the electron is so small that for many purposes it is ignored.

Chemistry, as we shall see, is controlled by the interactions of electrons from various atoms, and it is the number of electrons that determines the chemical nature of an element. Each element in its free, uncombined state has a unique number of electrons. The number of electrons is dictated by the number of positive charges on the nucleus which is equal to the number of protons in the nucleus. The number of protons, which is unique to each element, is referred to as the *atomic number*. The atomic number of each element is listed in most tables or lists of elements. (See the table inside the cover of this book.)

The other number that appears on most charts of elements is the *atomic weight*. As noted earlier, this number (in amu) is roughly equal to the number of neutrons plus the number of protons. For example, carbon, which has 6 protons and 6 neutrons, has an atomic weight of about 12, and oxygen with 8 protons and 8 neutrons has an atomic weight of about 16. A tremendous range of combinations is possible. These combinations range from one proton and no neutrons in hydrogen to 92 protons and 146 neutrons in uranium.

One puzzling factor is that with such an arrangement one would expect that all atomic weights would be simple whole numbers. A simple inspection of a table of elements would make it clear that this is not always the case. The atomic weight listed for chlorine is 35.5 and that for neon is 20.2. How can this be possible?

Remember that the atomic weight is calculated by adding the number of protons and the number of neutrons, and the atomic number is the number of protons. With this in mind, let us consider hydrogen. A lone proton with no other protons or neutrons has an atomic number of one and is a hydrogen nucleus; the atomic weight would be one. It turns out that there is a perfectly stable nucleus with one proton and one neutron. This nucleus would have an atomic number of one and an atomic weight of two. Because there is a one-to-one correspondence between an atomic number and an element, an atomic number of one means that this is also hydrogen, but this nucleus has an atomic weight of two. This material, which is often called heavy hydrogen or deuterium, accounts for about 1% of all naturally occurring hydrogen. Chemically it is still hydrogen, but heavier. These two forms of hydrogen are referred to as *isotopes*. An **isotope** (derived from the Greek word meaning "same place") is a form of an element that differs from the other forms of the same element in the atomic weight of the atoms. But, how does this explain the fractional atomic weights on the charts of elements?

To understand fractional weights let us consider neon, which is listed with an atomic weight of 20.2. It turns out that naturally occurring neon is a mixture of three isotopes with atomic weights of 20, 21, and 22. The atomic weight of 20.2 represents the weighted average of the naturally occurring mixture of the three isotopes. Obviously, the isotope with a weight of 20 is the predominant isotope.

PERIODIC LAW

As soon as the concept of elements became widespread in the latter part of the 18th century, scientists began trying to bring some order to the ever-increasing number of elements. The first published classification of the elements was by Lavoisier in 1789. An early chart prepared by John Dalton is shown in Figure 3.4. By 1850, there were 58 known elements, and many scientists were trying to develop a system to explain why some elements seemed to be similar to one another. In 1862, a Frenchman, A.E. Beguyer de Chancourtois, arranged the elements in ascending order of atomic weights. In his chart, de Chancourtois noticed that certain elemental properties (reactivity to water, low melting point, electrical conductivity, etc.) reoccurred at certain intervals. Properties that recur in some sort of a continuously repeating fashion can be referred to as *periodic*. John Newlands of England noted

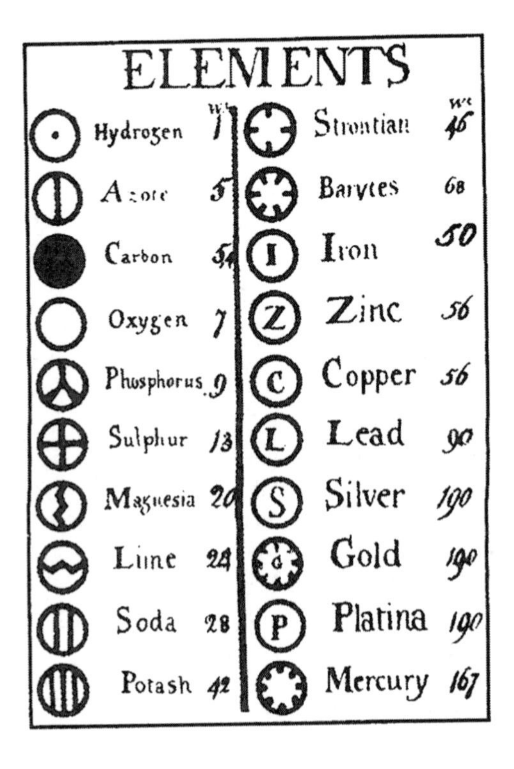

Figure 3.4 Chart of the elements produced by John Dalton in the early 1800s.

No.		No.		No.		No.		No.		No.		No.		No.	
H	1	F	8	Cl	15	Co & Ni	22	Br	29	Pd	36	I	42	Pt & Ir	50
Li	2	Na	9	K	16	Cu	23	Rb	30	Ag	37	Cs	44	Os	51
G	3	Mg	10	Ca	17	Zn	24	Sr	31	Cd	38	Ba & V	45	Hg	52
Bo	4	Al	11	Cr	19	Y	25	Ce & La	33	U	40	Ta	46	Tl	53
C	5	Si	12	Ti	18	In	26	Zr	32	Sn	39	W	47	Pb	54
N	6	P	13	Mn	20	As	27	Di & Mo	34	Sb	41	Nb	48	Ei	55
O	7	S	14	Fe	21	Se	28	Ro & Ru	35	Te	43	Au	49	Th	56

Figure 3.5 John Newlands' 1866 table of the elements.

that similar types of elements seemed to reoccur with every eighth element and, using an analogy to the octave in music, called his generalization the *law of octaves*. Although this law did not work as well with the heavier elements, in 1866 he made some adjustments and presented a table (Figure 3.5) based on his law. The charts of Newlands and de Chancourtois can be seen as attempts at the development of a periodic law. Such periodic ideas were not very well accepted in some circles. Once, at a meeting of the Chemical Society in England, Newlands was asked facetiously if he had ever tried to classify the elements by arranging them in alphabetical order.

The **periodic law** as we know it today was developed by the Russian chemist **Dmitri Ivanovitch Mendeleev** in 1869. The periodic table inside the front cover of this book is a direct result of the work of Mendeleev. The periodic table we are using today, although different in many ways, still bears some resemblance to Mendeleev's revised chart of 1871, shown in Figure 3.6. The chart is in German, as all major scientific work of that era was published in German, French, or English.

Some of the key differences between today's periodic table and Mendeleev's include the fact that today's chart has 18 columns, whereas his had only 8; today's chart has extra elements listed at the bottom, whereas his did not; and today we use atomic numbers to arrange the chart, whereas Mendeleev used atomic weights. The

Reihen	Gruppe I — R^2O	Gruppe II — RO	Gruppe III — R^2O^3	Gruppe IV RH^4 RO^2	Gruppe V RH^3 R^2O^5	Gruppe VI RH^2 RO^3	Gruppe VII RH R^2O^7	Gruppe VIII — RO^4
1	H = 1							
2	Li = 7	Be = 9, 4	B = 11	C = 12	N = 14	O = 16	F = 19	
3	Na = 23	Mg = 24	Al = 27, 3	Si = 28	P = 31	S = 32	Cl = 35, 5	
4	K = 39	Ca = 40	– = 44	Ti = 48	V = 51	Cr = 52	Mn = 55	Fe = 56, Co = 59, Ni = 59, Cu = 63.
5	(Cu = 63)	Zn = 65	– = 68	– = 72	As = 75	Se = 78	Br = 80	
6	Rb = 85	Sr = 87	?Yt = 88	Zr = 90	Nb = 94	Mo = 96	– = 100	Ru = 104, Rh = 104, Pd = 106, Ag = 108.
7	(Ag = 108)	Cd = 112	In = 113	Sn = 118	Sb = 122	Te = 125	J = 127	
8	Cs = 133	Ba = 137	?Di = 138	?Ce = 140	–	–	–	– – – –
9	(–)	–	–	–	–	–	–	
10	–	–	?Er = 178	?La = 180	Ta = 182	W = 184	–	Os = 195, Ir = 197, Pt = 198, Au = 199.
11	(Au = 199)	Hg = 200	Ti = 204	Pb = 207	Bi = 208	–	–	
12	–	–	–	Th = 231	–	U = 240	–	– – – –

*Spaces are left for the unknown elements with atomic masses 44, 68, 72, and 100.

Figure 3.6 Mendeleev's 1871 periodic table.

extra elements that modern charts list at the bottom were not known in 1871, and Mendeleev had to use atomic weights because atomic numbers had not yet been developed. Mendeleev demonstrated the power of the periodic law by accurately predicting the properties of three unknown elements that were later discovered and purified.

Our purpose in presenting the periodic law is not to explain it, but rather to use it. A more advanced chemistry text would endeavor to explain why the chart is laid out as it is and use this information to help the student learn more chemistry. The purpose of this chapter is to use the periodic table to help us understand the underlying principles of chemistry that will be needed in the remainder of the book; however, before we look at how chemicals react we need to explore some other fundamental concepts.

MOLE

In the section on atomic theory, we discussed the concept of atomic weights. The hydrogen atom was given a weight of 1, the carbon atom a weight of 12, and the oxygen atom a weight of 16 for the most abundant isotope of each element. As noted earlier, these are arbitrary weights set up so that the weight of the most common isotope of hydrogen would be very close to 1.* The problem with the atomic mass unit is that it is not very useful for most activities except with sophisticated instrumentation. Very few people have the equipment that handles matter one atom at a time; therefore, the concept of the **mole** was developed for those individuals who handle matter several grams, ounces, or pounds at a time.

The mole is a way of counting elementary particles and is similar to other multiple counting units. Everyone knows what a dozen eggs means. A dozen is 12. A dozen is, was, and always will be 12 of something. A gross is 144 of something; for example, a gross of pencils is 144 pencils. Well, a mole (mol) of carbon atoms is 602,200,000,000,000,000,000,000 atoms of carbon. (For those who are familiar with scientific notation the number is usually written as 6.022×10^{23}.) The only real difference between a dozen and a mole is the incredibly large size of the number of particles in a mole.

To give one a feel for how big the number of particles in a mole is, let us try some analogies. In his article "How to Visualize Avogadro's Number," Henk van Lubeck suggested that if one had a mole of dollars and put them in the bank at 5% annual interest for one second, the amount of interest would be enough to give every man, woman, and child on the planet $200,000. If one had a mole of red ants, they would completely cover 1000 planets the size of the Earth. If one wanted a mole of grains of sand, it would be necessary to skim off the top 6 feet (2 meters) of sand from the entire Sahara Desert.† As can be seen, this is an almost unimaginably large number. Why did we make it so big?

* The exact definition of the atomic mass unit is 1/12th the mass of one atom of the most abundant isotope of carbon.

† Analogies for the mole are from van Lubeck, H., How to visualize Avogadro's number, *J. Chem. Educ.*, 66, 762, 1989.

One of the most common units of mass used by scientists is the gram (g). In English units, this is equivalent to 0.0352 ounces (oz) (1 oz contains 28.4 g). Most researchers, therefore, were most likely to weigh out chemical material in the convenient unit of grams. Furthermore, it is helpful to use the atomic weights that were already developed for atomic-level discussions when weighing ordinary quantities of matter. As a result, 12 g of the most abundant isotope of carbon is defined to be 1 mol. This is the isotope where each carbon atom weighs 12 amu. Thus, a table of atomic weights can tell us the weight of one atom in atomic mass units or the weight of a mole of that element in grams. It turns out that 12 g of the most abundant isotope of carbon contains 6.022×10^{23} atoms. In fact, 1 mol of anything will contain 6.022×10^{23} elementary particles.

This concept can be easily extended to molecules as well as atoms. Consider water as an example. A water molecule (H_2O) contains two hydrogen atoms and one oxygen atom. The average atomic weight of each atom is 1 amu for hydrogen and 16 amu for oxygen. To get the weight of a water molecule, we add up the weights of all of the atoms:

Hydrogen	1 amu
Hydrogen	1 amu
Oxygen	16 amu
Total	18 amu

In a similar way as with elements, the molecular weight of one water molecule is 18 amu, 1 mol of water weighs 18 g, and 1 mol of water contains 6.022×10^{23} molecules.

SYMBOLS, FORMULAS, AND EQUATIONS

Chemists use *chemical symbols* for many of the same reasons that people in other professions use shorthand of various kinds. When discussing the same elements repeatedly, symbols are simply quicker and easier than writing out the name of the elements repeatedly. Some of the more commonly used elemental symbols are given in Table 3.2. The remaining symbols can be found in the table inside the front cover of this book. The symbols are based on the first letter of the Latin name of the element; this explains why potassium has the symbol K, as it is from the Latin word *kalium*. The symbols are generally single uppercase letters; if two or more elements begin with the same letter, then a second lowercase letter is added. Chemical symbols, of course, are used to represent elements, and elements combine to form compounds that are made up of molecules. These molecules can be represented by *chemical formulas*. Some

Table 3.2 Common Chemical Symbols

H	Hydrogen	Na	Sodium
He	Helium	P	Phosphorus
C	Carbon	S	Sulfur
N	Nitrogen	Cl	Chlorine
O	Oxygen	K	Potassium

examples of chemical formulas include H_2O for water and H_2SO_4 for sulfuric acid. The symbols represent one atom of that element unless it is followed by a subscript. The subscript indicates how many of a particular atom is present in the molecule. The formula H_2O indicates that a water molecule is made up of two atoms of hydrogen and one atom of oxygen. The formula H_2SO_4 indicates that a sulfuric acid molecule is made up of two atoms of hydrogen, one atom of sulfur, and four atoms of oxygen.

Chemical formulas can be used to represent chemical change by including them in *chemical equations*. Chemical equations spell out a statement in symbols that can be stated in words; however, stating it in words takes considerably more space. For example, we could say that carbon can react with oxygen to give carbon monoxide. In chemical symbols, this could be shown as

$$C + O_2 \rightarrow CO$$

Oxygen has to be written as O_2 because normal oxygen exists as a molecule containing two oxygen atoms. A careful examination of this equation shows that it violates the law of conservation of matter. The equation starts with one carbon atom and two oxygen atoms and ends up with one carbon atom and one oxygen atom. One just cannot lose an oxygen atom. The loss of one oxygen atom could be corrected by removing the subscript two, but this is not allowed because it would change the description of the oxygen molecule. This equation can be made to conform to the law of conservation of matter by the use of coefficients placed before some of the chemical formulas in the equation:

$$2C + O_2 \rightarrow 2CO$$

The equation can now be read that two atoms of carbon react with one molecule of oxygen to give two molecules of carbon monoxide. It is also perfectly correct to interpret the equation to say that 2 mol of carbon react with 1 mol of oxygen to give 2 mol of carbon monoxide.

CHEMICAL BONDING

What is it that holds molecules together? Why do these atoms stick together? We may not know in an absolute sense, but we have various models that explain it reasonably well. We are going to use a fairly simple model that, although it does not explain everything, will help us understand why certain compounds are formed and others are not. To develop our model we need to look at the periodic table inside the front cover of the book. The periodic table is designed in such a way that similar elements will appear directly below one another in the chart. For example, we would expect lithium, sodium, potassium, rubidium, cesium, and francium in column 1, or group 1 as it is usually called,* to be chemically similar materials, and they are.

* When using other periodic tables one should be aware that there are other systems for numbering the groups, usually involving a Roman numeral and the letter A or B.

Similarly, we would expect the elements in group 17 to be similar to one another in their chemistry. Fluorine, chlorine, bromine, iodine, and astatine undergo similar types of chemical reactions. Only the elements in groups 3 to 12, called the *transition metals*, and those listed at the bottom of the periodic table are similar to one another in the horizontal direction; for example, iron (26), cobalt (27), and nickel (28) show many similarities with one another.

To begin with, we will focus on the first rows or periods of the table. In the first three periods, the atomic numbers increase from 1 to 18. The atomic number, as noted earlier, corresponds to the number of positive charges on the nucleus and also the number of electrons around the nucleus. The bonding properties are related to the number of electrons. It is clear that after we pass the first two elements, hydrogen and helium, every ninth element will be similar. In other words, after the eight elements that are different, the ninth element will be similar to the first; that is, there are eight groups of similar elements.

Bonding occurs not only because of the interaction of electrons between one atom and another but also specifically because of the interaction of the outer electrons. In atoms except hydrogen and helium, we find that the inner electrons are too far from the edge of the atom to interact with other atoms. These outer electrons are usually referred to as **valence electrons**. How many valence electrons are there in any given atom? There generally seems to be eight or fewer for the smaller atoms. To understand this, we need to take a close look at group 18, specifically helium, neon, and argon.

These three gases are included in what are known as the *noble gases*, so named because these gases do not interact with any of the other elements. These gases do not form compounds with any element. Helium, neon, and argon have 2, 10, and 18 electrons, respectively. The model that we are using states that helium has two valence electrons and no inner electrons. Neon has eight valence electrons and two inner electrons. Argon has eight valence electrons and 10 inner electrons. The idea is that there is something very special about having eight valence electrons, and atoms that already have eight valence electrons have no reason to interact further with any other atoms. This brings us to the key number of eight electrons that can be used to understand most of the bonding found in chemical compounds. Atoms that have fewer than eight valence electrons undergo a reaction to enter into an arrangement with other atoms where they will have eight valence electrons; this is known as the **octet rule**. Hydrogen, helium, lithium, and beryllium are exceptions because for these very small atoms the desired number is two valence electrons. The octet rule is based on fundamental principles that have not been discussed in this chapter; however, this rule is a useful model and can be used to make many valid predictions about chemical bonding.

Figure 3.7 indicates how many valence electrons each atom has in this model. Note that the number of valence electrons generally corresponds to the group number or to the group number minus 10 with the exception of helium. From period 4 onward, one will note that there are 18 columns. The number of valence electrons varies in many of these columns; hence, the number of valence electrons in groups 3 to 11 is not discussed. For the remaining columns, the valence electrons can be predicted from the group, but there can be some variations. We will be referring to the

1	*2*	*3*	*4*	*5*	*6*	*7*	*8*	*9*	*10*	*11*	*12*	*13*	*14*	*15*	*16*	*17*	*18*
1 **H** **1**																	2 He **2**
3 **Li** **1**	4 **Be** **2**											5 **B** **3**	6 **C** **4**	7 **N** **5**	8 **O** **6**	9 **F** **7**	10 **Ne** **8**
11 **Na** **1**	12 **Mg** **2**											13 **Al** **3**	14 **Si** **4**	15 **P** **5**	16 **S** **6**	17 **Cl** **7**	18 **Ar** **8**
19 **K** **1**	20 **Ca** **2**	21 **Sc**	22 **Ti**	23 **V**	24 **Cr**	25 **Mn**	26 **Fe**	27 **Co**	28 **Ni**	29 **Cu**	30 **Zn** **2**	31 **Ga** **3**	32 **Ge** **4**	33 **As** **5**	34 **Se** **6**	35 **Br** **7**	36 **Kr** **8**
37 **Rb** **1**	38 **Sr** **2**	39 **Y**	40 **Zr**	41 **Nb**	42 **Mo**	43 **Tc**	44 **Ru**	45 **Rh**	46 **Pd**	47 **Ag**	48 **Cd** **2**	49 **In** **3**	50 **Sn** **4**	51 **Sb** **5**	52 **Te** **6**	53 **I** **7**	54 **Xe** **8**
55 **Cs** **1**	56 **Ba** **2**	71 **Lu**	72 **Hf**	73 **Ta**	74 **W**	75 **Re**	76 **Os**	77 **Ir**	78 **Pt**	79 **Au**	80 **Hg** **2**	81 **Tl** **3**	82 **Pb** **4**	83 **Bi** **5**	84 **Po** **6**	85 **At** **7**	86 **Rn** **8**
87 **Fr** **1**	88 **Ra** **2**																

Figure 3.7 Chart of elements and number of valence electrons.

usual values given in Figure 3.7. This model also does not work for the lower regions of the periodic table. Despite these problems, the model is still useful where numbers are given. How can we use this information to understand bonding?

Ionic Bonding

Ionic bonding is the simplest type of bonding. To explain this type of bonding, sodium chloride (table salt) can be used as an example. The chlorine atom has seven valence electrons. The addition of one electron would give it the eight valence electrons that correspond to the number of valence electrons of the noble gases (see later discussion on chemical bonds). Sodium, on the other hand, has only one valence electron. If sodium loses this electron, then the atom has 10 total electrons like neon. If the sodium becomes like neon, it will have a new set of eight valence electrons. Sodium and chlorine do, in fact, react by transferring one electron from sodium to chlorine. This transfer can be shown symbolically using Lewis dot structures, which were devised in 1916 by the influential American chemist G.N. Lewis. The Lewis dot structures show the valence electrons around an atom using a dot for each electron. The formation of sodium chloride from a sodium atom and a chlorine atom can be shown as follows:

$$\mathrm{Na}\cdot + \cdot\ddot{\underset{\cdot\cdot}{\mathrm{Cl}}}\!: \rightarrow \mathrm{Na}^{+} + :\ddot{\underset{\cdot\cdot}{\mathrm{Cl}}}\!:^{-}$$

It may still be unclear as to how transferring this electron leads to bonding and the formation of sodium chloride. When the sodium atom loses the electron, it does not have enough electrons to balance the positive charge on the nucleus and as a result has a net charge of +1. The chlorine, in contrast, gets an extra electron which is one too many for its nuclear charge and it develops a charge of −1. Because opposite charges attract each other, the sodium and the chorine will be attracted to each other, producing the ionic bond of sodium chloride. The charged particles that are formed in this process are known as **ions**. The names of positive ions are usually the same as the name of the element. The names of negative ions are derived from the name of the element, but the name is modified to end in "-ide."

Many atoms are able to bond in this way. Elements in group 17, such as chlorine, tend to accept one electron to get a total of eight, but elements in group 16 tend to accept two electrons to get eight. In contrast, elements such as sodium in group 1 tend to lose one electron and those in group 2 generally lose two electrons. Although this information can be deduced from Figure 3.7, the number of electrons an atom will lose in most reactions is shown more clearly in Figure 3.8. Figure 3.8 shows the number of electrons an atom generally loses in forming an ionic bond. If it would gain electrons, this is shown as a negative number. These numbers are the same as the charge on the ion that is formed and are known as **oxidation numbers**. Some other features of Figure 3.8 include elements for which our model does not work (i.e., Sb-51 and Po-84), indication of elements that usually do not bond ionically (i.e., B-5, C-6, Si-14, and Ge-32), and also oxidation numbers for some of the more common transition metals (i.e., Cr-24, Fe-26, Ni-28, and Cu-29).

1	*2*	*3*	*4*	*5*	*6*	*7*	*8*	*9*	*10*	*11*	*12*	*13*	*14*	*15*	*16*	*17*	*18*
1 **H**																	2 He **X**
3 **Li** **+1**	4 **Be**											5 **B** **X**	6 **C** **X**	7 **N** **–3**	8 **O** **–2**	9 **F** **–1**	10 **Ne** **X**
11 **Na** **+1**	12 **Mg** **+2**											13 **Al** **+3**	14 **Si** **X**	15 **P** **–3**	16 **S** **–2**	17 **Cl** **–1**	18 **Ar** **X**
19 **K** **+1**	20 **Ca** **+2**	21 **Sc**	22 **Ti**	23 **V**	24 **Cr** **+2, +3**	25 **Mn**	26 **Fe** **+2, +3**	27 **Co**	28 **Ni** **+2, +3**	29 **Cu** **+1, +2**	30 **Zn** **+2**	31 **Ga** **+3**	32 **Ge** **X**	33 **As** **–3**	34 **Se** **–2**	35 **Br** **–1**	36 **Kr** **X**
37 **Rb** **+1**	38 **Sr** **+2**	39 **Y**	40 **Zr**	41 **Nb**	42 **Mo**	43 **Tc**	44 **Ru**	45 **Rh**	46 **Pd**	47 **Ag** **+1**	48 **Cd** **+2**	49 **In** **+3**	50 **Sn** **+2, +4**	51 **Sb** **+3**	52 **Te** **–2**	53 **I** **–1**	54 **Xe** **X**
55 **Cs** **+1**	56 **Ba** **+2**	71 **Lu**	72 **Hf**	73 **Ta**	74 **W**	75 **Re**	76 **Os**	77 **Ir**	78 **Pt** **+2**	79 **Au** **+1, +3**	80 **Hg** **+1, +2**	81 **Tl** **+1, +3**	82 **Pb** **+2, +4**	83 **Bi** **+3**	84 **Po** **+2**	85 **At** **–1**	86 **Rn** **X**
87 **Fr** **+1**	88 **Ra** **+2**																

X = Do not bond ionically.

Figure 3.8 Oxidation numbers in ionic compounds.

This chart can be used to come up with the formulas of simple two-atom (binary) ionic compounds. To illustrate how to write these formulas let us consider a few examples. First, let us consider a compound made up of magnesium and chlorine. From Figure 3.8, it can be seen that magnesium needs to lose two electrons and chlorine needs to gain one:

$$Mg \rightarrow Mg^{2+} + 2e^-$$

$$Cl + e^- \rightarrow Cl^-$$

If magnesium has to lose two electrons, then these electrons have to go somewhere. The electrons are accepted by chlorine, but because each chlorine only accepts one electron, it will take twice as many chlorines as magnesium to make it even. The formula of magnesium chloride is $MgCl_2$.

The next example is sodium sulfide:

$$Na \rightarrow Na^+ + e^-$$

$$S + 2e^- \rightarrow S^{2-}$$

Sulfur needs to accept two electrons and will need two sodium atoms to do so. The formula of sodium sulfide is Na_2S.

The next example is calcium oxide:

$$Ca \rightarrow Ca^{2+} + 2e^-$$

$$O + 2e^- \rightarrow O^{2-}$$

Because one atom accepts the same number of electrons that the other atom releases, the compound would have the same number of calcium ions as the oxide ions. The formula of calcium oxide is CaO.

Aluminum oxide is a more complicated example:

$$Al \rightarrow Al^{3+} + 3e^-$$

$$O + 2e^- \rightarrow O^{2-}$$

To have one atom lose three electrons and the other accept only two, adjustments have to be made in the numbers of both ions. Two aluminums would lose six electrons and three oxygens would accept six electrons:

$$2Al \rightarrow 2Al^{3+} + 6e^-$$

$$3O + 6e^- \rightarrow 3O^{2-}$$

The formula of aluminum oxide is Al_2O_3.

Another example is iron chloride; however, there is some problem in determining its formula. Figure 3.8 indicates that there are two oxidation numbers for iron (Fe-26). To solve the problem one must know the oxidation number of iron in the compound in question. The oxidation number is given in a Roman numeral following the ion in question. Let us examine iron(III) chloride. The "III" indicates an oxidation state of +3:

$$Fe \rightarrow Fe^{3+} + 3e^-$$

$$Cl + e^- \rightarrow Cl^-$$

Hence, the formula of iron(III) chloride is $FeCl_3$.

Covalent Bonding

It is not always possible for atoms to remove electrons from another atom to achieve the octet of valence electrons. Consider the example of the formation of water from hydrogen and oxygen. Referring to Figure 3.7, hydrogen would like to accept one electron to obtain two and oxygen would like to accept two electrons to obtain eight, but neither has the ability to remove electrons from the other. The result is an arrangement whereby the valence electrons are shared. This arrangement is known as **covalent bonding**. If two hydrogen atoms provide two electrons, this will give oxygen a total of eight, and if the oxygen will provide each hydrogen with one electron, this will give each hydrogen a total of two. In other words, by sharing, oxygen can have its eight and each hydrogen can have its two electrons. This is illustrated by Lewis dot structures as follows:

H· ·Ö: → H:Ö:
H· (below O) ; H (below O in product)

Such a system allows atoms of the same element to share electrons to achieve the octet. Chlorine is one such element. Each chlorine atom has seven valence electrons and needs only one more to achieve the octet. If each of the two chlorine atoms provides one electron to the other, then by sharing each can have its octet shown as follows:

:C̤̈l· ·C̤̈l: → :C̤̈l:C̤̈l:

This type of diatomic molecule occurs in several gaseous elements such as hydrogen, nitrogen, fluorine, and oxygen.

A Lewis dot structure can be drawn every time one wishes to come up with the formula of a covalent compound, but in this chapter we will introduce the old concept of **valence** as an alternative approach. Valence can be thought of as the combining power of an atom. The higher the valence the more "other" atoms with which an atom can combine. The valences for elements that commonly bond covalently are shown in Figure 3.9. Most of these numbers could be deduced from Figure 3.7,

1	2	3	4	5	6	7	8	9	10	11	12	13	14	15	16	17	18
1 **H** 1																	2 He
3 **Li**	4 **Be** 2											5 **B** 3	6 **C** 4	7 **N** 3	8 **O** 2	9 **F** 1	10 **Ne**
11 **Na**	12 **Mg**											13 **Al** 3	14 **Si** 4	15 **P** 3, 5	16 **S** 2, 4, 6	17 **Cl** 1	18 **Ar**
19 **K**	20 **Ca**	21 **Sc**	22 **Ti**	23 **V**	24 **Cr**	25 **Mn**	26 **Fe**	27 **Co**	28 **Ni**	29 **Cu**	30 **Zn**	31 **Ga**	32 **Ge** 4	33 **As** 3	34 **Se** 2, 4	35 **Br** 1	36 **Kr**
37 **Rb**	38 **Sr**	39 **Y**	40 **Zr**	41 **Nb**	42 **Mo**	43 **Tc**	44 **Ru**	45 **Rh**	46 **Pd**	47 **Ag**	48 **Cd**	49 **In**	50 **Sn**	51 **Sb** 3	52 **Te** 2, 4	53 **I** 1	54 **Xe**
55 **Cs**	56 **Ba**	71 **Lu**	72 **Hf**	73 **Ta**	74 **W**	75 **Re**	76 **Os**	77 **Ir**	78 **Pt**	79 **Au**	80 **Hg**	81 **Tl**	82 **Pb**	83 **Bi**	84 **Po** 2	85 **At** 1	86 **Rn**
87 **Fr**	88 **Ra**																

Figure 3.9 Valence for covalent bonding atoms.

but not all of them. Also, as we saw with oxidation numbers, many of the elements have more than one valence. In fact, some of these elements have so many valences that only the major ones are shown in Figure 3.9. For example, nitrogen has so many valences or combining powers that the element has at least five different compounds with oxygen alone, including N_2O, NO, N_2O_3, NO_2, and N_2O_5. It should be noted that this valence method can predict the formulas of compounds that should exist but may not reveal all of the possible combinations.

To understand how to use the valence chart to predict the formula of binary (two-element) compounds, let us consider some examples. The first exercise is to find the formula for hydrogen fluoride. Each element has a valence of one; therefore, each atom can form one bond. Thus, the formula must be HF:

$$H- + -F \rightarrow HF$$

A second example is a compound made up of phosphorus and chlorine. Phosphorus has a valence of three or five, whereas chlorine has a valence of only one. Again, as with multiple oxidation numbers, one needs to know in advance which valence to use. In this case, the valence of five has been used. It will take five chlorines to equal the combining power of one phosphorus. The formula is PCl_5:

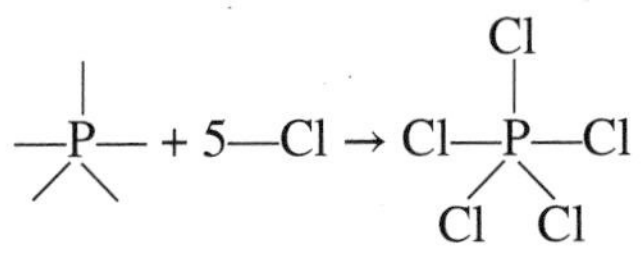

$$—\overset{|}{\underset{/\ \backslash}{P}}— + 5—Cl \rightarrow Cl—\overset{\overset{Cl}{|}}{\underset{\underset{Cl\quad Cl}{/\ \backslash}}{P}}—Cl$$

A very common simple compound of carbon and hydrogen is methane, a major constituent of natural gas. According to Figure 3.9, carbon can bond four times, whereas hydrogen forms only one bond, leading to a formula of CH_4:

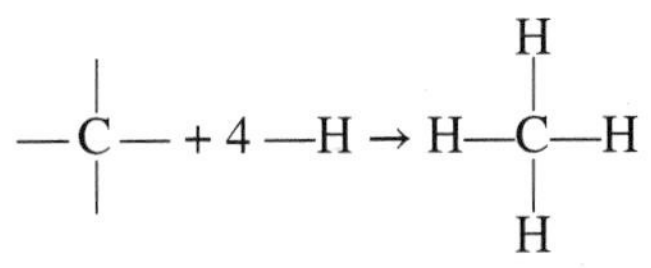

$$—\overset{|}{\underset{|}{C}}— + 4—H \rightarrow H—\overset{\overset{H}{|}}{\underset{\underset{H}{|}}{C}}—H$$

A compound can be formed between oxygen and fluorine. Because the valence of oxygen is two and the valence of fluorine is one, this compound would have the formula OF_2:

$$—O— + 2—F \rightarrow F—O—F$$

Another compound would be a gas made from carbon and oxygen. With a valence of four for carbon and two for oxygen, the compound would have a formula of CO_2:

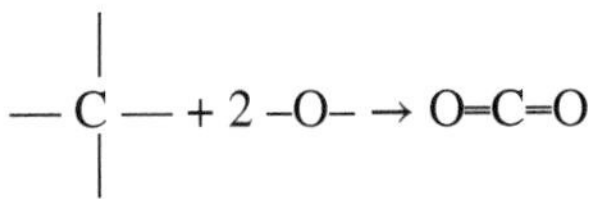

$$—\overset{|}{\underset{|}{C}}— + 2\ -O- \rightarrow O{=}C{=}O$$

Note the double bonds between the carbon and oxygen. The concept of multiple bonds is discussed in Chapter 4 in the section on organic chemistry.

Polyatomic Ions

There are many compounds that contain both ionic and covalent bonds. Consider, for example, a compound made up of sodium, oxygen, and hydrogen. Sodium has a strong tendency to form ionic bonds, and hydrogen has a strong tendency to form covalent bonds. Oxygen, however, can do both and does. The oxygen bonds ionically to the sodium and covalently to hydrogen. The product formed, NaOH, is both an ionic and a covalent compound:

$$Na^+ \; O\text{–}H^-$$

The negatively charged part of the molecule, OH^-, is known as the hydroxide ion and is a polyatomic ion. There are many different polyatomic ions. A list of a few of these ions is given in Table 3.3.

The process of writing formulas for these compounds is very similar to the case for the ionic compounds. In this case, one needs to refer Figures 3.8 and 3.9. Some examples are as follows:

- Potassium phosphate

$$K \rightarrow K^+ + e^-$$

$$3K^+ + PO_4^{3-} \rightarrow K_3PO_4$$

- Ammonium sulfide

$$S + 2e^- \rightarrow S^{2-}$$

$$2NH_4^+ + S^{2-} \rightarrow (NH_4)_2S$$

- Calcium carbonate

$$Ca \rightarrow Ca^{2+} + 2e^-$$

$$Ca^{2+} + CO_3^{2-} \rightarrow CaCO_3$$

Table 3.3 Common Polyatomic Ions

OH^-	Hydroxide	NO_3^-	Nitrate	NH_4^+	Ammonium
SO_4^{2-}	Sulfate	NO_2^-	Nitrite	CO_3^{2-}	Carbonate
SO_3^{2-}	Sulfite	PO_4^{3-}	Phosphate	HCO_3^-	Bicarbonate

DISCUSSION QUESTIONS

1. Outline the four points of Dalton's atomic theory. What was the importance of Dalton's theory in the evolution of chemical thought?
2. What part of Dalton's atomic theory was invalidated by the discovery of radioactivity? Why was the observation of radioactive emissions as particles important in this regard?
3. What is the role of chance observations in the advancement of science? Was A.H. Becquerel's discovery of radioactivity just by chance or was there more to it than that?
4. State the results of Rutherford's gold foil experiment in general terms and explain how the experiment outcome shows the general structure of the atom.
5. Fill in the missing information.
   a. Atomic wt. 1, atomic no. 1, no. of protons ______, no. of neutrons ______, and no. of electrons ______
   b. Atomic wt. ______, atomic no. ______, no. of protons 6, no. of neutrons 6, and no. of electrons ______
   c. Atomic wt. 56, atomic no. ______, no. of protons ______, no. of neutrons 30, and no. of electrons ______
   d. Atomic wt. 238, atomic no. ______, no. of protons 92, no. of neutrons ______, and no. of electrons ______
   e. Atomic wt. 32, atomic no. ______, no. of protons ______, no. of neutrons ______, and no. of electrons 16
6. Because the atomic mass unit is defined so that the atomic weight of an atom would be very close to a whole number, then explain how chlorine could have an atomic weight of 35.5.
7. What part of Dalton's atomic theory was invalidated by the discovery of isotopes and why?
8. Explain how Mendeleev was able to predict the existence of elements that had not been discovered when he developed his periodic table. Why was Mendeleev so surprised by the discovery of the elements helium, neon, argon, krypton, and xenon, none of which was known when he developed the table?
9. Answer the following questions:
   a. How many pencils are there in a dozen?
   b. How many pencils are there in a mole of pencils?
   c. If each pencil weighed 2 billion billion amu, how much would a mole of pencils weigh in grams?
   d. If you had a dozen dogs, then how many dog legs would there be?
   e. If you had a mole of dogs, then how many dog legs would there be?
   f. If you had a mole of water molecules (H_2O), then how many hydrogen atoms would there be?
10. For each of the following equations point out the coefficients and the subscripts:
    a. $C + O_2 \rightarrow CO_2$
    b. $2C + O_2 \rightarrow 2CO$
    c. $2H_2 + O_2 \rightarrow 2H_2O$
    d. $3H_2 + N_2 \rightarrow 2NH_3$
    e. $4C + S_8 \rightarrow 4CS_2$

11. Give the formula for the compound made from each of the following pairs of elements:
    a. Ionic
        i. Potassium and chlorine
        ii. Silver and sulfur
        iii. Iron(II) and iodine
        iv. Aluminum and sulfur
        v. Copper(I) and nitrogen
        vi. Sodium and bromine
        vii. Copper(II) and nitrogen
        viii. Calcium and chlorine
        ix. Magnesium and iodine
    b. Covalent
        i. Phosphorus(V) and iodine
        ii. Nitrogen and hydrogen
        iii. Silicon and chlorine
        iv. Iodine and chlorine
        v. Arsenic and fluorine
        vi. Boron and bromine
        vii. Hydrogen and sulfur(II)
        viii. Carbon and sulfur(II)
12. Give the formula for the compound made from the following elements and polyatomic ions:
    a. Sodium and sulfate
    b. Magnesium and bicarbonate
    c. Ammonium and chlorine
    d. Calcium and phosphate
    e. Iron(II) and nitrate

BIBLIOGRAPHY

Chang, R., *Chemistry*, 4th ed., McGraw-Hill, New York, 1991, p. 307.
Hill, J.W., *Chemistry for Changing Times*, 6th ed., Macmillan, New York, 1992, p. 63.
Hudson, J., *The History of Chemistry*, Chapman & Hall, New York, 1992.

CHAPTER 4

Types of Chemical Compounds and Their Reactions

The world of chemistry contains millions of compounds and a huge number of different types of chemical reactions. In order to keep track of all of this information it is necessary to categorize the material in some way. In this chapter, we are going to break down the field of chemistry into roughly five areas. The first three areas are based on types of chemical reactions that account for a vast majority of all chemical reactions. We will then look at organic chemistry, which is based on a particular type of chemical compound, and, finally, we will look at certain nuclear processes.

ACIDS AND BASES

Acids

Although there are a large number of chemical reactions that could be discussed, only three general classes of reactions—acid–base, precipitation, and oxidation–reduction—are discussed in this chapter. These three classes of reactions explain the chemical behavior of a large variety of substances.

There are three major theories of acids and bases. This chapter focuses only on one of them, the simplest and the oldest. This theory was developed by a Swedish chemist, **Svante Arrhenius**, in the latter part of the 19th century. Let us first consider Arrhenius' definition of acids. Arrhenius said that acids are substances that, when dissolved in water, produce hydrogen ions (H^+). This definition works fairly well with one minor modification. The 20th-century chemists fairly quickly came to the conclusion that H^+ ions do not exist as such, but rather H^+ attaches itself to a water molecule to form the hydronium ion (H_3O^+). An **acid** can be defined as a substance that, when dissolved in water, produces hydronium ions (H_3O^+). The properties associated with acids should be the properties of the H_3O^+ ion. What is the behavior that the H_3O^+ ion exhibits?

Characteristic properties of acids, and therefore of the H_3O^+ ion, include the following:

1. Sour taste due to the H_3O^+ ion
2. Ability to affect the color of acid–base indicators (specifically, acids will change blue litmus paper to red)
3. Ability to dissolve certain metals (e.g., iron, zinc, tin) with the evolution of hydrogen gas
4. Ability to undergo reaction with bases

The sour taste is a property of acids that is familiar to nearly all of us. Anything that tastes sour contains an acid. Vinegar contains acetic acid; orange juice, grapefruit juice, and lemon juice contain citric acid; and soft drinks contain phosphoric and carbonic acids. Acids are also found in many places; for example, the gastric juice in our stomach is a dilute solution of hydrochloric acid. Many common toilet bowl cleaners are also a solution of hydrochloric acid.

All of these properties are associated with the H_3O^+ ion, but few acids contain this ion when the acid is in the pure form. This ion is produced when the acid is dissolved in water, as suggested in the definition given earlier. For example, when hydrogen chloride gas is dissolved in water, it reacts with the water to produce a hydrochloric acid solution:

$$HCl + H_2O \rightarrow H_3O^+ + Cl^-$$

Many substances can produce H_3O^+ ions and are, therefore, acids. Some of the most common acids are sulfuric acid (H_2SO_4), nitric acid (HNO_3), hydrochloric acid (HCl), phosphoric acid (H_3PO_4), and acetic acid ($HC_2H_3O_2$). Of these, sulfuric, nitric, and hydrochloric acids are really strong and are widely used.

The reaction of acids with bases is discussed in the later discussion on neutralization, but the reaction of acids with metals is outlined here. Two hydrogen atoms, one each from the two H_3O^+ ions, can be released to form hydrogen gas in the presence of certain metals. In this process, the metal will dissolve, releasing metal ions into the solution. The following is an example for zinc:

$$2H_3O^+ + Zn \rightarrow Zn^{2+} + H_2\uparrow + 2H_2O$$

To summarize, an acid is a substance that, when dissolved in water, produces hydronium ions (H_3O^+).

Bases

The other half of the acid–base discussion is, of course, about bases. Base is also sometimes referred to as **alkali**. A **base** can be defined as a substance that, when dissolved in water, produces hydroxide ions (OH^-). The properties of the bases are associated with the OH^- ions and therefore, include the following:

1. Bitter taste due to the OH^- ion
2. Ability to affect the color of acid–base indicators (specifically, bases will change red litmus paper to blue)
3. Slippery or soapy feel on the skin as a result of the ability of OH^- to dissolve the surface layers of the skin
4. Ability to undergo reaction with acids

Many household products contain bases. One example is from the world of medicine. As noted earlier, gastric juice contains hydrochloric acid, the overproduction of which can lead to what is known as heartburn or acid indigestion. Antacids that combat this disorder are slurries of insoluble bases. Insoluble bases are used because they do not react with body tissue but will react with gastric acid. Milk of magnesia, for example, is a slurry of magnesium hydroxide. Other basic household products containing bases include washing soda, lye, ammonia-type household cleaners, oven cleaners, and drain cleaners. Lye, oven cleaner, and drain cleaner are particularly caustic and should be used with extreme caution.

The properties of bases are associated with the OH^- ion. Common bases differ from most common acids in that most acids produce the H_3O^+ ion in solution, whereas most bases already contain the OH^- ion as a solid. One of the most common bases is sodium hydroxide, which is made up of a sodium ion and a hydroxide ion. The hydroxide ion is released when the solid sodium hydroxide is dissolved in water:

$$Na^+OH^- \text{ (solid)} \rightarrow Na^+ \text{ (solution)} + OH^- \text{ (solution)}$$

The common name of sodium hydroxide is lye. The other really strong, common base is potassium hydroxide (KOH).

Ammonia is the one commonly used base that does not contain OH^- ions, but rather produces them by reaction with water in solution. Pure ammonia is a gas that is very soluble in water. When the ammonia dissolves, a chemical reaction occurs that produces OH^- ions:

$$NH_3 + H_2O \rightarrow NH_4^+ + OH^-$$

To summarize, a base is a substance that, when dissolved in water, produces hydroxide ions (OH^-).

Acid and Base Strength

Hydrochloric acid is considered a strong acid, whereas formic acid, which is found in bee stings, is considered a weak acid. How do these two acids differ from each other? They differ in the number of H_3O^+ ions a given number of acid molecules will produce. Consider the following chemical equations:

$$HCl \text{ (10,000)} + H_2O \rightarrow H_3O^+ \text{ (10,000)} + Cl^-$$

$$HCHO_2 \text{ (10,000)} + H_2O \rightarrow H_3O^+ \text{ (500)} + CHO_2^-$$

If one were to start with 10,000 molecules of HCl and place them in water, every one of them would react with the water and produce 10,000 H_3O^+ ions; however, the weak acids do not react completely. If one were to start with 10,000 molecules of formic acid ($HCHO_2$), only 500 of them would react with the water to produce H_3O^+ ions. The other 9500 $HCHO_2$ molecules would remain as they are. These acids are said to be weak because they have only a weak ability to produce H_3O^+ ions.

The difference between strong and weak bases works very much the same way as for the acids. If one takes 10,000 NaOH units and dissolves them in water, then 10,000 OH^- ions will be released into the solution. However, the 10,000 ammonia (NH_3) molecules will produce only 100 OH^- ions. The other 9900 NH_3 molecules would remain as they are, as follows:

$$Na^+ OH^- (10,000) + H_2O \rightarrow Na^+ + OH^- (10,000) + H_2O$$

$$NH_3 (10,000) + H_2O \rightarrow NH_4^+ + OH^- (100)$$

Neutralization

One of the properties of acids and bases is that they react with each other. If the amounts of the acid and the base are measured carefully, the resulting solution after reaction will be neither acidic nor basic; that is, it will be neutral. Such a reaction is said to be a **neutralization reaction**. The examination of the products of a neutralization reaction usually will lead to the observation that a salt is produced. A **salt** is an ionic compound where the negative ion is something other than the hydroxide ion (OH^-). Neutralization is usually explained by considering the reaction in terms of the acid and base involved. For example,

$$NaOH + HCl \rightarrow H_2O + NaCl$$

$$2KOH + H_2SO_4 \rightarrow 2H_2O + K_2SO_4$$

In these two cases, the salts are NaCl and K_2SO_4. Both the reactions produce water. If these two reactions are analyzed very carefully, one will find that the essentials of the two reactions are the same. Let us analyze the reaction between NaOH and HCl first by considering what happens when they are put in water, and then see how the water and the salt are produced.

$$NaOH \text{ (with } H_2O) \rightarrow Na^+ + OH^-$$

$$HCl + H_2O \rightarrow H_3O^+ + Cl^-$$

$$H_3O^+ + OH^- \rightarrow 2H_2O$$

$$Na^+ + Cl^- \rightarrow NaCl$$

Most neutralization reactions when analyzed in this manner include the reaction shown above between H_3O^+ and OH^-. This reaction represents the essence of the neutralization reaction. The last reaction that produces the salt does not really occur unless the water is evaporated.

pH Scale

We now know that H_3O^+ ions are associated with acids and OH^- ions are associated with bases; however, the variety of concentrations of these ions that one finds in solution is really incredible. The concentrations range from the very high to the very low. To discuss this phenomenon, it is necessary to describe how chemists measure the concentration of solutions.

The concentration unit that is important for this discussion is moles per liter (mol/L), referred to as **molarity** (*M*). A 1-*M* solution would be one that has 1 mol of material dissolved in 1 L of solution. It would also be 1 *M* if 1/2 mol were dissolved in 1/2 L of solution; however, if 1/2 mol were dissolved in 1 L, then the solution would be 0.5 *M*.

The tremendous variation in concentrations of H_3O^+ ions that can occur in solution is shown in Table 4.1. The first column shows a wide range of common concentrations written in the normal decimal notation. It is worth noting that writing some of these concentrations is very cumbersome. This situation can be improved considerably by writing the concentrations in scientific notation, where the negative exponent corresponds to the number of zeros to the right of the decimal point plus one. This notation can be shortened even more. One can see that the only change down the chart is the exponent, and it is always negative. The pH scale is based on these numbers and is shown in the last column.

Table 4.1 Variation in the Concentrations of H_3O^+ Ions

Decimal Notation (*M*)	Scientific Notation	pH
1	1×10^{0}	0
0.1	1×10^{-1}	1
0.01	1×10^{-2}	2
0.001	1×10^{-3}	3
0.0001	1×10^{-4}	4
0.00001	1×10^{-5}	5
0.000001	1×10^{-6}	6
0.0000001	1×10^{-7}	7
0.00000001	1×10^{-8}	8
0.000000001	1×10^{-9}	9
0.0000000001	1×10^{-10}	10
0.00000000001	1×10^{-11}	11
0.000000000001	1×10^{-12}	12
0.0000000000001	1×10^{-13}	13
0.00000000000001	1×10^{-14}	14

Table 4.2 Corresponding Concentrations for H_3O^+ and OH^- Ions

H_3O^+ Concentration	OH^- Concentration	pH
1×10^{0}	1×10^{-14}	0
1×10^{-1}	1×10^{-13}	1
1×10^{-2}	1×10^{-12}	2
1×10^{-3}	1×10^{-11}	3
1×10^{-4}	1×10^{-10}	4
1×10^{-5}	1×10^{-9}	5
1×10^{-6}	1×10^{-8}	6
1×10^{-7}	1×10^{-7}	7
1×10^{-8}	1×10^{-6}	8
1×10^{-9}	1×10^{-5}	9
1×10^{-10}	1×10^{-4}	10
1×10^{-11}	1×10^{-3}	11
1×10^{-12}	1×10^{-2}	12
1×10^{-13}	1×10^{-1}	13
1×10^{-14}	1×10^{0}	14

Two things need to be noted. First, the scale is inverse; that is, the pH increases as the H_3O^+ ion concentration decreases. Second, the scale is not linear. When the pH changes by one unit, the H_3O^+ ion concentration changes by a factor of 10. Despite some of these problems, this scale is very useful. The scale usually uses numbers from 0 to 14 to express an incredible range of concentrations.

It turns out that in water solutions there is a relationship between the H_3O^+ and OH^- ion concentrations. If the H_3O^+ ion concentration is known, the OH^- ion concentration can be calculated and *vice versa*. Table 4.2 shows several sets of corresponding concentrations for H_3O^+ and OH^- ions. On close examination it can be noted that the very low concentrations of H_3O^+ ion correspond to a basic solution because there is more OH^- ion present than H_3O^+ ion. These factors lead to the interpretation of the pH scale. When pH values are less than 7, the solution is acidic; when they are greater than 7, the solution is basic or alkaline. A pH of exactly 7 corresponds to neutrality, as the H_3O^+ concentration equals the OH^- concentration.

To get a better sense about pH values, let us consider the pH values of some common solutions. The pH of gastric juice is roughly 1.6 to 1.8, lemon juice is about 2.1, vinegar is about 2.5, and soft drinks range from 2 to 4. At the other end of the scale are fresh egg whites at pH 7.6 to 8.0, milk of magnesia at 10.5, and bleach at 12.2.

PRECIPITATION REACTIONS

We noted earlier that ionic bonding occurs because in the electron transfer process positive and negative ions are produced, and these oppositely charged ions attract each other. As will be noted in Chapter 6, water is a *polar molecule*, meaning that one end of the molecule has a relatively positive charge when compared to the relatively negative charge on the other end. Water is good at separating positive ions from negative ions by inserting itself between the oppositely charge ions. The net result is that some ionic compounds dissolve in water. For dissolved compounds, the ions float around in the water independent of each other. In some cases, however, the ions of an ionic compound attract each other too strongly and cannot be pulled apart by the water molecules. These compounds are water insoluble. For example, $NaNO_3$ is water soluble; therefore, when placed in water, the Na^+ ions move around separately from the NO_3^- ions. In contrast, silver chloride, AgCl, is not water soluble; therefore, when placed in water, the material remains separate from the water.

It turns out that both NaCl (table salt) and $AgNO_3$ are water soluble. Let us consider what happens if we mix a solution of NaCl with a solution of $AgNO_3$. Because both salts are water soluble, the ions would all be floating around independently. Once the solutions are mixed, the new solution will contain the following ions: Na^+, Cl^-, Ag^+, and NO_3^-. However, the Ag^+ ion and Cl^- ion cannot be together in the solution because AgCl is not water soluble. The AgCl will immediately settle out as a solid, usually referred to as a *precipitate*. The reaction can be shown in the following manner, where (aq) represents a water solution and (s) an undissolved solid:

$$NaCl\ (aq) + AgNO_3\ (aq) \rightarrow AgCl\ (s) + NaNO_3\ (aq)$$

Because nothing really happens to the Na^+ and NO_3^- ions, they do not need to be included in the preceding equation. The equation can be simplified to what is known as the *net ionic equation* by not including the extraneous ions:

$$Cl^- + Ag^+ \rightarrow AgCl$$

To be able to predict precipitation reactions, one needs to have some basic information about the solubility of ionic compounds in water. Some rules for compounds of general environmental interest are as follows:

1. The compounds of Na^+, K^+, and NH_4^+ are all water soluble.
2. The compounds of NO_3^- are all water soluble.
3. Most compounds of Cl^- are water soluble; exceptions include AgCl, mercury(I) chloride (Hg_2Cl_2), and lead chloride ($PbCl_2$). Lead chloride is slightly soluble.
4. Most sulfates are water soluble; exceptions include $PbSO_4$, Hg_2SO_4, Ag_2SO_4, and $CaSO_4$. Calcium sulfate is slightly soluble.
5. Most hydroxides are water insoluble; exceptions include NaOH, KOH, and $Ca(OH)_2$. Calcium hydroxide is slightly soluble.
6. Most carbonates and phosphates are water insoluble; exceptions include compounds of Na^+, K^+, and NH_4^+.

To illustrate the use of these rules to predict precipitation reactions, consider mixing a solution of Na_3PO_4 with a solution of $Ca(NO_3)_2$. The following ions would be present in the solution: Na^+, PO_4^{3-}, Ca^{2+}, and NO_3^-. Rule 6 would suggest that Ca^{2+} and PO_4^{3-} are an insoluble pair. Therefore, this mixture would precipitate as

$Ca_3(PO_4)_2$. In contrast, mixing the solutions of NH_4Cl and $NaNO_3$ would not produce a precipitation reaction because there are no insoluble pairs among the ions in these solutions. Generally, any solution produced in which a significant concentration of an insoluble set of ions is present will give rise to a precipitation reaction.

OXIDATION–REDUCTION

The third very large class of reactions is oxidation and reduction. Starting from a very simple idea, this topic can develop into a very complex one. In this chapter, we have tried to keep the concept as simple as possible. For the purpose of simplicity, we have used a definition in each case that is as simple and easy to use as possible. Three separate definitions are provided. Each one is useful in certain situations, whereas another definition is better in other situations.

First Definition

In the 18th century, as mentioned in Chapter 2, one of the phenomena of the natural world that scholars were trying to explain was combustion. It appeared that something was lost when most things were burned; for example, one often starts with a fireplace full of wood and ends up with a smaller amount of ashes. This conclusion seemed to be so obvious that nearly everyone accepted the idea. It was not until the discovery of oxygen independently by Priestly and Scheele in the early 1770s that the theory of combustion began to change drastically. Scientists came to the conclusion that burning was the combination of the material with oxygen; hence, the term **oxidation**. The first definition of oxidation is related to this new understanding of the combination of materials with oxygen. Oxidation is defined as the chemical combination of a substance with oxygen. Some examples include the combustion of charcoal (carbon), sulfur, and natural gas (methane, CH_4):

$$C + O_2 \rightarrow CO_2$$

$$S + O_2 \rightarrow SO_2$$

$$CH_4 + 2O_2 \rightarrow CO_2 + 2H_2O$$

Another example of the addition of oxygen to a substance would be the so-called burning of food in our bodies. This can be shown with the addition of oxygen to glucose:

$$C_6H_{12}O_6 + 6O_2 \rightarrow 6CO_2 + 6H_2O$$

The corrosion of metals, such as the rusting of iron, is another example of oxidation:

$$4Fe + 3O_2 \rightarrow 2Fe_2O_3$$

The examples shown so far all show the direct addition of oxygen from the air; however, oxygen can also be added by being transferred from another substance. Copper oxide can be converted to copper metal by heating it in the presence of hydrogen gas. Hydrogen is oxidized by the oxygen released by copper:

$$CuO + H_2 \rightarrow Cu + H_2O$$

If we focus on copper in this reaction, it can be noted that the oxygen has been removed from copper. These types of reactions have been carried out since the beginning of the Bronze Age, although the ancients did not know the chemistry behind the process. When a metal ore (often an oxide) was converted into the pure metal, it was referred to as being *reduced*. This observation gives rise to the other half of our first definition. **Reduction** is the chemical removal of oxygen from a substance—for example, the reaction of tin dioxide with carbon:

$$SnO_2 + C \rightarrow Sn + CO_2$$

In this reaction, we would say that the tin has been reduced and the carbon has been oxidized. Reduction can also involve the simple release of elemental oxygen gas, as in the heating of mercuric oxide:

$$2HgO + Heat \rightarrow 2Hg + O_2$$

In this reaction, we have the simple reduction of mercury by driving off oxygen through heating.

Second Definition

Over the years, many chemists began to notice that when oxygen was added hydrogen was often lost. This process was particularly true for compounds of carbon. This observation can be illustrated by the combustion of methane:

$$CH_4 + 2O_2 \rightarrow CO_2 + 2H_2O$$

The carbon is oxidized in as much as oxygen is added to it, but the carbon also loses hydrogen. This leads to a second set of definitions that are particularly useful in organic chemistry. **Oxidation** is the chemical removal of hydrogen from a substance. **Reduction** is the chemical combination of a substance with hydrogen. Such a set of definitions means that oxidation–reduction reactions can occur where oxygen is neither added to nor subtracted from the compound in question.

- Reduction of ethylene (C_2H_4)

$$C_2H_4 + H_2 \rightarrow C_2H_6$$

- Oxidation of isopropyl alcohol (rubbing alcohol, C_3H_8O)

$$3C_3H_8O + Cr_2O_7^{2-} + 8H_3O^+ \rightarrow 2Cr^{3+} + 3C_3H_6O + 15H_2O$$

Third Definition

The third definition is the most general and the most useful for chemists, but it is also the most abstract. One can learn more about this definition by noting the similarities between the two following reactions:

$$2Mg + O_2 \rightarrow 2Mg^{2+}O^{2-}$$

$$Mg + Cl_2 \rightarrow Mg^{2+}(Cl^-)_2$$

(The charges were included in the products to illustrate the fact that these products are ionic.) From a chemist's point of view, the same thing is happening to the magnesium in the two cases. The magnesium atom is losing two electrons to become a magnesium ion. Because the first case is clearly an oxidation of magnesium, the second is defined to be an oxidation of magnesium as well, although the reaction involves neither oxygen nor hydrogen. **Oxidation** is defined as the loss of electrons from a chemical species.

If oxidation involves the loss of electrons, then the electrons must go somewhere; therefore, it may be useful to define the chemical species accepting these electrons as being reduced. For example, both gold trioxide (Au_2O_3) and gold trichloride ($AuCl_3$) can be decomposed by heating:

$$2(Au^{3+})_2(O^{2-})_3 + \text{heat} \rightarrow 4Au + 3O_2$$

$$2Au^{3+}(Cl^-)_3 + \text{heat} \rightarrow 2Au + 3Cl_2$$

The gold ions are experiencing the same chemical change in both cases, which is the gain of electrons. Because the first reaction is clearly reduction (i.e., the gold species loses oxygen), it seems reasonable to call the second reaction reduction as well. Therefore, we define **reduction** as the gain of electrons by a chemical species.

Using this third definition, oxidation and reduction always occur together because the released electrons during oxidation have to go somewhere, which causes the species that accepts them to be reduced. In the magnesium reactions, for example, in the first reaction oxygen gains two electrons to become O^{2-} and is therefore reduced, whereas magnesium loses two electrons to become Mg^{2+} and is therefore oxidized. In the second reaction, the magnesium is once again oxidized, but here chlorine gains one electron to give Cl^- and is thereby reduced.

If one can cause the electrons to flow through a wire to go from one species to another, electricity can be produced. Such an arrangement is the basis of the chemical battery. One example would be the lead storage battery found in cars. These batteries produce electrons at one place and consume them at another:

$$Pb + HSO_4^- \rightarrow PbSO_4 + H^+ + 2e^-$$

$$PbO_2 + HSO_4^- + 3H^+ + 2e^- \rightarrow PbSO_4 + 2H_2O$$

One plate is made of lead and the other of lead impregnated with lead dioxide. Because lead and the lead oxide are on different plates, the electrons must flow through a wire, motor, light, etc., to go from one plate to another.

Oxidation–Reduction Summary

- Oxidation is the chemical combination of a substance with oxygen.
- Oxidation is the chemical removal of hydrogen from a substance.
- Oxidation is defined as the loss of electrons from a chemical species.
- Reduction is the chemical removal of oxygen from a substance.
- Reduction is the chemical combination of a substance with hydrogen.
- Reduction is defined as the gain of electrons by a chemical species.

ORGANIC CHEMISTRY

An entire field of chemistry is devoted to the study of one element, carbon. This field, known as **organic chemistry**, involves a huge number of compounds and is related to everything from living organisms to medicines to plastics to paper to paint, and so on. But, how can one element be responsible for so many compounds? Carbon has the unique ability to bond to itself to form chains of indefinite length. In addition, given that carbon has a valence of four, it not only can make chains but also can produce branches to the chain, as well:

```
     H   H   H   H   H   H   H   H
     |   |   |   |   |   |   |   |
 H — C — C — C — C — C — C — C — C — H
     |   |   |   |   |   |   |   |
     H   H   H   H   |   H   H   H
                 H — C — H
                     |
                 H — C — H
                     |
                     H
```

As one might be able to imagine, an incredible number of compounds can be made from only carbon and hydrogen. For example, using only 20 carbon and 42 hydrogen atoms, one can theoretically come up with 366,319 different compounds.

Formulas and Structural Drawings in Organic Chemistry

The preceding drawing is an example of a *structural formula*. This drawing shows exactly which atoms are bonded to each other and the exact order of this bonding. Such drawings are very clear in describing the structure of the molecule at hand; however, they are very laborious to draw. As a result, organic chemists over the years have come up with simplified ways of drawing organic compounds. One way involves the grouping of carbon and hydrogen atoms that are attached to each other to give rise to what are known as *condensed structural formulas*. The compound whose structural formula is given earlier would appear as follows when drawn as a condensed structural formula:

$$\begin{array}{l} CH_3CH_2CH_2CH_2\underset{\displaystyle |\atop \displaystyle CH_2CH_3}{C}HCH_2CH_2CH_3 \end{array}$$

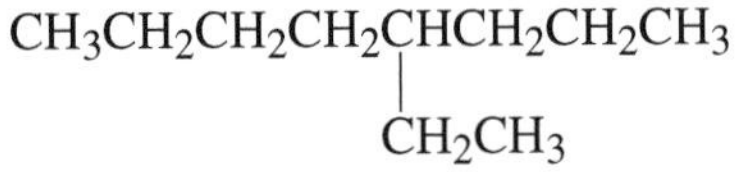

To take the simplification process one step further, chemists have developed what are known as *line-bond formulas*. In these drawings, many of the carbon atoms are shown simply as bends in a line (or in some cases as the end of a line), and many of the hydrogen atoms are omitted. The formula that has been considered would appear as follows in a line-bond drawing:

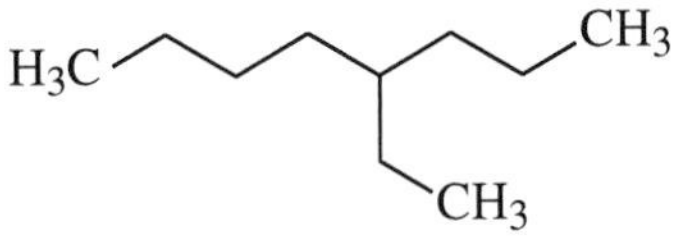

H_3C CH_3 CH_3

Hydrocarbons

The compound that has been considered is an example of a **hydrocarbon**, a compound containing only carbon and hydrogen. This compound is also an example of a group of compounds known as alkanes. **Alkanes** are organic compounds containing only carbon, hydrogen, and single bonds. A single bond is the typical covalent bond created by sharing one pair of electrons between two atoms. The simplest example of an alkane is methane (CH_4). Methane is the principal component of natural gas and is one of the final products of the breakdown of biological material in the absence of oxygen. If we consider a two-carbon molecule, we have ethane (CH_3CH_3), a gas that can be obtained from natural gas and petroleum. The organic alkane with three carbons is propane ($CH_3CH_2CH_3$). Propane is commonly used as a compressed fuel in tanks in situations where natural gas is not available. Propane can also be obtained from natural gas and petroleum.

An interesting situation arises when an alkane with four carbon atoms is considered. One can write two structural formulas for an alkane molecule with four carbons:

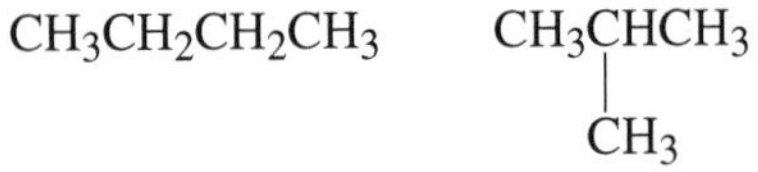

$$CH_3CH_2CH_2CH_3 \qquad CH_3\underset{\displaystyle |\atop \displaystyle CH_3}{C}HCH_3$$

(1) (2)

Because we can write two different structural formulas, these are two different compounds. These compounds are said to be **isomers**. Isomers are compounds with the same number of atoms of each element in the molecule but with different structures. Compound 1 is *n*-butane and compound 2 is isobutane.

Hydrocarbons can also contain double bonds, where some atoms share two pairs of electrons. Hydrocarbons that contain at least one double bond are known as **alkenes**. The simplest alkene is ethylene ($H_2C{=}CH_2$). Ethylene is one of the major organic chemicals used in the world today. Many things can be made from it, including polyethylene milk jugs. It is obtained from petroleum. There are many alkenes; for example, propylene ($H_2C{=}CHCH_3$) is an alkene with three carbons.

By sharing three pairs of electrons between atoms, one can form a triple bond. Hydrocarbons with at least one triple bond are known as **alkynes**. The simplest of the alkynes is acetylene, $HC{\equiv}CH$, which is a common fuel in welding torches. One possibility for hydrocarbons that we have not mentioned up to this point is to attach the two ends of carbon chain to each other, thereby making a ring. This arrangement gives rise to whole new classes of compounds such as *cycloalkanes* and *cycloalkenes*. (There are also cycloalkynes, but they are fairly uncommon.) Some examples are shown as follows:

CH_3

cyclohexane methylcyclopentane cyclohexene

Ring compounds are nearly always drawn using line-bond formulas.

The last type of hydrocarbons would be the **aromatic compounds**. It turns out that rings containing three double bonds and six carbon atoms have unusual chemical properties. These rings have usual stability and undergo a unique set of reactions. The simplest of these aromatic compounds is benzene:

benzene

Other aromatic compounds are created by adding various groups to the benzene ring. Another approach to making more aromatic compounds is by adding more benzene-type rings. One could also do both. Some examples are as follows:

CH_3 CH_2

toluene naphthylene styrene

Toluene is a common solvent associated with glue and paints, and it has many other uses. Naphthylene is found in mothballs, and styrene is used to make the common plastic polystyrene that is found in Styrofoam™.

Organic Chemistry Including Other Elements

A very large segment of organic chemistry includes compounds that contain elements other than carbon and hydrogen. The most common other elements found in organic compounds are oxygen, nitrogen, sulfur, phosphorus, fluorine, chlorine, bromine, and iodine. The inclusion of any of these other elements produces compounds with new sets of properties. These other compounds are found in nearly every area where organic chemicals are found. Some examples are given in Table 4.3.

NUCLEAR CHEMISTRY

Fundamentals of Nuclear Processes

To discuss nuclear chemistry, we need to focus on the nucleus of the atom. Recall that the nucleus of an atom is made up of protons and neutrons that have an extremely small volume. Each proton and neutron has roughly the same mass, 1 amu. The

Table 4.3 Some Examples of Organic Compounds Containing Oxygen, Nitrogen, Fluorine, and Chlorine

Structure	Name	Comments
CH_3CH_2OH	Ethyl alcohol	Solvent, used in intoxicating beverages
$CH_3C(=O)OH$	Acetic acid	Found in vinegar
$CF_2ClCFCl_2$	Trifluorotrichoroethane	One of the CFCs, now banned
$HC(=O)H$	Formaldehyde	Tissue preservative
$CH_3C(=O)CH_3$	Acetone	Solvent, found in fingernail polish remover
$CH_3CH_2CH_2C(=O)OCH_2CH_3$	Ethyl butyrate	Pineapple scent
$C_6H_5CH_2CH(CH_3)NH_2$ (benzene ring–CH_2–C with H_3C, NH_2, H)	Amphetamine	Drug of abuse, stimulant
$CH_2(OH)CH_2(OH)$	Ethylene glycol	Antifreeze

number of protons corresponds to the atomic number of the element, whereas the sum of the number of protons plus the number of neutrons corresponds to the atomic mass of the isotope. The standard symbol ${}^{m}_{n}X$ is used to indicate the isotope under consideration. The n is the atomic number, which is equal to the number of protons, and m is the atomic mass, which is equal to the number of protons plus the number of neutrons ($m - n$ = the number of neutrons). Some examples are ${}^{12}_{6}C$, ${}^{1}_{1}H$, ${}^{16}_{8}O$, and ${}^{35}_{17}Cl$.

Although the nuclear binding force that holds the nucleus together is not well understood, the ratio of protons to neutrons in the nucleus seems to be related to nuclear stability. For the lighter atoms, the number of neutrons should be approximately equal to the number of protons to produce a stable nucleus, whereas for the heavier atoms the number of neutrons should exceed the number of protons for maximum stability.

Some nuclei are unstable, and although these may exist for a period of time, they will sooner or later eject a particle and become a different nucleus. Some of the radioactive isotopes are isotopes of perfectly stable elements, such as carbon, for example, which has the isotope ${}^{14}_{6}C$. Other radioactive isotopes come from elements that are entirely radioactive. All uranium isotopes are radioactive, including ${}^{238}_{92}U$. Some elements would not exist were it not for artificial creation. Technetium and plutonium are manmade elements giving rise to isotopes such as ${}^{239}_{94}Pu$ and ${}^{98}_{43}Tc$.

When these nuclei disintegrate, they give off either small particles or packets of energy or both. The matter or energy that is given off is known as radiation, and the common types were identified by Rutherford in the early years of the 20th century. He gave the radiation the labels of α, β, and γ. The **α particle** is essentially the nucleus of a helium atom. It has a mass of 4 amu and a charge of +2. Radium-226 emits an α particle when it undergoes radioactive decay:

$${}^{226}_{88}Ra \rightarrow {}^{222}_{86}Rn + {}^{4}_{2}He$$

Thorium-234, in contrast, emits a **β particle**, which is an electron. The electron with a charge of −1 may seem like an odd particle to be emitted from a positively charged nucleus, but clearly experimental evidence suggests that the nucleus does in fact in certain cases eject an electron. Because a negatively charged particle is lost, the positive charge on the nucleus will increase by one. Because the electron is so light (1/1837 amu), its weight is usually rounded off to zero, hence the weight of the nucleus does not change:

$${}^{234}_{90}Th \rightarrow {}^{234}_{91}Pa + {}^{0}_{-1}e$$

The case of **γ radiation** is slightly different. Gamma radiation is electromagnetic radiation similar to visible radiation. The main difference is that it is a much higher energy, potentially destructive type of emission. Because γ radiation is pure energy, it has no mass or charge; therefore, reactions giving off γ radiation would not give rise to a new isotope. This type of emission can occur when an isotope goes from an unstable state to a stable state; however, γ radiation can also be given off along with other types of radiation:

$$^{238}_{92}U \rightarrow ^{234}_{90}Th + ^{4}_{2}He + \gamma$$

These three types of radiation vary considerably in their penetrating power and ability to do damage. The relatively large α particles can do considerable damage because of their size, but they have relatively less penetrating power. They are much more likely than the smaller β particles to hit a nucleus of an atom. This situation is analogous to fastening bowling pins firmly to the floor of a bowling alley and throwing various things at them. If you throw a large bowling ball at the pins, it will not get far; however, if you throw a golf ball at them, it just might find its way through them. The bowling ball represents the α particle and the golf ball represents the β particle. A bullet fired from a gun might be a good representation of γ radiation.

The α particles can generally be stopped by a piece of paper but can do considerable amount of damage within the range of their penetration. These factors bear directly on the health risks of isotopes that emit alpha radiation. External exposure is not very dangerous because the radiation cannot penetrate into the vital organs. However, if such isotopes are taken internally, the situation instantly becomes very serious because the radiation can be emitted near vital organs and penetration is not necessary.

The β particles have intermediate penetrating power and can generally be stopped by a block of wood or aluminum foil; however, γ radiation with its high penetrating potential requires several centimeters of lead or a concrete wall to stop it. Interestingly, γ radiation is not as dangerous internally as it is externally. When taken internally it tends to penetrate all the way to the exterior, whereas external γ radiation can easily penetrate into vital organs.

All radioisotopes are converted into new isotopes at a predictable rate. Each radioisotope possesses a **half-life ($t_{1/2}$)**, which is a property of that isotope and represents the time required for half of the radioisotope to be converted. Half-lives can vary from less than a second to several billion years, but for each isotope the half-life is fixed. No matter how much of the isotope one starts with at the beginning of the half-life period, there will be half of it left at the end. This can be illustrated using the isotope uranium-238 (U-238), which has a half-life of 4.5 billion years. This long half-life explains why there is still so much uranium left around after the formation of the Earth about 4 billion years ago. For the sake of argument, let us assume that we start with 8 billion atoms of U-238. After one half-life or 4.5 billion years, there would be one-half or 4 billion atoms left. For each half-life of 4.5 billion years, the amount would be reduced to half, as shown in Table 4.4. The time spans of course are too long to be observed for U-238, but many other radioisotopes decay much more quickly. For example, consider polonium-218 (Po-218), an isotope that has a half-life of 3.05 minutes. Table 4.5 illustrates 10 half-lives (30.5 minutes); at the end of this time, only 0.098% of the atoms remain. As a rule of thumb, 10 half-lives are usually considered sufficient for a radioisotope to disappear; however, in a theoretical sense there will always be some left.

Table 4.4 Illustration of the Half-Life of Uranium-238 (U-238)

Number of Atoms of U-238 at the Beginning of the Time Period (billions)	Time Elapsed (billion years)	Number of Atoms of U-238 at the End of the Time Period (billions)
8	4.5	4
4	4.5	2
2	4.5	1
1	4.5	0.5

Table 4.5 Illustration of the Half-Life of Polonium-218 (Po-218)

Number of Atoms of Po-218 at the Beginning of the Time Period (billions)	Time Elapsed (minutes)	Number of Atoms of Po-218 at the End of the Time Period (billions)
8	3.05	4
4	3.05	2
2	3.05	1
1	3.05	0.5
0.5	3.05	0.25
0.25	3.05	0.125
0.125	3.05	0.0625
0.0625	3.05	0.03125
0.03125	3.05	0.015625
0.015625	3.05	0.0078125

Nuclear Fission

In 1939, two German scientists, Hahn and Strassman, found that when U-235 atoms are struck by relatively slow-moving neutrons, they would split into a variety of smaller nuclei. The splitting of one atomic nucleus into two or more smaller atomic nuclei is known as **nuclear fission**. During this fragmentation, considerable amount of energy as well as several neutrons are released. Because the released neutrons could conceivably split other U-235 atoms, scientists were very quick to understand that, once started, this process could suddenly split a large number of atoms and produce huge amounts of energy. This idea can be shown through an example of one of the fission reactions of U-235 (there are several reactions, but they are all similar):

$$^{235}_{92}\text{U} + ^{1}_{0}\text{n} \rightarrow ^{236}_{92}\text{U} \rightarrow ^{141}_{56}\text{Ba} + ^{92}_{36}\text{Kr} + 3^{1}_{0}\text{n} + \textit{energy}$$

When the neutron strikes the U-235 nucleus, it is absorbed, creating the very unstable U-236 nucleus that promptly falls apart, producing barium-141 (Ba-141) and krypton-92 (Kr-92), along with three neutrons and lots of energy. If the conditions were just right these three neutrons could conceivably split other U-235 nuclei, setting

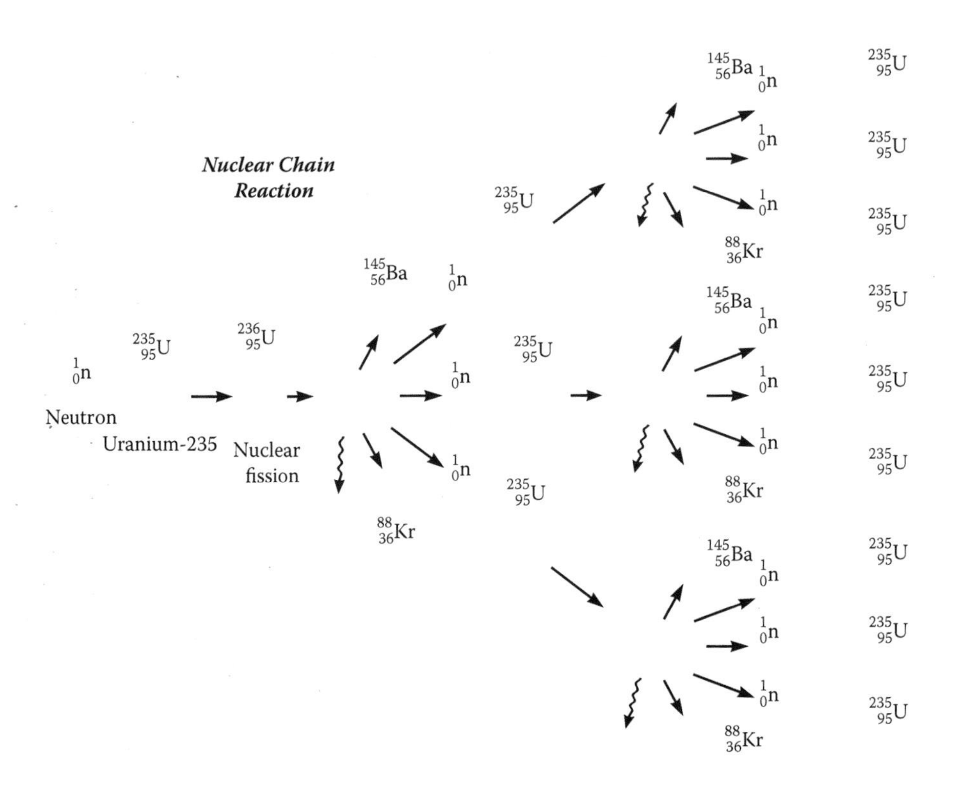

Figure 4.1 Diagram of a nuclear fission chain reaction.

up what is known as a chain reaction (see Figure 4.1). If on average one neutron from each reaction splits at least one additional U-235 atom, the reaction will be sustained in a controlled fashion. If the average is significantly greater than one additional U-235 atom split for each reaction, then the reaction will accelerate. These reactions are, of course, the basis of the nuclear energy industry that is discussed in Chapter 7.

Nuclear Fusion

The reason energy is released in splitting very large nuclei, such as U-235, is that smaller nuclei are more stable. Generally, the most stable nuclei are the medium-sized ones—for example, iron. It stands to reason, therefore, that combining nuclei smaller than medium size together would also release energy. This nucleus-combining process is exactly what occurs in the stars of the universe to produce energy. The sun, being a star, produces its energy in this way. Small nuclei such as hydrogen are combined to make larger nuclei and release large quantities of energy.

This process is known as **nuclear fusion** and involves the combining of small atomic nuclei to make larger nuclei. The biggest drawback to this approach is that it takes a temperature of several million degrees to get the process started. In Chapter 7, the potential of using this process to produce energy commercially will be discussed.

DISCUSSION QUESTIONS

1. List some of the key properties of acids. Which ion is responsible for these properties?
2. List some of the key properties of bases. Which ion is responsible for these properties?
3. Classify each of the following as an acid or a base:
   a. Vinegar
   b. Lemon juice
   c. Oven cleaner
   d. Soft drinks
   e. Lye
   f. Household cleaners (ammonia type)
   g. Gastric juice
   h. Drain cleaner
   i. Milk of magnesia
   j. Toilet bowl cleaner
4. Explain why one might put vinegar, which contains at least 4% acetic acid, on a salad and eat it; yet, it would not be recommended to put 4% sulfuric acid on your salad and eat it.
5. In a neutralization reaction, an acid and a base will react to produce water and a salt. In many cases it will not be obvious that anything changed upon reaction. How would you demonstrate that the neutralization reaction has occurred?
6. What is the relationship between the number of H_3O^+ ions at pH 4 and 5 (1/2, 1/3, ..., 1/10, ..., same, ..., double, triple, 4 times, ..., 10 times, etc.)? Be sure to note which is greater.
7. For each of the following sets of solutions determine whether a reaction will occur if the two solutions are mixed. If no reaction will occur, then indicate no reaction. If a reaction will occur indicate the products.
   a. KNO_3 (aq) + NaCl (aq) →
   b. Na_2CO_3 (aq) + $CaCl_2$ (aq) →
   c. $(NH_4)_2SO_4$ (aq) + $Pb(NO_3)_2$ (aq) →
   d. Na_3PO_4 (aq) + KCl (aq) →
   e. KOH (aq) + $Fe(NO_3)_3$ (aq) →
   f. $MgCl_2$ (aq) + Na_3PO_4 (aq) →
8. State if the indicated compound, ion, or element undergoes oxidation, reduction, or neither.
   a. $C_3H_6 + H_2 \rightarrow C_3H_8$
      C_3H_6 __________
   b. $CuS(Cu^{2+} + S^{2-}) + O_2 \rightarrow Cu + SO_2$
      Cu __________
      S __________
   c. $Ca + F_2 \rightarrow CaF_2(Ca^{2+} + 2F^-)$
      Ca __________
      F __________
   d. $2Fe_2O_3 + 3C \rightarrow 3CO_2 + 4Fe$
      Fe __________
      C __________

e. $Ag^+ + NaCl \rightarrow AgCl(Ag^+ + Cl^-) + Na^+$
   Ag ________

f. $3CH_4O + Cr_2O_7^{2-} + 8H_3O^+ \rightarrow 2Cr^{3+} + 3CH_2O + 15H_2O$
   CH_4O ________

g. $Zn + 2H_3O^+ \rightarrow Zn^{2+} + H_2 + 2H_2O$
   Zn ________

h. $S + O_2 \rightarrow SO_2$
   S ________

i. $Fe_2O_3 + 2Al \rightarrow Al_2O_3 + 2Fe$
   Fe ________
   Al ________

9. Which of the following pair of compounds represent isomers and which do not?

a. 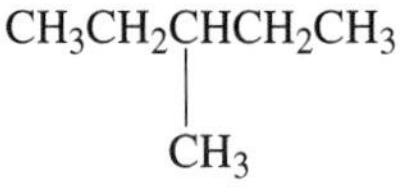
```
CH3CH2CHCH2CH3  and  CH3CH2CH2CHCH3
        |                      |
       CH3                    CH3
```

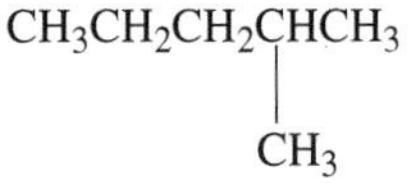

b.  and

c. $CH_3CH_2CH_2CH{=}CH_2$ and $CH_3CH_2CH_2CH_2CH_3$

d.
```
                      S
                      ||
NH4+ SCN-  and  H2NCNH2
```

e. KCN and NaCN

f.  and CH_3

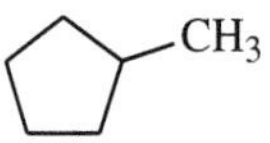

10. Name and distinguish between the various types of hydrocarbons.
11. Give the condensed structural formula for the following compounds:

a. 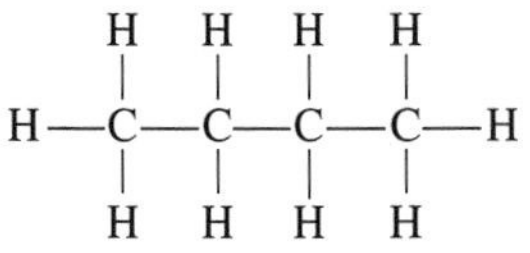
```
    H   H   H   H
    |   |   |   |
H—C—C—C—C—H
    |   |   |   |
    H   H   H   H
```

b. 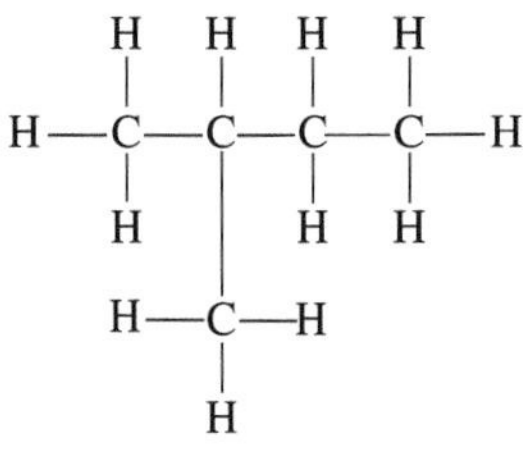
```
    H   H   H   H
    |   |   |   |
H—C—C—C—C—H
    |   |   |   |
    H   |   H   H
        |
    H—C—H
        |
        H
```

12. Give the line-bond formulas for the following structures:

a.

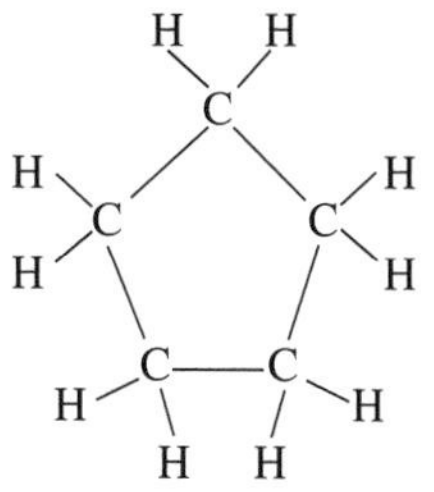

b.

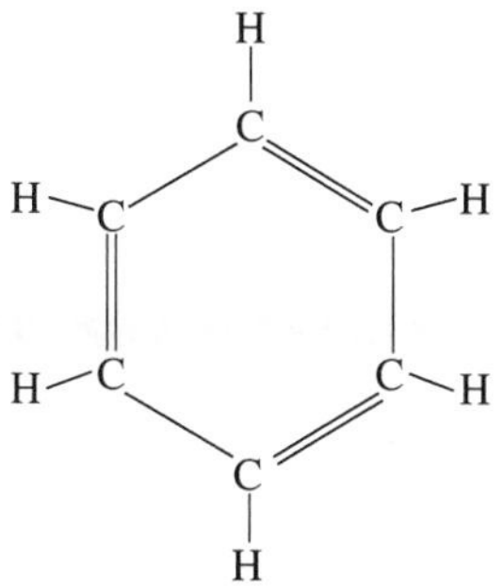

13. If we had 16 mg of a radioactive isotope with a half-life of 10 seconds, how much would be left after 10, 20, 30, and 40 seconds?
14. Explain the advantages of using an α-emitter such as radium-233 to implant into a cancerous tumor to target the tumor cells for destruction. Why would you use a γ-emitter such as technetium-99 as a tracer that is injected into the human body to highlight certain organs?
15. Explain why the fission of uranium-235 nuclei can be carried out in such a way as to lead to a chain reaction.
16. Which are the most common places in the universe where nuclear fusion occurs? Why?

BIBLIOGRAPHY

Hudson, J., *The History of Chemistry*, Chapman & Hall, New York, 1992.

Joesten, M.D., Netterville, J.T., and Wood, J.L., *World of Chemistry: Essentials*, Saunders College Publishing, New York, 1993, p. 118.

CHAPTER 5

Element Cycles

The need to discuss **element cycles** comes directly from the law of conservation of matter. All the things that are produced in any manner must be produced by the transformation of previously existing matter. This law puts restraints not only on human activities but also on natural systems, as well. Unfortunately, humans often try to ignore these restraints; however, nature is designed to be in harmony with the law of conservation of matter. Natural processes have been operating for thousands, perhaps millions, of years with little change. Such continuous processes are said to be **sustainable**. To be sustainable over a long period of time, the same atoms of the elements have to be used over and over again. The path an element takes in going through various combined states, finally returning to an earlier state, is referred to as an element cycle. The purpose of this chapter is to discuss only the natural cycles of the elements. The remainder of this book, to some extent, discusses how humans have interfered with these processes, leading to many of our environmental problems.

COMPARTMENTS

For the purpose of our study, we have divided our environment into **compartments**. The fact, however, is that no part of the environment can be totally isolated from the whole; therefore, one must be careful not to overstate the separate nature of each compartment. The definition of a compartment is ultimately arbitrary. Any compartment can be defined in a manner convenient to the discussion at hand.

The planet on which we live is commonly divided into four compartments, usually referred to as *spheres*. They are the **lithosphere**, **hydrosphere**, **atmosphere**, and **biosphere**. The first three spheres are collectively known as the **geosphere**. Each of these spheres exists in distinct places, whereas the biosphere is intertwined with the geosphere.

The lithosphere is the crust of the Earth which includes soil, rocks, and all the other materials on which we walk. The hydrosphere includes all of the Earth's water, including liquid, frozen, and vapor forms. This sphere includes the oceans, lakes, rivers, and glaciers. The atmosphere consists of the air above the planet. Finally, the biosphere is comprised of the biological matter on the planet, both living and dead.

Various elements move through each of these compartments at different rates. Some elements may remain in a certain compartment for a few days, whereas some may remain for thousands of years. The average amount of time that an atom of a given element, or a molecule of a given substance, remains in a compartment before it is removed in some way is referred to as the **residence time**.

CARBON CYCLE

How plants appear to grow out of nothing is often a mystery to many people. To construct buildings, one has to dig up stones or clay and make bricks, and animals grow by eating food, but from where do plants get their food? Johann Baptista van Helmont, an early scientist, was curious about such observations. He performed an interesting experiment. He planted a 5-lb willow sapling in 200 lb of dried soil. Then, for a period of 5 years, he added only pure water to the plant. At the end of the 5 years, he weighed the willow tree and the soil separately. When he dried the soil and weighed it, he found that the soil weighed only 2 oz less than it did at the beginning of the experiment. Because the plant weighed 169 lb, van Helmont concluded that everything in a plant is made up of water.

The factor that van Helmont did not consider was air. Air contains a small amount of the carbon dioxide (CO_2) gas. This gas, although not a major component of the atmosphere (currently approximately 0.039%), is nonetheless a very important one. The carbon dioxide in the atmosphere represents a huge atmospheric reservoir of carbon. This factor is important because the key element in living things is carbon. Van Helmont was not completely wrong, though, because plants are partly made up of water but the key element is carbon.

The **carbon cycle** has several features, which are illustrated in Figure 5.1. First, the biological aspects of this cycle will be discussed. Carbon is removed from the atmosphere by green plants through a process known as **photosynthesis**. This process, which is outlined in the following equation, requires sunlight, water, and carbon dioxide:

$$H_2O + CO_2 + \text{Sunlight} \rightarrow \text{Carbohydrates} + O_2$$

Photosynthesis is a process through which carbon dioxide and water are converted into carbohydrates, which are a key food for plants. As discussed in the later parts of this chapter, these carbohydrates are not the only materials required for a plant to survive, but they are certainly very important.

Photosynthesis can also take place under water. Many aquatic green plants extract the carbon dioxide directly from the water in which it is dissolved. The oceans, in fact, have great quantities of carbon dioxide dissolved in them. The amount of nonbiological carbon present in the ocean is about 65 times that present in the atmosphere.

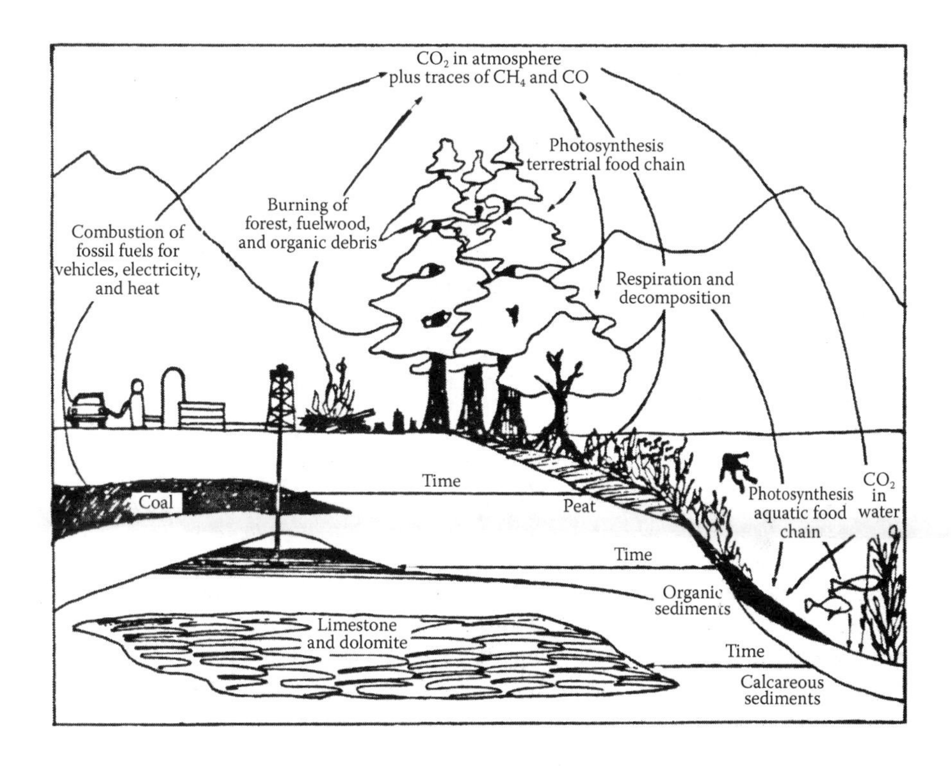

Figure 5.1 The carbon cycle. (Courtesy of Amy Beard.)

The carbon that is incorporated into these plants is eventually converted into other forms. Some of the plant carbon becomes animal carbon when these plants are consumed by animals. In living plants and animals, some of the carbon-containing material is used as an energy source in a process called **respiration**. Respiration is actually the reverse of photosynthesis, except that in photosynthesis light energy is required and in respiration energy is released as heat or motion:

$$\text{Carbohydrates} + O_2 \rightarrow H_2O + CO_2 + \text{Heat/motion}$$

As a result of respiration, the carbon is released back into the atmosphere or the water and the cycle is complete. No organism converts all of its carbon back into carbon dioxide during its lifetime. Some carbon is needed as part of the structure of the organism, and other carbon compounds are excreted as solid or liquid wastes. These waste products and the organism itself when it dies are attacked by decomposing microorganisms. These microorganisms convert the remaining carbon to carbon dioxide through their respiration.

Sometimes dead biological matter burns intentionally or accidentally. This could happen when one burns wood in a fireplace, burns leaves in the fall, or is the victim of an accidental house fire. Although these vary in intensity, the results of respiration and combustion are chemically the same. During combustion, carbon dioxide and water are released; however, all of the energy is released immediately as heat. Whether the process is respiration or combustion, the cycle is completed and carbon dioxide is returned to the atmosphere or to the hydrosphere. Atmospheric carbon dioxide has a residence time of about 2 years before the cycle restarts.

A small amount of biological carbon is removed from the carbon cycle and trapped underground. This organic material is eventually converted into coal, oil, or natural gas—that is, fossil fuels. The process is slow, taking millions of years, and has been going on for a very long time, thus resulting in the build-up of fairly large deposits. Beginning in the middle of the 19th century, fossil fuels have been exploited at an ever-increasing rate. Burning of these fossil fuels represents the completion of a very long biological carbon cycle by which the carbon is returned to the atmosphere:

$$\text{Fossil fuels} + O_2 \rightarrow CO_2 + H_2O + \text{Heat}$$

A very slow nonbiological carbon cycle that has been operating over millions of years involves the formation of sedimentary rocks such as limestone. Most of the nonbiological carbon in the ocean is in the form of bicarbonate ion (HCO_3^-) rather than carbon dioxide. Additionally, some of the carbon is in the form of carbonate ion (CO_3^{2-}). All of this is the result of the fact that carbon dioxide reacts with water, as is illustrated in the following three reactions:

$$CO_2 + H_2O \rightleftarrows H_2CO_3$$

$$H_2CO_3 + H_2O \rightleftarrows HCO_3^- + H_3O^+$$

$$HCO_3^- + H_2O \rightleftarrows CO_3^{2-} + H_3O^+$$

Note that all of these reactions are reversible. The production of limestone is due to the fact that calcium ions gradually find their way into the ocean. Calcium carbonate is not very water soluble; therefore, it forms a solid sediment and sinks to the bottom of the ocean. Over time, this sediment becomes limestone rock. Limestone is a solid, compressed form of calcium carbonate:

$$Ca^{2+} + CO_3^{2-} \rightarrow CaCO_3 \text{ (solid)}$$

Actually, some marine biological activity also adds to this process. Many marine organisms build shells and skeletons out of calcium carbonate and related compounds. When these organisms die, these shells and skeletons sink to the bottom of the ocean and become a part of the calcium carbonate sediment.

This carbon is eventually recycled through volcanic activity. When exposed to volcanic heat, limestone decomposes into lime and carbon dioxide:

$$CaCO_3 + \text{Heat} \rightarrow CaO + CO_2$$

OXYGEN CYCLE

Oxygen is a very reactive gas. (In fact, it is so reactive that one would not expect to find it in the atmosphere.) No other planet or moon in the solar system has an atmosphere containing significant quantities of free oxygen. Yet, the atmosphere of the Earth contains huge quantities of the gas, almost 21% of all of the gases to be specific. So, what is oxygen doing in our atmosphere and how did it get there?

It is generally agreed that when the atmosphere of the Earth came into existence, it contained no oxygen for all practical purposes. This early atmosphere was probably made up of nitrogen, water vapor, carbon dioxide, ammonia (NH_3), and methane (CH_4). The first forms of life on the Earth evolved without oxygen. The first production of oxygen in the atmosphere is thought to have been carried out by some photosynthetic bacteria. The photosynthesis reaction indicates that the process involves the consumption of carbon dioxide and water and the release of free oxygen. As these bacteria and other new species began to push oxygen into the atmosphere, some species began to die, as oxygen was toxic to them. Eventually, when oxygen levels reached modern atmospheric concentrations, the vast majority of species not only were compatible with the oxygen atmosphere but were also dependent on such an atmosphere.

The oxygen and carbon cycles are very similar in that they share many of the same key reactions. A quick review of the reactions from the carbon cycle will show that all of the reactions in the biological part of the cycle also involve oxygen. Generally, when oxygen is consumed, carbon dioxide is released and *vice versa*. To a large extent, oxygen can be seen as a part of the same cycle with carbon except that it moves in the opposite direction. Such a situation arises because when green plants draw carbon dioxide from the atmosphere and water from their surroundings, to carry out photosynthesis, the carbohydrates created contain less oxygen. This leftover oxygen is returned to the atmosphere. The **oxygen cycle** continues with the breakdown of carbohydrates and similar materials. This process releases carbon dioxide and water and consumes oxygen.

The only significant way in which oxygen enters the atmosphere is through photosynthesis. Without photosynthesis, all of the oxygen in the atmosphere would eventually react and be removed from the air ... so be kind to your green friends. It is important to keep these observations about photosynthesis in perspective. There is an incredible amount of oxygen in the atmosphere. If the process of photosynthesis stopped, it would take millions of years to remove all of the oxygen from the atmosphere. For example, the burning of fossil fuels will not affect the amount of oxygen in the atmosphere significantly, although combustion consumes oxygen. If all fossil fuels in the world were burned at one time, such a fire would have little effect on the atmospheric oxygen level.

Oxygen is removed from the atmosphere in the same way as carbon dioxide is released into it, through the respiration of plants and animals, decomposition of dead biological matter, and decomposition of waste products. A small amount of oxygen is consumed in a variety of other processes, but the amount is small when compared to respiration, decomposition, and combustion.

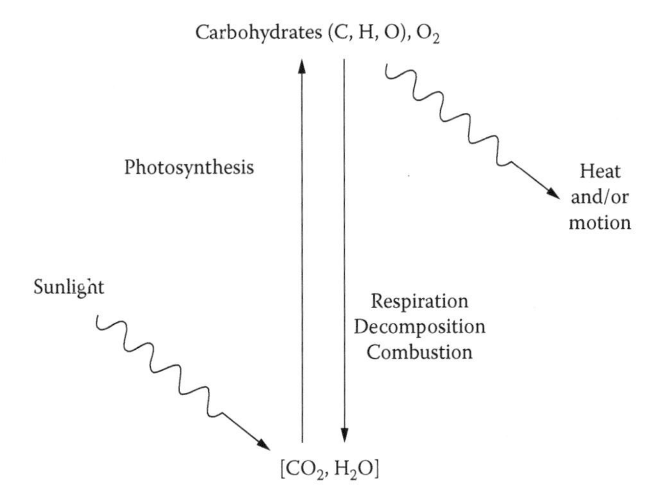

Figure 5.2 The biological cycling of carbon, hydrogen, and oxygen.

HYDROGEN CYCLE

Hydrogen is one of the most common elements in organic material. Although the backbone of organic compounds is carbon, nearly every biological molecule contains hydrogen. The source of this hydrogen is water. Because every water molecule contains two hydrogen atoms and because there is such an incredible amount of water in the world, the supply of hydrogen is nearly inexhaustible. A quick check of the carbon cycle shows that all of the reactions in the biological part of the cycle involve the consumption or production of water. In photosynthesis, carbon dioxide and water are consumed and oxygen is produced. The hydrogen is clearly incorporated into the carbohydrate products. Conversely, in respiration, decomposition, and combustion, the organic molecule is broken up and its hydrogen is released into the water. The biological parts of the carbon cycle and **hydrogen cycle** involve the same reactions and cycle in the same direction. The biological cycling of carbon, hydrogen, and oxygen is shown in Figure 5.2.

NITROGEN CYCLE

In contrast to oxygen, nitrogen is the gas that one would expect to find in the atmosphere. Nitrogen gas is relatively unreactive. In fact, nitrogen is so unreactive that pure nitrogen is often used in place of air around very reactive chemicals. One should not get the idea that nitrogen does not form compounds because there are countless compounds containing this element. The nonreactivity of nitrogen gas is a result of the diatomic nature of the nitrogen molecule (N_2) that is found in pure nitrogen gas.

The two nitrogen atoms are bonded very tightly to one another. An individual nitrogen atom would readily react with many chemicals, but a considerable quantity of energy is required to break the bond between two nitrogen atoms in nitrogen gas.

As a result of its nonreactivity, the nitrogen in the atmosphere tends to be present for a long time. As an example, the residence time of oxygen in the atmosphere is about 5000 years, which is a long time, but that of nitrogen is about 10 million years. This long residence time is directly related to its nonreactivity.

Life on the planet requires nitrogen because many of the compounds that sustain life, including the important proteins and nucleic acids, contain nitrogen. There is no shortage of the element nitrogen, but it is rather difficult to release it from the air and convert it into usable nitrogen compounds. The process of converting free nitrogen in the air into chemically combined nitrogen is known as **nitrogen fixation**.

Nonbiologically, nitrogen can be fixed through lightning discharges, but this source of nitrogen fixation is minor compared to biological processes. The lightning discharge provides the energy to cleave the nitrogen molecule, which allows the nitrogen atoms to react with the oxygen in the air to give nitric oxide. The nitric oxide is slowly converted into other nitrogen-containing compounds, which usually end up in the water or the soil as nitrate (NO_3^-):

$$N_2 + O_2 \rightarrow 2NO \rightarrow \rightarrow NO_3^-$$

This pathway alone does not provide sufficient nitrogen fixation to support life. To make up for the difference, certain bacteria evolved that have the ability to fix nitrogen from the atmosphere. The roots of legumes are important places where one of these bacteria, *Rhizobium*, is found. Legumes are a class of plants that includes peas, beans, clover, and alfalfa. The bacteria in these plants can fix enough nitrogen for the host plant with some remaining. Other types of plants that do not have this ability slowly deplete the soil of nitrogen. Several methods can be used to replenish soil nitrogen. A farmer could spread manure on a field, or the farmer might plow under grasses or crop residue. This approach works because decomposing bacteria release nitrogen compounds into the soil. Any type of animal wastes or dead organisms release nitrogen compounds. Another option is to take turns by first planting legumes and then nonlegumes. This practice, which is known as *crop rotation*, allows the nonlegumes to use the excess nitrogen in the soil from the legumes.

Modern farmers usually turn to fertilizers, which are produced using the Haber process (see Chapter 1). This process represents a manmade method for fixing nitrogen from the atmosphere. The process is so widely used that it is competing with natural fixation pathways. (In excess of 30% of all fixed nitrogen is generated using the Haber process.) The method involves heating hydrogen and nitrogen under pressure to produce ammonia. Ammonia can be applied directly to the soil, where it is naturally converted to nitrates, or it can be artificially converted to nitrates and packaged as nitrate fertilizer:

$$N_2 + 3H_2 \rightarrow 2NH_3 \rightarrow \rightarrow NO_3^-$$

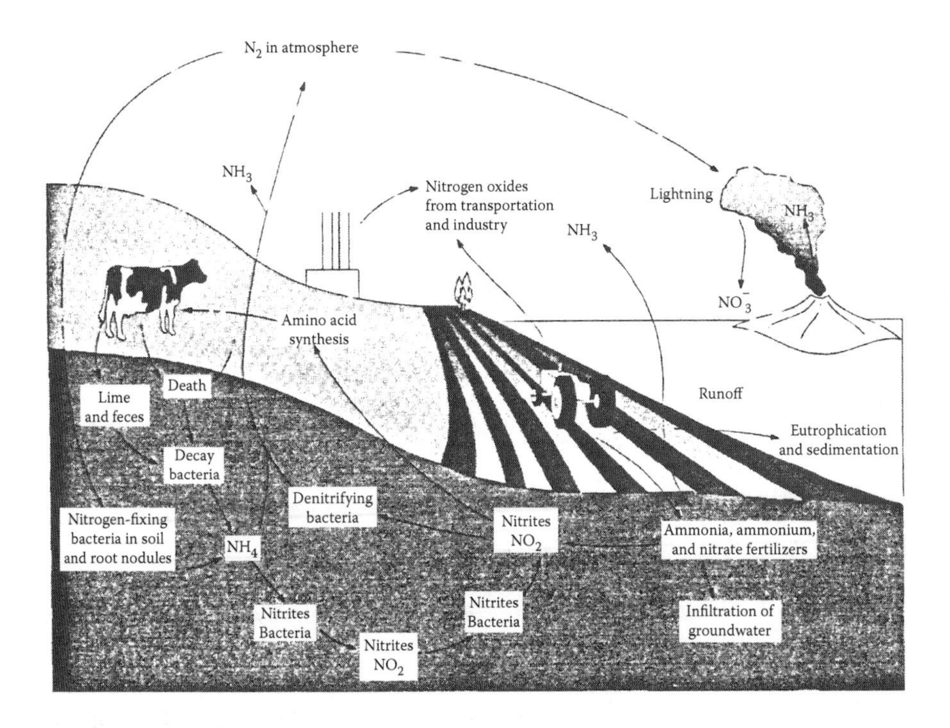

Figure 5.3 The nitrogen cycle. (From Cunningham, W.P. and Saigo, B.W., *Environmental Science, A Global Concern,* William C. Brown, Dubuque, IA, 1990. With permission of The McGraw-Hill Companies.)

The processes mentioned earlier represent the ways in which nitrogen is chemically combined; however, if the fixed nitrogen is never returned to the atmosphere, theoretically there would never be a need to fix it any more. Two factors make continued fixation necessary. First, nitrates are water soluble and are constantly being washed into the oceans. Second, a class of bacteria that exist in water-logged soils returns nitrogen to the atmosphere as nitrogen and nitrous oxide (N_2O); these are referred to as *denitrifying bacteria*. The nitrous oxide is slowly converted in the atmosphere into free nitrogen and oxygen. The processes of the nitrogen cycle are summarized in Figure 5.3.

PHOSPHORUS CYCLE

Phosphorus is a very important element in the biological world. This element is an important component of the genetic material of life—deoxyribonucleic acid (DNA). DNA and ribonucleic acid (RNA) are known as nucleic acids and are essential to the functioning of any living cell. Phosphorus is also an important component of the bones of vertebrates. In many ecosystems, phosphorus is the least available of the key elements; for example, phosphorus is usually the limiting element in aquatic ecosystems. The restricted availability of phosphorus may limit the growth of plants in many situations. In addition to nitrogen and potassium, the other major component of fertilizers is phosphorus.

The **phosphorus cycle** is very different from the cycles mentioned earlier. The phosphorus cycle does not have an atmospheric component as the carbon, oxygen, and nitrogen cycles do, and phosphorus is not available in huge amounts in the oceans as is the case for hydrogen. The biological phosphorus cycle is short. All dead organisms and the waste products of various organisms contain phosphorus; when acted upon by decomposing bacteria, they produce inorganic phosphorus, generally in the form of phosphate (PO_4^{3-}). This inorganic phosphorus in the soil is reabsorbed by plants and converted back into biological phosphorus. In most terrestrial ecosystems the phosphorus is recycled in this way an average of 46 times. Eventually the phosphorus will be removed through runoff because the inorganic phosphorus is slightly water soluble and can be slowly washed away.

The phosphorus that is washed away is slowly lost from the terrestrial biological cycle. Most of the phosphorus over a period of time slowly works its way into the ocean. In the ocean, the phosphorus can be used by marine organisms and is used over and over again, as many as 800 times. The inorganic phosphorus is only partially water soluble, so gradually some of the phosphorus settles down as sediment at the bottom of the ocean. This phosphorus is eventually trapped in sedimentary rock. Only after some geological event exposes this rock is the phosphorus available again for use in the biological world. The release of this phosphorus, again as phosphate (PO_4^{3-}), occurs very slowly as the rock weathers and breaks down. The entire cycle takes millions of years.

We humans, in our need to grow ever-increasing volumes of crops, must increase the amount of phosphorus in the soil. The only way to do so is to dig up phosphorus-bearing rocks, crush them, and extract the phosphorus. Fertilizer containing phosphorus is then spread on the fields but eventually washes back into the oceans. Most of the readily available phosphorus-bearing rock is found in China, Morocco, South Africa, and the United States. Although the total amount of phosphorus in the world is probably sufficient for agricultural needs for the foreseeable future, the amount of inexpensive phosphorus is probably limited. If agricultural demands increase, the supplies of inexpensive phosphorus could be depleted within 40 years.

Human intervention has taken a very slow process of rock weathering, which leads to slow fertilization of the soil, and sped it up. This has led to large amounts of phosphates ending up in the waterways, encouraging algae growth that leads to problems discussed further in Chapter 13.

SULFUR CYCLE

The **sulfur cycle** has many features that are common to the phosphorus cycle. When operating without human interference, the sulfur cycle is geologically very slow and has only a negligible atmospheric component. The residence time for sulfur in the ocean is 40 million years and in rock it is estimated to be 9.7 billion years. (This estimate suggests that most rock sulfur is permanently fixed, as Earth is thought to be only 4.5 billion years old.) Small atmospheric emissions of sulfur are produced in the form of hydrogen sulfide (H_2S, "rotten egg" gas) from sources such as geysers, hot springs, volcanoes, anaerobic microorganisms, and other natural sources. The slowness of this cycle is not as serious a problem for sulfur as it is for phosphorus because the biological demand for sulfur is not as high. Most soils contain a sufficient reserve of sulfur to satisfy the needs of most plants. The large-scale burning of fossil fuels by humans has drastically altered the sulfur cycle. Fossil fuels contain some sulfur, and when they are burned sulfur dioxide gas is produced. The major form of sulfur in the atmosphere today is sulfur dioxide, 99% of which is due to humans. This sulfur is eventually returned to the ground in the form of rain, which is discussed in Chapter 9.

OTHER CYCLES

There are many other cycles that could be discussed. Many elements move around in our environment. Some of them move as a result of biological forces, others move due to natural geological forces, and still others move because of human interference. With most elements, the reality is a combination of all three of these factors. Element cycles could be discussed for mercury, lead, tin, cadmium, aluminum, iron, zinc, chromium, copper, and others. The law of the conservation of matter states that matter can neither be created nor destroyed; therefore, in a biological system, elements must be cycled or they will eventually be depleted.

DISCUSSION QUESTIONS

1. Much effort is expended by many nations and corporations to secure and protect various sources of natural resources. In some situations, continually using new natural resources makes sense, whereas in other situations such an approach creates serious problems. Discuss some of the limitations of using new natural resources vs. finding ways to reuse material over and over again. How do these limitations relate to element cycles?
2. Name and define each of the compartments that make up the geosphere.
3. Define the biosphere and indicate a fundamental difference between the various spheres of the geosphere and the biosphere in terms of the ease with which they may be located.
4. Because compartments can be defined arbitrarily based on the usefulness of the definition, suggest some other compartments of the natural world that you think might be useful in studying the environment.
5. Why might the residence time of a substance in a compartment (e.g., a lake) be important?
6. Could you grow a plant in a sealed glass container containing only fertile soil, water, oxygen, and nitrogen if it is sitting in the sunlight?

7. Which of the element cycles requires photosynthesis and why?
8. Why is oxygen not an element one would normally find in the atmosphere of a planet? Why is nitrogen an element one would expect to find in the atmosphere of a planet?
9. What would be the implications for the world's food supply if the Haber process had not been developed? Are there any other alternatives for the nitrate fertilization of crops on a massive scale?
10. Compare and contrast the various features of the following element cycles: nitrogen, phosphorus, and sulfur.

BIBLIOGRAPHY

Cunningham, W.P. and Saigo, B.W., *Environmental Science, A Global Concern*, William. C. Brown, Dubuque, IA, 1990, p. 38.

Garrels, R.M., Mackenzie, F.T., and Hunt, C., *Chemical Cycles and the Global Environment: Assessing Human Influences*, W. Kaufmann, Los Altos, CA, 1975.

Mooney, H.A., Vitousek, P.M., and Matson, P.A., Exchange of materials between terrestrial ecosystems and the atmosphere, *Science*, 238, 926–932, 1987.

Paasivirta, J., *Chemical Ecotoxicology*, Lewis Publishers, Chelsea, MI, 1991.

Sarmiento, J.L., Ocean carbon cycle, *Chemical and Engineering News*, 71(22), 30–43, 1993.

Vaccari, D.A., Phosphorus: a looming crisis, *Scientific American*, 300(6), 54–59, 2009.

Vitousek, P.M., Aber, J., Howarth, R.W., Likens, G.E., Matson, P.A., Schindler, D.W., Schlesinger, W.H., and Tilman, G.D., Human alteration of the global nitrogen cycle: causes and consequences, *Issues in Ecology*, August 30, 2000, http://esa.sdsc.edu/tilman.htm.

CHAPTER 6

Toxicology

One major concern in any study of the environment is that many of the changes we humans bring to the environment may also be harmful to ourselves and the other organisms that inhabit our planet. Some changes may make the planet a physically less desirable place to live in, some may spread infectious diseases, and some may bring us in contact with dangerous chemical substances. As this is a chemistry text, we will focus on the last concern.

It is difficult to understand to what extent chemicals are dangerous to the environment. Much of popular culture believes that "chemicals" are harmful. There is a widespread belief that all things that are "chemical" should be kept out of our lives. But what is a chemical? Often the materials being referred to are synthetic chemicals (i.e., man-made), because in some sense everything is a chemical as everything is made up of atoms and molecules. The fact is that not all "man-made" chemicals are dangerous nor are all "natural" chemicals safe. For example, consider the large number of synthetic plastics and synthetic pharmaceuticals. Our life span has increased because of these man-made chemicals. As citizens, with the help of experts, we must be able to distinguish between dangerous and safe chemicals.

What do we mean by dangerous chemicals? Some chemicals may blow up, some may kill plant life, and others may make us ill or, perhaps, kill us. As a broad class, these substances are known as **hazardous chemicals**. Hazardous chemicals are those substances that are flammable, explosive, corrosive, allergenic, or toxic. This umbrella definition includes nearly everything that would be directly harmful to life regardless of the concentration or the amount. Nearly anything can be poisonous in high enough concentration. This chapter focuses on chemicals that are dangerous at fairly low concentrations or in small amounts. Generally, these substances are more classically what we think of as poisons. The more technical term usually associated with a poison is **toxin**. A toxin is defined as a substance that can cause death in test animals at a low concentration.

Toxicology is defined as the study of toxins or poisons. Poisons have a fascinating and colorful history. A well-known incident of ancient history was the poisoning of the Greek philosopher Socrates by the forced ingestion of hemlock. As a culture, we find stories about poisoning fascinating, and as a result poisoning has been the plot of many a mystery novel.

HISTORY OF TOXICOLOGY

The point when humans became aware that some substances are poisonous is lost in antiquity. The earliest writing dealing with poisonous substances and their pharmacological effects was probably the Egyptian papyrus entitled *Ebers* produced about 1500 BCE. About a thousand years later, the writings of Hippocrates, Aristotle, and Theophrastus included some references to poisons. The origins of modern toxicology can probably be traced back to Paracelsus of the early 16th century. It was Paracelsus who was the first to state the dose–response relationship in his famous quote, "All substances are poisons; there is none that is not a poison. The right dose differentiates a poison and a remedy." In 1700, Bernardino Ramazini set the foundation for occupational medicine with his book, *Diseases of Workers*, which was followed by the work of Dr. Percivall Pott regarding chimney sweeps in England. About the same time, Dr. John Hill made observations about the connection between nasal cancer and the use of snuff.

In 1815, the first comprehensive work focused exclusively on toxicology was produced by the Spaniard Mathieu Orfila at the University of Paris. From this point onward, the field of toxicology began to grow. The use of toxicological principles in the study of the environment was spurred on in 1962 by the publication of **Rachel Carson**'s book, *Silent Spring*. Carson's book brought attention to the widespread use of pesticides as well as other chemicals throughout the environment. This book was partly responsible for the later ban on dichlorodiphenyltrichloroethane (DDT) in the United States and spurred the environmental movement. After this point, the application of toxicology to environmental issues expanded rapidly.

ENVIRONMENTAL TOXICOLOGY

Of course, the focus of this book is not on poisons in general, but rather on those that show up in the environment as a result of the activities of our society. When the study of toxic substances is focused on in this way, it is known as **environmental toxicology**. Environmental toxicology is a complex field because we must consider not only the amount of toxins affecting various organisms but also their source and mechanism of distribution throughout the environment. Often toxic effects are present in the environment before the poisonous substance or substances have been identified.

Environmental toxins can be considered in one of two ways based on their effects. The study of the effect of toxins on human health is known as **environmental health toxicology**; however, many of the toxins in the environment can have effects on an ecosystem. The study of these effects is known as **ecotoxicology**.

When considering environmental toxins there is a tendency to think of them as anthropogenic; however, many of the toxic materials found in the environment are produced or transported to a location naturally. Naturally occurring arsenic or selenium is sometimes found in water supplies. Many organisms produce toxic

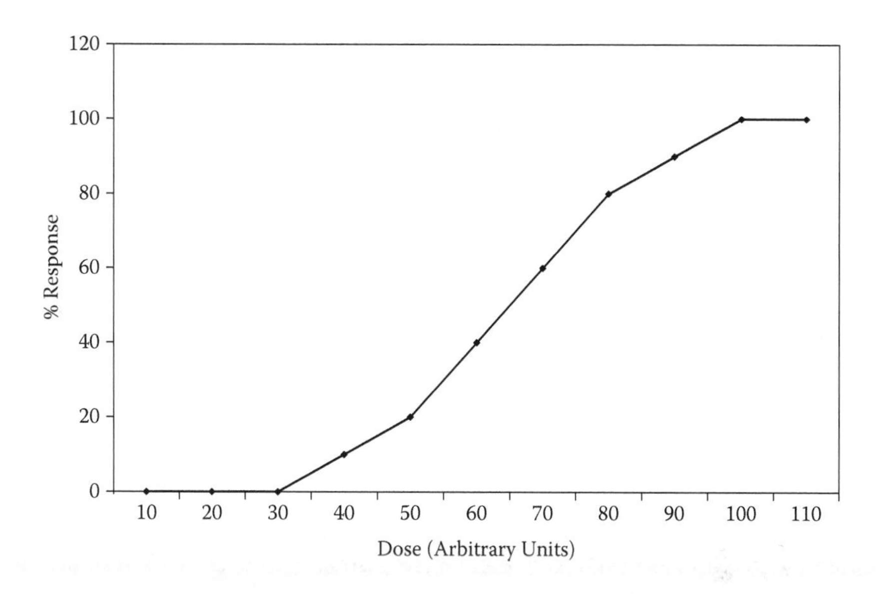

Figure 6.1 A typical dose–response curve.

substances for their defense or other reasons, and these substances are referred to as *biotoxins*, a term developed by the toxicologist B.W. Halstead. Some examples include nicotine, bee venom, snake venom, and aflatoxin from mold.

TOXICITY MEASUREMENTS

To establish just how toxic a chemical is, it is necessary to find a way to measure the toxic effects of the chemical. This testing is usually done using a number of individuals from a target species. These individuals are then exposed to a dose of the chemical in question, and the percentage of individuals exhibiting a response such as death is determined. By doing repeated tests using various doses, one can construct a dose–response curve similar to the one shown in Figure 6.1. For most toxins and with most species there is a level below which there is no observable effect, as can be seen in the hypothetical curve. This level is known as the **threshold dose**, which in Figure 6.1 is 30 units. As the dose increases from the threshold dose, the percentage of the individuals exhibiting a response also increases until there is a 100% response.

Generally, toxicologists are interested in the dose at which 50% of the individuals show the effect. If the effect is death, then this dose is known as the $\mathbf{LD_{50}}$, whereas if the effect is some other response then the dose is called the $\mathbf{ED_{50}}$. To compensate for the variations in the size of individuals, the dosage is often expressed in milligrams of toxin per kilogram of body weight (mg/kg).

The type of testing described earlier is most easily carried out by what is called *acute toxicity testing.* **Acute toxicity** usually manifests itself in a short period of time such as hours or days. Although there is no exact limit, the time period is usually less than 2 weeks. Toxicity that manifests itself over a longer period of time is known as **chronic toxicity**. Chronic toxicity testing is carried out in a manner similar to acute testing but requires a lot of time. The time spans for chronic toxicity testing are usually a year or more. Often, in chronic testing, the concern is the safe use of a chemical that has some chronic toxic properties. Toxicity testing is used to determine the highest dose at which no effect is observed and that is known as the **no-observed-effect level (NOEL)**. Obviously, the NOEL is very closely associated with the threshold dose.

These are the most commonly used methods for assessing the toxicity of chemicals; however, these approaches are not without controversy. First, there are humanitarian concerns. Procedures such as these usually involve thousands of laboratory animals. Many of these test animals are killed, as one would expect, and many of those that do not die endure considerable pain. Additionally, animal tests are very expensive and time consuming. The complete testing of a toxin might take several years and might cost hundreds of thousands of dollars.

From a scientific point of view, the most difficult aspect of animal testing is interspecies variation. Which species should be used as the test population? Many species react very differently from others when exposed to a specific toxin. Dioxin, which is found in the herbicide Agent Orange from the Vietnam era, has been widely studied for its toxic effects. Actually, dioxin is a class of compounds, and the individual dioxin most commonly studied is called 2,3,7,8-tetrachlorodibenzo-*p*-dioxin (TCDD). The LD_{50} varies greatly from species to species. For example, the LD_{50} of TCDD using the guinea pig is about 0.0006 mg/kg, whereas that using the hamster is about 1.1 mg/kg. Clearly, the kind of toxicity numbers one gets often depends on which species is used. Furthermore, the reason this testing is usually done is to assess the risk of these chemicals to humans. Humans clearly are not guinea pigs or hamsters. It is not altogether clear whether toxicity data collected on animals can be directly applied to humans.

Of course, getting toxicity data on humans is difficult. Data can be collected from accidental poisoning cases. For some common poisons, this information may be plentiful, whereas for less common ones the data may be spotty to nonexistent. In evaluating toxic substances, it is usually best to utilize human data if they happen to be available, but if not then data from animal studies are often used.

One approach used when human data are not available and when for various reasons one may not want to do animal testing is the **quantitative structure–activity relationship (QSAR)**. This approach is based on the structure of the chemical molecule in question. Each part of the molecule is assigned a value to indicate its expected contribution to the toxicity of the overall molecule. The value is based on the toxicity of other molecules where the part can be found. By analyzing the parts, a toxicity of an entire molecule can be assigned. Other approaches using cell cultures are also an alternative.

In the environment, it is quite common to have more than one toxin present at a given time. These toxins can interact in such a way as to cause the mixture of toxins to be more or less toxic than the sum of the effects of each toxin individually. For example, the air pollutants ozone and sulfur dioxide are more toxic to plants together than one would expect them to be individually. When the effect of a combination of toxins is greater than would be expected from the sum of the toxins individually, the effect is known as **synergism** (*synergistic effect*). The metal cadmium, however, is less toxic in the presence of zinc. This effect of decreasing the toxicity is known as **antagonism** (*antagonistic effect*).

ROUTES OF EXPOSURE

For a toxin to have an effect on an individual organism, it must be able to interact in a meaningful way with the organism in question. This section discusses some of the ways in which humans are exposed to toxins. By analogy, one can project the likely path of exposure for some other species; however, we will try not to cover all of the species that might be of interest. Without injection, the toxin must cross some type of membrane from the exterior of the body to the interior. The most obvious tissue that covers the body is the skin, but, in fact, the body has two other major areas where external materials come in contact with membranes that protect the interior of the body. These major areas are the gastrointestinal (GI) tract and the lungs. The GI tract is basically an open tube that runs through the body, and the lungs are basically air sacs inside the body. Let us begin by considering the GI tract.

Gastrointestinal Absorption

The GI tract provides the major route for the absorption of nutrients and water into the body. The absorption of nutrients and water occurs in the small intestine and to a lesser extent in the stomach. In the large intestine only water is absorbed. Most substances that are absorbed in the GI tract simply pass through the membrane. A few substances such as blood sugar are actively carried across the membrane. Most toxins are similar to the nutrients that the body needs, and they cross the membranes in the GI track along with everything else.

Dermal Absorption

The GI tract is designed to allow many materials to pass into the bloodstream, whereas the skin is generally designed to prevent most things from entering the body. However, the skin is not an impervious wall; some things pass through it. Many of the substances that pass through the skin are either harmless or even beneficial. If the substance that can pass through happens to be a toxin then a situation arises where simple contact with that poison can cause symptoms to develop. Sometimes certain solvents or agents can help a substance pass through the skin. Many drugs now are delivered via a transdermal patch, which contains the necessary agents to help the drug penetrate the skin. If a toxin happens to be mixed with certain solvents or agents, it might be able to penetrate the skin.

Respiratory Inhalation

The lungs represent the third area through which external substances can enter the body by crossing a membrane. The membranes in the lungs are much thinner than those of the skin or those in the GI tract. They are designed to accommodate the rapid movement of gases into the bloodstream. Unfortunately, if there are poisonous gases in the air that are inhaled, then these gases will enter the bloodstream. In addition to gases, fine particles can also get into the lungs. The likelihood of these particles causing lung damage is related to the size of the particles. Larger particles are not prone to cause damage to the lungs because these particles tend to be removed by the **cilia** (fine hairlike fibers) in the nasal passages or the upper respiratory tract. Very fine particles, however, may go right in through these areas and find their way into the **alveoli**, the air sacs where gases pass in and out of the bloodstream. The alveoli have no mechanism to remove these fine particles. Particles such as these, when respired deep into the lungs, are trapped. These particles remain there and irritate the cells in the alveoli and the bronchioles leading to the alveoli. The lungs respond to the foreign particles by sealing off the damaged area with scar tissue. This process is called **fibrosis**. The sealed area is, of course, no longer able to provide for the exchange of gases between the blood and the air. With continued exposure, the lungs accumulate more and more scar tissue, which at some point can severely impair their functioning.

CLASSIFICATION OF TOXINS

Toxic substances can be classified based on their mode and site of action. Not all of the possible classifications are considered in this book. The following are discussed: **respiratory toxins**, **general metabolic toxins**, **neurotoxins**, **endocrine system toxins**, **carcinogens**, **mutagens**, and **teratogens**. Some of the environmentally important toxins in each of these groups are discussed in this chapter.

RESPIRATORY TOXINS

As noted earlier in the discussion on the routes of exposure, toxins that cause problems in the lungs are generally either gases or particulates. The most important of the gaseous respiratory toxins are sulfur dioxide, nitrogen dioxide, and ozone. First, these gases are discussed followed by the various particulates.

Sulfur Dioxide

Sulfur dioxide (SO_2) plays an important role in Chapter 9, where it is noted to be a common constituent of one type of air pollution. The gas is a respiratory tract irritant that acts by directly damaging lung tissue and by increasing airway resistance. The increased airway resistance is produced because the gas causes the airway to constrict. In addition, irritation of lung tissue causes the lungs to produce mucous secretions that interfere with the flow of air. All of these factors can make breathing difficult and painful.

Sulfur dioxide is lethal at approximately 500 ppm.* Such a level is rarely encountered in environmental situations. Except in very unusual situations, the level of sulfur dioxide in the atmosphere rarely exceeds 10 ppm; however, long-term exposure to low levels of a toxin such as sulfur dioxide may increase the risk of health problems. Such a risk of chronic exposure can be a problem for the elderly and the infirm. Serious medical problems can be made worse by the presence of low levels of sulfur dioxide in the air.

Generally, plants are much more sensitive to sulfur dioxide pollution than animals. Exposure of some strains of wheat and barley to sulfur dioxide levels of 0.15 ppm for 72 hours can reduce grain yields. Depending on the level of exposure, sulfur dioxide kills leaf tissue and bleaches chlorophyll. Loss of chlorophyll decreases the plant's ability to carry out photosynthesis. Soybeans exposed to 0.8 ppm sulfur dioxide for 24 hours show extensive chlorophyll loss. If the levels are high enough, the death of the plant is certain. The plant life in some areas around which smelting operations are carried out has been absolutely devastated by sulfur dioxide. The sulfur dioxide is produced when sulfide metal ores are roasted to recover the pure metal. The sulfur in the ore is oxidized to sulfur dioxide. In the past, the sulfur dioxide was often released directly into the air. Some of the most well-known examples occurred at Sudbury, Ontario, and Copperhill, Tennessee. The areas around these plants were reduced to virtual deserts. In recent years, the sulfur dioxide emissions in these areas have been reduced or stopped all together; however, the recovery has been very slow and is still not complete.

Nitrogen Dioxide

This gas has effects similar to those of sulfur dioxide in that it causes inflammation of lung tissue. These effects are quite pronounced at high levels. Exposure at a level of 50 to 100 ppm for a few minutes to an hour can cause lung inflammation that may last for 6 to 8 weeks. Exposure to levels of 150 to 200 ppm causes a condition known as *bronchiolitis fibrosa obliterans*, which might lead to death in 3 to 5 weeks. At a level of 500 ppm, a condition known as *silo-filler's disease* develops and

* The concentration unit of parts per million is similar to percent composition, which can be thought of as parts per hundred (percent). Using this model, a substance present at 0.0001% is present at a concentration of 1 ppm.

might lead to death in 2 to 10 days. (The disease is so named because silage, which is chopped-up damp plant material, can produce the gas in a closed silo. When a worker enters the silo to work with the silage, he may be exposed to the gas and die.)

Fortunately, environmental levels of nitrogen dioxide are generally fairly low, in the range of 1 to 3 ppm. At this level, the gas is much less dangerous to humans. Chronic exposure at 10 to 25 ppm for several months is necessary to produce severe symptoms in animals. There is some evidence that the gas may have long-term effects on the ability of the body to fight against lung infections. Nitrogen dioxide is also not as toxic to plants as sulfur dioxide. Therefore, it would appear that one has less to fear from nitrogen dioxide as an air pollutant than from sulfur dioxide. Unfortunately, nitrogen dioxide does its damage indirectly, as will be seen in Chapter 9.

Ozone

Ozone (O_3) is another form of the element oxygen. Diatomic oxygen (O_2) is the more stable form. Both materials contain only oxygen but are different molecules, one being O_2 and the other O_3. Two different molecular forms of the same element are known as **allotropes**.

Ozone is a pungent-smelling, bluish gas. The smell of ozone gas can usually be encountered around arc welding or a model train transformer. In very low concentrations for short periods it is not particularly harmful, but in slightly higher concentrations for longer periods this gas becomes harmful to humans and particularly to plant life. The potency of this air pollutant is illustrated by the fact that, although commonly encountered levels of this gas are low (generally only 0.1 to 0.5 ppm for typical pollution levels), it is considered as one of the more serious air pollutants.

Usually, the threshold of toxic effect in healthy, exercising individuals is at about 0.15 ppm. Above this level individuals begin to experience coughing, shortness of breath, and nose and throat irritation. Ozone is very irritating to the lungs. The lungs respond by producing fluid, which may begin to fill up the alveoli (microscopic air sacks). Ozone is one of the strongest oxidizing agents known and as such attacks the lung tissue by oxidizing it. The damage can be severe, particularly in people who are already compromised by illness or age.

The damage done by ozone to plants is even greater than the damage done to humans and other animals. Ozone attacks plants, producing yellow spots on the green leaves. Plants are so sensitive that a brief exposure at 0.06 ppm may cut the photosynthesis rate in half. It has been estimated that crop losses in sensitive crops such as peanuts, soybeans, and wheat range from 5 to 20% because of regional ozone pollution.

Particulates

Many indoor and outdoor air pollution sources fill the air with fine particulates that can cause severe lung disease. There are several sources of fine particles that can lead to lung disease. One source, which is discussed in detail in Chapter 9, is the industrial combustion of coal that leads to the production of smoke containing relatively small particles of soot and fly ash. Another common source of airborne particles is cigarette smoke. These particles are inhaled not only by the smoker but also by other people in the vicinity as well.

Some specific types of particulates have been known for a long time to cause disease, and these diseases have come to have specific names associated with them. One such condition is associated with coal mining. Coal miners are exposed to a significant amount of coal dust, and if proper precautions are not taken the miners develop a condition known as **black lung disease**. Other conditions are silicosis, caused by silica dust; asbestosis, caused by asbestos fibers; brown lung, caused by cotton fibers; and farmer's lung, caused by hay mold. The common factor in each case is the small size of the particles.

RATES OF CHEMICAL REACTIONS, CATALYSIS, AND ENZYMES

Several of the groups of toxins to be discussed in the following sections require some understanding of rates of reaction. In particular, this information will be required when discussing heavy metals and neurotoxins. As you are probably aware, not all chemical reactions occur at the same rate. Some reactions proceed at a very slow rate—for example, the rusting of iron or steel, which may proceed gradually over a period of years. Some reactions, though, are quite rapid, such as the explosion of a stick of dynamite. The speed of a reaction is usually controlled by three factors:

1. Temperature
2. Concentration
3. Catalyst

Generally, if one provides heat to a reaction, it will speed up. For example, if one is trying to prepare a roast in the oven, it will cook faster at 375°F (190°C) than at 325°F (165°C). This observation is generally true of most reactions.

Increasing concentration plays a role by bringing more molecules closer together that are available to undergo a reaction. This can be illustrated by placing a burning piece of wood into a bottle of pure oxygen. The wood will begin to burn much more brightly and faster. Because combustion requires oxygen, using pure oxygen makes more oxygen atoms available than would be the case in air, which is only 20% oxygen.

Another way to speed up a reaction is to add a **catalyst**. A catalyst is a substance that speeds up a reaction without being consumed in the reaction. An example would be the production of sulfuric acid. To produce sulfuric acid, sulfur dioxide is converted into sulfur trioxide:

$$2SO_2 + O_2 \rightarrow 2SO_3$$

This reaction, however, is too slow to be useful. To speed up the reaction, nitric oxide and nitrogen dioxide are added to the mixture. These nitrogen oxides do participate in the reaction but are not consumed.

Catalysts are very important to biological systems. The reason is that an organism can use a catalyst, but temperature and concentration are usually beyond the organism's control. If a reaction is too slow to occur at body temperature, the body cannot just turn on a tiny oven. Raising the temperature would interfere with other reactions. The response of living organisms is that they build their own special catalysts called **enzymes**. Enzymes are biological catalysts. Each of these enzymes speeds up one or perhaps a few chemical reactions. Enzymes are composed of protein, as is much of the material of which the body is composed. Proteins are very large molecules principally composed of carbon, hydrogen, oxygen, and nitrogen. Without enzymes biological organisms just cannot survive.

GENERAL METABOLIC TOXINS

Metabolic toxins are those toxins that act by interfering with some essential biochemical process. The effect can be massive, causing immediate death, or the effect can be slow and cumulative, leading to slow decline and eventually death. Cyanide is an example of the first type, whereas arsenic is of the second type. The toxins in this section are referred to as general metabolic toxins because they include those toxins that have multiple modes of action, including interfering with an essential biochemical process. Toxins of the general type that are being discussed include the heavy metals, organochlorine pesticides, polychlorinated biphenyls (PCBs), and dioxins.

Carbon Monoxide

Carbon monoxide is a well-known poison. This colorless gas can be found coming out from the tailpipe of any car, and most people know that running a car in a closed garage can fairly quickly lead to death. Carbon monoxide is a potent and fairly easy-to-understand metabolic poison. Blood contains a very important substance called hemoglobin (Hb), the function of which is to carry oxygen to the tissues. In the lungs, the hemoglobin combines with oxygen and is converted into oxyhemoglobin (HbO_2):

$$Hb + O_2 \rightarrow HbO_2$$

The oxyhemoglobin is carried by the blood to the tissues, where it gives up its oxygen and reverts back to hemoglobin:

$$HbO_2 \rightarrow Hb + O_2$$

The effect of carbon monoxide comes from the fact that hemoglobin has over 200 times greater affinity for carbon monoxide than oxygen. Therefore, in the presence of carbon monoxide, the hemoglobin combines with carbon monoxide to form carboxyhemoglobin (HbCO) in preference to the reaction with oxygen:

$$Hb + CO \rightarrow HbCO$$

The carboxyhemoglobin is carried to the tissues, where it is incapable of providing oxygen. Not only does the HbCO not have an oxygen molecule to release, but it is also very stable and will circulate in the blood for some time without losing the carbon monoxide. The result is that the hemoglobin becomes incapable of carrying oxygen for quite some time. The hemoglobin will eventually release the carbon monoxide if the victim is removed from the source of the gas. The inhalation of pure oxygen will help drive the carbon monoxide off the hemoglobin.

Generally, carbon monoxide levels of up to about 100 ppm can be tolerated for several hours without ill effect; however, after about 1 hour at about 600 to 700 ppm, noticeable effects begin to appear. Above 1000 ppm, a mild headache will develop and the skin will appear reddish. As the concentration increases, the symptoms will get more severe until about 2000 ppm, when death may occur. Death is almost certain at 4000 ppm or higher for 1 hour. The concentration in relation to the time of exposure is important because, for example, 1000 ppm is not fatal after 1 hour but probably would be after 4 hours.

Carbon monoxide has numerous sources in nature, the home, and industry. This gas can be produced whenever something is burned. The production of carbon monoxide is the result of the incomplete combustion of materials containing carbon. (A more complete discussion of carbon monoxide production is found in Chapter 9.) Some of the sources of this poisonous gas are the internal combustion engine, coal-fired industrial plants, cigarettes, charcoal burners, burning leaves, burning timber from cleared forestlands, and the ever-popular home fireplace. Over time, carbon monoxide is converted into carbon dioxide by natural processes in the atmosphere:

$$2CO + O_2 \rightarrow 2CO_2$$

In remote areas where very little carbon monoxide is produced, such a removal process leaves only a very low worldwide carbon monoxide background level of 0.1 ppm. Although carbon monoxide is removed slowly, this does not mean that the gas cannot build up to fairly high concentrations in some localities. Heavy traffic can produce urban levels of 20 to 50 ppm. Although such levels are not fatal and probably do not produce any obvious effects over several hours, it should not be assumed that these concentrations are harmless. Because carbon monoxide attaches itself to hemoglobin and reduces the oxygen-carrying capacity of the blood, one's powers of concentration and judgment may be diminished. These problems may increase with the length of the exposure. The U.S. government considers an average concentration of 35 ppm for 1 hour or 9 ppm for 8 hours as the upper limit acceptable for carbon monoxide. Many cities often exceed these levels. Extremely high levels approaching or exceeding 100 ppm may occur in underground traffic corridors and in tunnels.

Nitric Oxide

Nitric oxide (NO) is a gas that has some similarities to carbon monoxide. Both gases are toxic and the mode of action of each is similar. Each of the oxides reacts slowly with oxygen, but nitric oxide reacts slightly faster than carbon monoxide:

$$2NO + O_2 \rightarrow 2NO_2$$

As a result, nitric oxide is removed from the atmosphere much more quickly than carbon monoxide. The toxicity of nitric oxide is rarely a problem, although it is attracted to hemoglobin much as carbon monoxide is. The environmental concentrations are simply not high enough. Nitric oxide has been found to be involved in a number of biological processes in the body, and as a result one would not be surprised if it had toxic biological effects. This simple little diatomic molecule is involved in the dilation of blood vessels, the activation of certain immune responses, and the transmission of nerve impulses in certain activities in the brain. To date, no toxicity of nitric oxide from environmental exposure has been demonstrated.

Heavy Metals

At the bottom of the periodic table several metals are found that are toxic to varying degrees. These metals include mercury, thallium, and lead. The toxic effects of these metals occur in somewhat similar ways. Heavy metal poisons act by interfering with various enzymes. One of the ways to inhibit an enzyme involves sulfur. In most enzymes sulfur is present, often as a sulfhydryl group (–SH). Although there are usually not many of these groups, the sulfhydryl group is often important for the enzyme to function. Heavy metals often tie up two of these groups by attaching themselves to the sulfur atoms:

$$\text{Active Enzyme}\begin{matrix}SH\\SH\end{matrix} + Hg^{2+} \longrightarrow \text{Inactive Enzyme}\begin{matrix}S\\|\\Hg\\|\\S\end{matrix} + 2H^+$$

As can be seen from the reaction, it is the metal ion of mercury and not the neutral metal atom that causes the problem. Hence, it may be said that the neutral metal atom is usually not toxic.

This section discusses mercury, lead, cadmium, and arsenic. There are other toxic heavy metals, but they behave in similar ways to the metals we will discuss. It is worth noting that arsenic is not a metal but is instead considered to be a metalloid. Metalloids are elements that are metal like but have some properties of nonmetals.

Mercury

Ever hear the expression, "Mad as a hatter?" This expression is a reference to mercury poisoning. In the 19th century, mercury(I) nitrate solutions were used to stiffen the felt in hats. The continuous exposure to these solutions led to mercury poisoning. Mercury poisoning leads initially to sore gums and loose teeth but progresses to a condition known as *erethism*. This condition is characterized by jerking, irritability, emotional instability, and mental disturbances. Unfortunately, by the time these symptoms are obvious a lot of damage has already been done to the brain and the nervous system.

There is an antidote for mercury, but it must be used quickly before irreversible damage is done. The antidote is a compound developed by the British during World War I to counter the effects of an arsenic-containing poisonous gas called *Lewisite*. This compound came to be known as **British anti-Lewisite (BAL)**. BAL acts by tying up the mercury in a fairly nonreactive compound known as a chelate (from the Greek word *chela* meaning "claw"):

$$\text{(BAL)}\quad \begin{array}{l} CH_2-OH \\ | \\ CH\ -SH \\ | \\ CH_2-SH \end{array} + Hg^{2+} \longrightarrow \begin{array}{l} CH_2OH \\ | \\ CH\ -S \searrow \\ |\qquad\qquad Hg \\ CH_2-S \nearrow \end{array}$$

The chemistry involved in the chelation of mercury using BAL is very similar to the chemistry of the attack of mercury on some enzymes. Both are reactions of mercury ions with sulfhydryl groups.

One of the worst incidents of mercury poisoning occurred in Iraq in the winter of 1971–1972, when wheat treated with a mercury fungicide was used to make flour, which in turn was used to make bread. The wheat was originally intended to be used for seed and was treated with a mercury compound to prevent the growth of mold during storage. By the time the incident was over, 6530 people had been hospitalized and 459 had died.

Elemental mercury has long been considered to be less toxic than mercury compounds because it is the mercury ions that cause the enzyme damage. This view seems to stand up fairly well based on some of the evidence. The poison control center of the Department of Health of New York City reported no serious complications with 18 separate accidental ingestions of liquid mercury over a 2-year period.

Mercury vapor, however, seems to be another story. Because mercury is a liquid with a moderate boiling point (675°F, 357°C), it produces a certain amount of mercury vapor at room temperature. The mercury vapor when inhaled or absorbed through the skin is somehow converted into the poisonous mercury ions. People chronically exposed to mercury vapor could suffer the effects.

Liquid mercury can also be a serious problem if it is not handled properly. Chapter 13 discusses the dumping of liquid mercury in places such as waterways, which may not be as harmless as one might expect.

Lead

Lead is another heavy metal that has been used extensively in our culture. The metal has been around at least since the Egyptian and Babylonian times. Lead is an attractive metal for many purposes because it is soft and malleable. The Romans used lead extensively for plumbing. The Latin word *plumbum* was the origin for the English word "plumbing" and the chemical symbol for lead, Pb. In the Middle Ages, lead was used as a roofing material on public buildings and cathedrals.

The toxicity of lead has been known for over 2000 years. Despite the danger, lead has found continued use in our culture in ever-increasing amounts. Lead is currently used in the manufacture of lead storage batteries and was historically used as an additive in gasoline to cut down on engine knock. The lead from gasoline found its way out of the car's tailpipe and into the air. Starting in the 1970s, cars were equipped with catalytic converters that were poisoned by heavy metals; therefore, it was necessary to ban the addition of lead to gasoline. Atmospheric deposition of lead in the United States from auto exhausts has largely disappeared.

We continue to use lead for many other purposes. Lead is used to make solder, to glaze pottery, to make ornamental leaded glass, and to undercoat steel products. Another important use of lead that has been abandoned is as a pigment for paints. The two major colors in which lead was used as a pigment were yellow and white. White was made from basic lead carbonate ($2PbCO_3.Pb(OH)_2$) and yellow was made from lead chromate ($PbCrO_4$).

At one time, lead was used to seal tin cans used for commercial canning. A small amount of lead can get into the food in this way. In the early years of the canning industry, this process was probably not well controlled. As a result, if one got their entire diet from canned food, it could result in lead poisoning. An extreme example of this type of lead poisoning occurred in 1845. After the Franklin expedition left England in search of the Northwest Passage, the party was never heard from again. In the 1980s, the graves of many of the crew were found in the Canadian Arctic. The bodies were well preserved from having been buried in the permafrost. Autopsies indicated that they had died of lead poisoning. The source of the lead was probably their canned food, sealed with lead solder.

Lead is encountered constantly in our everyday lives. This metal shows up in our food and in our water from lead-soldered water pipes, and some is even found in the air. It does not matter whether the lead is in the form of lead ions or neutral lead atoms because lead atoms are converted into toxic ions in the body. Because we are constantly taking lead into our bodies, why don't we all come down with lead poisoning? The reason is because the body can get rid of a certain amount of lead every day through urine, feces, and sweat. As long as this amount, approximately 230 micrograms (μg, 1 millionth of a gram), is not exceeded, the body will pass the lead right on through. If a person takes in more than 230 μg of lead per day, the lead will begin to build up in the body and cumulative lead poisoning will develop. Chronic lead poisoning is irreversible except in the early stages. Treatment must be started early if permanent effects are to be avoided.

Where would one encounter the extra lead to bring on lead poisoning? It could be as simple as handling large amounts of lead on a continuous basis, such as in a battery factory. This lead can be absorbed through the skin. One fascinating example of lead poisoning is a condition known as *pica*, which affects many children in poverty-stricken areas. Children who are ill fed and anemic develop an appetite for unusual things. A material the children are often attracted to is paint, which is found peeling off furniture, woodwork, and walls in older homes. The paint in these older homes is usually the older, lead-based type of paint. The children chew on or eat the paint chips and develop lead poisoning.

The tragedy of lead poisoning in children is that lead attacks the brain and nervous system of these developing young people. Unless it is caught in time, the result is retardation and hyperactivity. Lead is a cumulative poison that does continual damage over time. In adults, the nerve damage manifests itself as irritability, sleeplessness, and irrational behavior. Lead also usually results in the development of anemia. The lead interferes with an enzyme that is necessary to make hemoglobin, a key compound in blood. The hemoglobin levels drop and the person becomes anemic.

Lead can be removed from the body by BAL in much the same way as mercury. Another compound that can tie up lead ions and remove them from the body is the ethylenediaminetetraacetate (EDTA) ion. The reaction is as follows:

$$CaEDTA^{2-} + Pb^{2+} \rightarrow PbEDTA^{2-} + Ca^{2+}$$

Cadmium

Compared to mercury and lead, cadmium is a fairly recently observed toxic metal in the environment. The metal was discovered in Germany in 1817 as a byproduct of zinc refining. In the short history of cadmium the metal has become a fairly useful material. It is used in paint pigments and plastic additives and has also found various uses in the metal industry for processes such as electroplating and the production of alloys. The major use by far, however, is in the production of rechargeable nickel–cadmium batteries.

Environmental pollution is of major concern, as cadmium is removed from the body very slowly. The half-life of cadmium in the body is about 30 years. As a result, any continuous exposure to cadmium will gradually build up over time. The most well-known large-scale cadmium poisoning occurred in Japan when farmers ate rice contaminated with cadmium. The water source for the rice farms went near some zinc mines upstream and cadmium was an impurity in the zinc. The disease was given the name *itai-itai*, which literally means "painful-painful." There were a number of deaths from the poisoning.

Industrial cadmium pollution peaked in the 1960s and has declined since, with all industrial sources currently accounting for less than 25% of the exposure. The most common sources of cadmium pollution currently are burning fossil fuels, the use of phosphate fertilizers, and natural sources such as soils. One group of individuals in our society has a greater than normal exposure to cadmium: smokers. Plants tend to bioaccumulate cadmium from the soil, and tobacco plants do this as well as any. In addition, because tobacco is smoked, the cadmium-laden material from the plant ends up in the lungs as opposed to the GI tract, which would be the case for plants that are eaten. Cadmium is absorbed into the blood from the lungs much more efficiently than it would be from the GI tract.

Because cadmium is mainly used in nickel–cadmium batteries, recycling of these batteries is of prime importance to keep the cadmium out of the environment. These cadmium-type rechargeable batteries are being gradually replaced by nickel–metal hydride batteries that have no cadmium, but these hydride batteries are more expensive.

Arsenic

Of the world's poisons of which one might be aware, arsenic would certainly be one. The name of this element finds its way into all kinds of literature, such as the title of the play "Arsenic and Old Lace." Actually, one rarely encounters pure elemental arsenic. Arsenic is nearly always encountered in the form of a chemical compound. Most of these compounds are solids containing either other inorganic elements or organic groups. The most notable exception is the toxic gas arsine, AsH_3. Arsine is the most acutely toxic form of arsenic.

Microorganisms exposed to arsenic can produce a compound known as *dimethylarsenate*, which is easily taken up by shellfish and fish. Consumption of these aquatic items by humans can expose them to arsenic poisoning. The symptoms of acute arsenic poisoning include vomiting, diarrhea, abdominal pain, and esophageal pain. Eventually, vasodilation and heart-rate depression occur, followed eventually by circulatory failure and death. More insidious effects of arsenic are caused by chronic poisoning. Those of a criminal mind will often give low doses of arsenic over a long period of time because the symptoms are low grade and relatively nonspecific. The low-dose arsenic will give the impression of a long-term chronic illness with slow decline toward death. The arsenic will build up, particularly in the hair, but one needs to think to look for it. Arsenic can occasionally be an environmental poison. The main way in which one would come in contact with environmental arsenic is through food. Except for intentional poisoning, this situation is not common. There are a few locations where arsenic is found naturally in the water supply. In these places, residents may suffer from various skin disorders. More extensive exposure may lead to gangrene of the lower extremities, or "blackfoot disease." Natural water supply contamination has been observed in parts of Taiwan and South America.

Organochlorine Pesticides, Polychlorinated Biphenyls, and Dioxins

The members of this group of chemical toxins are related by their structures rather than by their uses or functions. This section could have been titled, "The Pesticides, the Insulators, and the Impurities." The structural similarities become very apparent when looking at the structures of some examples of each:

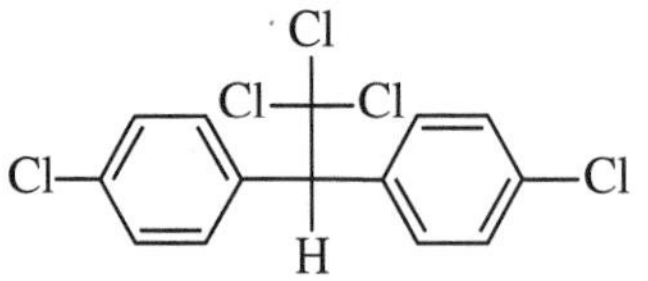

DDT (Organochlorine Pesticide) 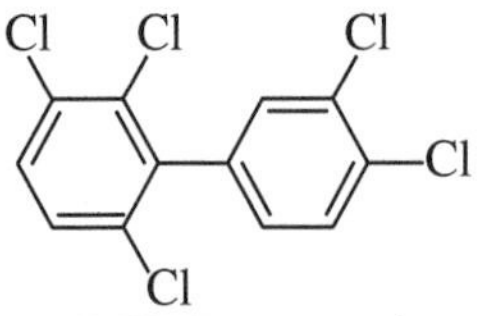 PCB Compound 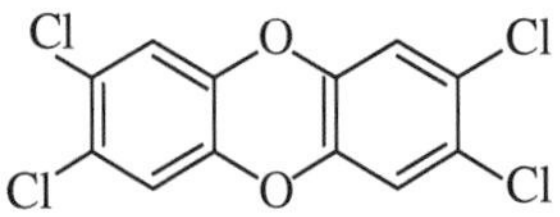 Chlorinated Dioxin

The related structures of this group of chemicals give them many similar properties. The importance of these properties becomes apparent in the latter portions of this chapter concerning environmental persistence and the environmental movement of toxins. First, let us consider the organochlorine pesticides.

The organochlorine pesticides are a group of substances that are designed to rid ourselves of various undesired organisms. **Pesticides** are substances that kill or otherwise control unwanted organisms; many types of pesticides kill various types of pests. **Insecticides** are substances that kill insects. **Herbicides** are compounds that kill plants. **Fungicides** are substances used to control the growth of various types of fungus. There are other pesticides, but these make up the bulk of the pesticides produced in North America. This section mostly deals with insecticides.

The earliest known insecticides probably date back to Greece around 1000 BCE, when sulfur dioxide was produced by burning sulfur. Sulfur dioxide is very toxic to insects, but unfortunately it is also toxic to humans. This process was used to kill insects in homes or other enclosed spaces. While the sulfur burned in the enclosed space, the occupants and their animals left. Later, after the sulfur dioxide has diffused out of the house, they would return. This process is still used occasionally today.

Over the ensuing centuries, many materials have been used as insecticides, including elemental sulfur, fluorides, boric acid, and oils of various types. In recent times, arsenic and its compounds were used. Given arsenic's generally toxic properties, which have been discussed, there was a desire to find an insecticide with less toxicity to humans. Such materials became available in the late 1930s and early 1940s. These materials were the organochlorine insecticides.

The most well known of the organochlorines is ***p*-dichlorodiphenyltrichloroethane**, or **DDT**. The properties of DDT as an insecticide were discovered by the Swiss chemist Paul Müller in 1939. Müller was able to demonstrate that DDT was an extremely effective insecticide with little apparent effect on humans and other animals. Although developed as an agricultural insecticide, the material proved quite effective in controlling many insect-borne diseases such as malaria, yellow fever, typhus, and plague. The Allies during World War II used DDT powder directly on the troops to combat lice-induced typhus, which is often a problem in the unsanitary conditions of battle. DDT worked well and was used extensively. Many considered DDT to be almost a "superinsecticide" until a few problems began to emerge. The first problem was that various species of insects began to develop resistance to the toxicity of the chemical. As much as this was disconcerting, there did not appear to be anything dangerous in the observation. Although DDT had very little toxicity on mammals or for the most part birds, its negative effects on fish, aquatic invertebrates, and many useful insects such as honeybees were fairly widespread. By the late 1950s another problem was becoming apparent. There appeared to be mounting evidence that DDT was causing a decline in bird populations, specifically carnivorous birds found on top of the food web. The chemical appeared to interfere with the ability of the birds to produce eggshells thick enough to be sat upon during nesting. The eggs cracked, leading to a severe decrease in reproduction, and bird populations began to decline. Rachel Carson's *Silent Spring* brought attention to this problem and others. In time, DDT was banned in the United States, but it is still used in certain parts of the world. The main argument for its continued use is to combat malaria in tropical areas of the world. DDT is very effective in controlling the mosquitoes that spread the disease, but there are other chemicals that could carry out the same function. DDT is still used in poorer countries of the world because it is inexpensive. Despite the environmental damage that DDT can cause, it is a powerful argument to note the thousands of lives that can be saved for a relatively small expenditure of funds.

Within this group of similar chemicals is a collection of closely related compounds known as **PCBs**, which were developed in the 1930s for use in a number of applications, the most notable of which included their use as insulators in large electrical transformers, as heat-exchange fluids, as an additive in plastics, and as a component in paints.

The toxicity of PCBs is certainly a complex issue. These compounds actually have little acute toxicity except for their effects on aquatic organisms. Chronic toxicity can be more of a problem, as it varies by species and some of the symptoms are suspected but not proven. Two large-scale human PCB exposure incidents were due to the accidental contamination of rice oil by PCBs. The first incident took place in Japan in 1968 and the second in Taiwan in 1979. A large number of symptoms were noted over the next weeks to years, but the only one conclusively associated with the PCBs was a condition known as **chloracne**, which is also associated with dioxins. The manufacture of these compounds was discontinued in the United States in 1976, and any use of PCBs was severely restricted.

The toxic materials we usually refer to as **dioxins** are actually polychlorinated dioxins. As can be seen from the structures given earlier, these polychlorinated dioxins have some similarity to the PCBs. The dioxins themselves have no known use. These materials are in fact byproducts of other processes. One of the more well-known processes is the manufacture of Agent Orange, the herbicide used in the Vietnam War. Dioxin was produced as a byproduct of the manufacture of the active ingredient in Agent Orange, 2,4,5-trichlorophenoxyacetic acid (2,4,5-T). After the war, many soldiers complained of symptoms that they felt were related to exposure to Agent Orange. Other common sources of dioxins include tobacco smoke, engine exhaust, bleached paper products, incineration of chlorine-containing plastics, and burning wood. One additional problem with dioxins is that they can be products of the high-temperature chemistry of burning wood and chlorinated plastics. The dioxins themselves are, therefore, difficult to destroy. If incineration is the method of destruction, then very high temperatures must be attained to break down the dioxin molecules.

Much alarm was raised with regard to dioxins when it was reported that these compounds were the most toxic substances known. It is true, as reported earlier, that the LD_{50} for TCDD is extremely low at 0.0006 mg/kg for the guinea pig, but is not nearly as low as substances such as botulinum toxin. Finally, it is worth noting that in most situations dioxin usually arises as a mixture of compounds for which most are much less toxic than TCDD.

Many animals seem to be more sensitive to dioxin than humans. Upon exposure to dioxin, these animals develop a wasting disease and simply waste away over time, eventually dying. In humans, dioxins have been suggested to be related to many disorders, but the data are not clear cut. The only certain disease linked to dioxin (and PCB) poisoning in humans is chloracne. Chloracne is generally nonfatal but manifests as disfigurement that usually affects the face. One well-known

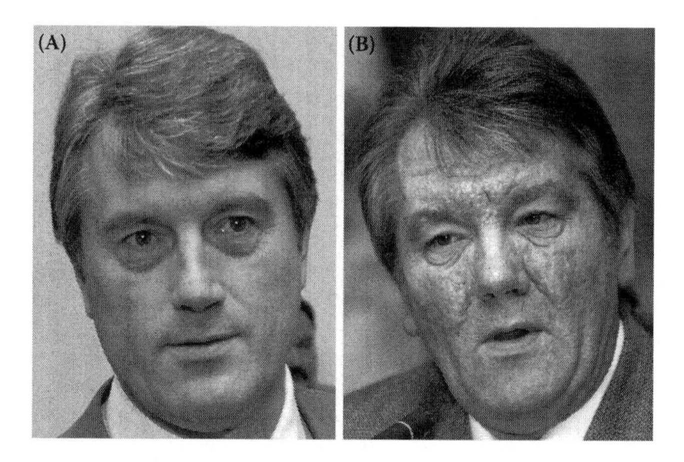

Figure 6.2 Victor Yushchenko (A) before (July 2004) and (B) after (December 2004) apparent dioxin poisoning. (Photographs by STF/AFP/Getty Images and used with permission.)

incident of alleged dioxin poisoning is the case of Ukrainian presidential candidate Victor Yushchenko. The poisoning was suspected when over a period of 5 months Yushchenko's facial appearance deteriorated severely, as shown in Figure 6.2.

Dioxins are definitely environmental toxins that need to be restricted, but the matter of exactly how dangerous they are to humans is unclear and continues to be debated.

NEUROTOXINS

There are many toxins whose principal site of action is the nervous system, and these toxins are known as *neurotoxins*. Nerves are the transmitters of messages both to and from the brain. The nerve fiber must be able to be "turned on" and "turned off" to transmit information. Nerve cells, or **neurons**, are very long, thin cells that are in essence microscopic fibers; however, a single nerve does not connect the brain to a given organ or muscle. The transfer of a nerve signal from one nerve to another occurs across what is known as a **synapse**, a microscopically narrow gap between the two nerve cells. The transmission of the nerve impulse across this space occurs chemically through **neurotransmitters**. Many neurotoxins disturb the chemical processes occurring at the synapse during nerve signal transmission.

Most neurotoxins affect the nerves that are known as cholinergic nerves. These nerves generally send commands to the heart, lungs, and skeletal muscles and several other sites. These nerves use *acetylcholine* as a neurotransmitter. The process by which the nerve impulse is transmitted across the synapse is shown in Figure 6.3. When the nerve impulse arrives at the nerve ending, acetylcholine is produced from acetic acid and choline by catalysis from the enzyme acetylase. The acetylcholine travels to the end of the other nerve and stimulates that nerve. The acetylcholine must be removed, or the nerve will be continuously stimulated. Under

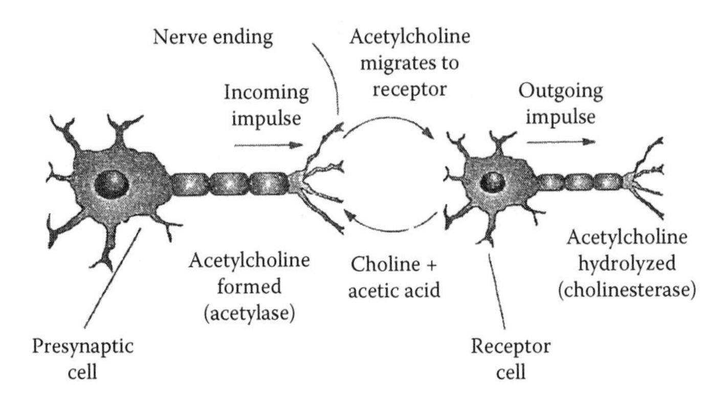

Figure 6.3 Nerve impulse transmission at a synapse in a cholinergic nerve. (From Hill, J.W., *Chemistry for Changing Times*, 6th ed., Macmillan, New York © 1992. With permission of Pearson Education, Inc., Upper Saddle River, NJ.)

continuous stimulation no further information will be transmitted. Remember that a nerve must be able to be turned on and off to be able to transmit information. Once acetylcholine has stimulated the nerve, it is normally destroyed by the action of the enzyme cholinesterase, which converts the acetylcholine back into acetic acid and choline:

$$\text{Acetic acid} + \text{Choline} \rightarrow \text{Acetylcholine} + H_2O$$

$$\text{Acetylcholine} + H_2O \rightarrow \text{Acetic acid} + \text{Choline}$$

Anything that interferes with this cycle will interfere with nerve transmission.

Neurotoxins generally interfere with the acetylcholine cycle in one of the three ways. The first two ways relate to ways to keep the nerve impulse from crossing the synapse. Certain bacteria such as *Botulinus* produce toxins that inhibit the synthesis of acetylcholine (Figure 6.4). If no acetylcholine is produced, then the impulse cannot be transmitted. The other way to stop transmission is to block the receptor nerve from receiving the impulse. Nicotine is one such poison. As a pure compound, nicotine is a potent nerve poison. The lethal dose for a 150-lb (70-kg) man has been estimated to be only about 75 mg (0.0026 oz.). Nicotine is, of course, associated with cigarette smoking. Luckily for the smoker, the quantities inhaled during cigarette smoking are well below the lethal dose. Other nerve poisons of this type include atropine, curare, morphine, codeine, and cocaine.

The third way to stop nerve impulse transmission is to permanently turn the nerve on. Remember that acetylcholine turns the receptor nerve on; if the acetylcholine cannot be destroyed, then the nerve cannot be turned off. One class of neurotoxins is known as **anticholinesterase poisons**. These poisons act by inhibiting the enzyme cholinesterase; without cholinesterase, acetylcholine cannot be broken down.

It is the last class of poisons that is probably most important environmentally. Many widely used insecticides fall into this class of toxins. These insecticides can be dangerous due to accidental contact or through their dispersion into the environment.

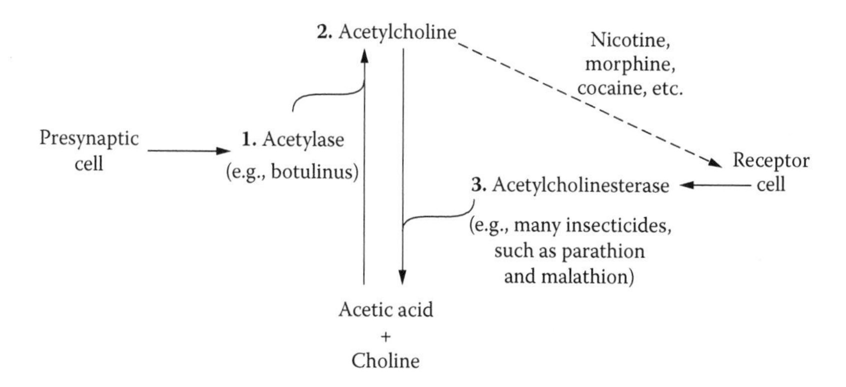

Figure 6.4 Neurotoxins typically interfere with acetylase (1), with reception of nerve impulse (2), or with acetylcholinesterase (3).

Although they would obviously be toxic if ingested, they can also be absorbed through the skin. The earliest and most toxic of this class of insecticides was parathion, which was first licensed for use in 1944. Parathion has a rat LD_{50} of 20 mg/kg. A dose of 120 mg has been known to kill an adult human, and as little as 2 mg has been known to be fatal to children. A relative of parathion, malathion, is considerably less toxic to mammals while maintaining its insect-killing ability. The rat LD_{50} for malathion is over 800 mg/kg.

A relative of these insecticides is the nerve gas sarin, which was first produced during World War II. This liquid releases vapors that can be absorbed through the skin. The material is so toxic that one drop of it can kill a person. The lethal dose is thought to be about 0.01 mg/kg.

ENDOCRINE TOXINS

In humans, as well as in numerous other species, many of the biochemical processes of the body are under the control of hormones. These hormones control our body's rate of metabolism, our blood sugar levels, our rate of growth as children, and our development and maintenance of various sexual and reproductive characteristics. Many of these hormones are related to the endocrine glands. **Endocrine hormones** are molecules that are released by an endocrine gland into the bloodstream. These hormones eventually attach to **receptors** on certain cells in the body to produce a response. For example, when insulin is released by the pancreas, it attaches itself to various cells of the body and causes the cells to allow blood sugar into the cell.

There is a large set of endocrine hormones that are produced from various glands and have numerous effects. This section focuses on the estrogens as they tend to be of utmost interest environmentally. The natural estrogen hormones include estradiol, estrone, and estriol, and these hormones are responsible for the development of female sexual characteristics and regulation of reproduction.

It turns out that a significant number of synthetic chemicals can mimic the estrogens, and these are known as **environmental estrogens**. As compared to other hormone receptors of the body, the estrogen receptors seem to be easily fooled into accepting other molecules as if they were estrogens. Because the biological effects of estrogens on an organism can be extensive, the environmental estrogens can produce fairly extensive effects, including feminization of males, abnormal female responses, disruption of reproduction, and effects on fetal development. Although these effects can be seen in humans, the greatest effects seem to occur in wildlife, including mammals, birds, fish, and reptiles.

The subject of environmental estrogens and endocrine toxins, in general, has become a very hotly debated area. Most of the studies involve situations where there are many uncontrolled variables. Often, studies involve low-level exposure leading to small effects over a large area, which can be very difficult to interpret. Alternatively, one can use laboratory studies, but they do not translate well into the real world. Occasionally, large amounts of chemicals are put into the ecosystem with fairly clear results. After a chemical spill in 1980 in Lake Apopka in Florida, the alligator population in the lake dropped by nearly 90% in the years following the spill. The problems persisted for a number of years after the spill was cleaned up. The effects included hatching problems and male alligators with unusually small penises. Unfortunately, most environmental estrogen insults produce much less clear-cut results.

Among the compounds that show estrogenic activity are PCBs, dioxins, a number of pesticides including a degradation product of DDT, and components used in the manufacture of plastics. Of particular interest recently is **bisphenol A (BPA)**, which is used in the manufacture of plastics used in reusable drink containers, DVDs, cellphones, eyeglass lenses, and as protective layers inside food and drink cans. The fact that BPA has estrogenic activity is not a surprise, as this fact was determined in the 1930s. The debate rages around the question of how significant the effect is at the very low levels likely to come from these plastic products. In spite of 25 years of scientific study, there appears to be no consensus on the danger of BPA. Because of the expected higher sensitivity of infants, BPA is no longer present in baby bottles and baby cups; however, decisions concerning exposure of other human populations have not been made.

Currently, the United States operates under the philosophy that chemicals can be used until they are proven unsafe. Under this approach, the U.S. government has allowed the use of BPA while debate regarding its safety continues. Many people advocate the **precautionary principle**, which states essentially that any potentially harmful substance should not be used until it is proven safe. Under this principle, BPA would be banned.

Many of these estrogenic compounds continue to find their way into the environment via commercial and industrial activities, but fish may also be exposed through human sewage. Treated human sewage often contains birth-control hormones, nonylphenol spermicide, and urinary human estrogens, among other things. All of these can have estrogenic effects on the fish in the water.

ALLERGENS

Allergens are a large class of compounds that activate the immune system. People vary widely in their response to allergens. Some individuals seem to be unaffected by a wide variety of allergens, whereas others need to take injections to counteract the effects of many of these substances. Allergens are a problem because they overstimulate the immune system. As a result, the immune system that is supposed to protect us becomes our enemy. Many allergens are naturally present in the environment, but pollution can add many more. Allergens can have a particularly potent effect when emitted indoors. One of the more potent and commonly encountered environmental allergens is formaldehyde. This compound, which is commonly used to preserve biological specimens, is often emitted from common indoor items such as particleboard, plywood, and foam insulation. Some people are so strongly affected that they cannot stay in rooms where formaldehyde vapors are present.

CARCINOGENS

Of all of the words that strike fear in our hearts, cancer is probably right at or near the top of the list. Considerable research has been carried out into the causes of cancer, and it would be impractical to review all of the material on the subject in this section. What is clear is that certain chemicals have the ability to cause cancers of various types. These chemicals are known as *carcinogens*. Cancer is a group of diseases related to cell growth and cell function. Once cancer develops, the cells affected are no longer under the control of growth regulation mechanisms. The cells reproduce at will at a rate that is usually faster than normal cells. These cells also lose functional identity and locational restriction; for example, if liver cancer develops, the cancerous liver cells no longer behave as liver cells, and such cells may be found at places other than the liver. Cancerous cells, by growing rapidly in places where they do not belong, can interrupt essential body functions and lead to death.

Cancer is widely believed by many scientists to be an environmental disease. The cascade of events leading to cancer is generally thought to start with some insult to the body by some external agent (chemicals, radiation, repeated injury, etc.). Although nearly all of us come into contact with potential cancer-causing insults, we do not necessarily develop cancer. For one thing, the process often takes 20 or 30 years, which tends to make it a disease associated with aging. Additionally, the carcinogens do not seem to cause cancer on their own, but instead seem to require repeated exposure to a promoter after initial exposure to the carcinogen. As a result, except in the case of a few cancer-related agents, not everyone who is exposed to a carcinogen develops cancer.

The history of environmental cancer-causing agents goes back to England in the late 1700s when two English physicians made the connection between external substances and cancer. As mentioned earlier, Dr. Percivall Pott noted that chimney

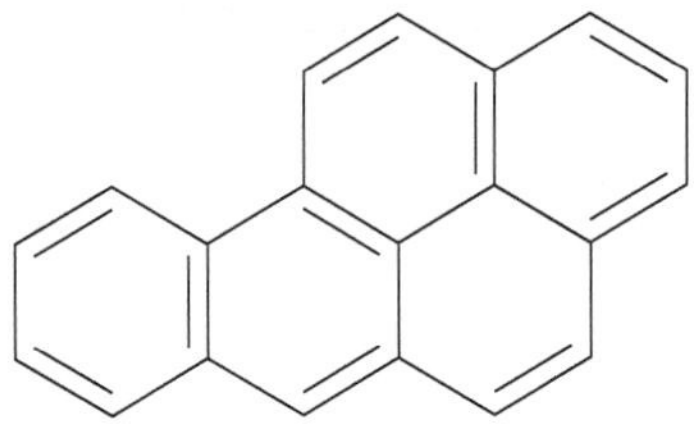

Figure 6.5 Benzo(α)pyrene.

sweeps had a higher incidence of skin cancer than the general population, and Dr. John Hill observed a correlation between snuff use and nasal cancer. In 1933, the carcinogen in chimney soot was identified as benzo(α)pyrene (see Figure 6.5).

Toward the end of the 19th century, the German physician Ludwig Rehn noted the increased incidence of bladder cancer in dye factory workers. Interestingly, 30 years went by, in many cases, between the beginning of employment and the onset of cancer, suggesting (as noted earlier) that carcinogens often take many years to show their effect. Initially, it was thought that aniline was the carcinogen, but, although it is very toxic, aniline was found to be noncarcinogenic. By 1937, it was confirmed that the culprit was 2-naphthylamine. The requirements for a compound to be a carcinogen are not obvious from its structure; for example, 2-naphthylamine is a carcinogen and 1-naphthylamine is not (see Figure 6.6).

Not all carcinogens are synthetic chemicals. Many substances exist in nature that are carcinogenic. These cancer-causing agents are often produced by plants, animals, and bacteria to protect themselves. Our food naturally contains a large collection of carcinogens. Varied items such as herbal teas, corn, peanuts, black pepper, celery, and strawberries all contain carcinogens. Obviously, we all have differing susceptibilities to these natural carcinogens. The point is not to make one fearful of eating most foods, but rather to help each of us to not overreact to the chemical carcinogens that are produced by humans. Due care should be taken with a known carcinogen, but it should be understood that carcinogens can never be totally isolated

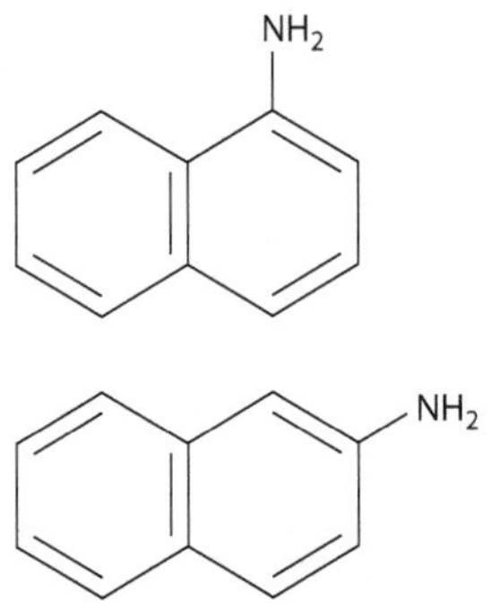

Figure 6.6 (Top) 1-Naphthylamine and (bottom) 2-naphthylamine.

from us. It is often tempting to insist that the government totally ban all carcinogens in the environment. Such a policy is not practical for anthropogenic carcinogens and impossible for carcinogens in general.

MUTAGENS

Mutagens are closely related to carcinogens. Mutagens are chemicals that damage or alter the genetic information in cells. Because the development of cancer is closely related to changes in the genes in cells, it is clear that mutagens and carcinogens may in many cases be the same substances. Mutagens are usually of the greatest concern when they interfere with the reproductive cycle. Any change in the genetic information in sperm cells or egg cells can cause changes in embryonic or fetal cells during pregnancy and can lead to birth defects. Many chemicals have been shown to be mutagenic to bacteria, various plants, animals, and, in some cases, human cell cultures. No one has ever proved that mutagens cause birth defects in humans because of the difficulty in demonstrating that a given birth defect is directly related to a specific chemical; however, some of these chemicals are strong enough mutagens in animals to be of a reasonable concern in humans. Caution is advised while using these chemicals.

Benzo(α)pyrene and ozone are two mutagenic substances of environmental importance. As discussed earlier, ozone was noted to be extremely toxic to plants, and it is in plants that ozone shows its mutagenic activity. Benzo(α)pyrene was discussed as being a carcinogen and not surprisingly is a mutagen, as well. The compound was demonstrated to be a mutagen to mice.

TERATOGENS

In the late 1950s and early 1960s in West Germany, a truly tragic story began to unfold. Large numbers of children were born with severely shortened or missing limbs. After some research, the problem was traced to a widely used medication, thalidomide. The drug was marketed as a remedy for morning sickness, nausea, and insomnia and was considered so safe in Germany that it was sold without a prescription. Unfortunately, if the pill was taken during the first 2 months of pregnancy, which was common because it was used to combat morning sickness, severe birth defects resulted. By the time this tragic episode had run its course, over 4000 cases were reported in West Germany and over 1000 cases in Great Britain. Only about 20 cases of thalidomide-induced birth defects occurred in the United States because an official of the Food and Drug Administration (FDA) was unconvinced of its safety. The drug was never approved for use in the United States.

Thalidomide is an example of a teratogen. Teratogens induce birth defects by causing direct damage to the developing embryo, but not by disturbing the genetic material in the cells. As a result, they cannot be classified as mutagens, as mutagens damage or alter the genetic information in cells. Teratogens are of most concern during the embryonic stage of pregnancy (18 to 55 days). It is during this period that the critical organs of the body are formed. Although damage can be caused during the fetal period (56 days to term), it is less likely and will probably be less severe if it occurs.

One of the more unusual teratogens was diethylstilbestrol (DES). DES was a synthetic estrogen-like molecule that could be used as an inexpensive, but effective, natural estrogen substitute. It was originally used to treat disorders such as vaginitis, gonorrhea, and menopausal problems. It was also used to suppress lactation. Eventually, the drug was believed to be useful to decrease the likelihood of miscarriages and was approved by the FDA for use in pregnant women in 1947. Initially, there was no apparent problem in the offspring from these pregnancies. It was only as the female children from these pregnancies approached and entered adulthood that the problem became apparent. A rare cancer known as clear cell adenocarcinoma (CCA), which almost never appeared in young women, began to show up in these DES offspring. The cancer was generally vaginal or cervical. In 1971, doctors at Massachusetts General Hospital demonstrated the connection between DES and CCA, and the FDA sent out an alert warning against using DES during pregnancy. Women had continued to use DES during pregnancy over the 20-year period between 1947 and 1971 because it took about 20 years to see the effect of this teratogen. Other conditions associated with DES offspring include infertility, tubal pregnancy, miscarriage, and premature delivery in females, as well as genital abnormalities and perhaps infertility in males.

Other teratogens to be noted include beverage alcohol and the anti-acne drug acutance. The most commonly encountered environmental teratogens are arsenic, lead, mercury, and PCBs, which were discussed earlier.

ENVIRONMENTAL DEGRADATION OF TOXINS

When considering the effect of toxins on the environment, two factors have to be considered other than the toxicity of the substance. First, we need to know how long it lasts in the environment and, second, how it moves around in the environment. Toxins that find their way into the environment at nontoxic levels will build up to toxic levels if they do not break down before more toxin is added. Toxins that do not break down in the environment are said to be **persistent**. If a toxin breaks down more slowly than the rate at which more toxin is added, it will eventually approach toxic levels and will continue to build up to relatively high levels. Persistence also implies that once a toxic level of an environmental toxin occurs it will continue to be a threat for a long period of time. For example, the half-life of DDT in soil is 10 years and the half-life of TCDD is 9 years. The result is that after 20 years 25% of the DDT on a field would remain.

Generally, most chemicals in the environment break down in one of three ways. One of the most common methods is reaction with water—that is, hydrolysis. Water is very common in the environment in most climates and is key to the degradation of many materials. Another important element in the degradation of environmental chemicals is light. Ultraviolet light, which is a small component of the light received from the sun, has the ability to break chemical bonds. This interaction can start a series of events leading to the breakdown of a molecule.

Finally, microorganisms can be involved in the conversion of environmental toxins to nontoxic materials. A very large variety of microbiological organisms consume a wide variety of substances as food. Many toxins are consumed by microorganisms and broken down that way.

ENVIRONMENTAL MOVEMENT OF TOXINS

If toxins released into the environment simply stayed in one place, they would probably not cause much of a problem. The toxic material would remain where it is placed and perhaps cause a problem at that point. If there was a desire to clean up the material, it would all be right there. Unfortunately, toxins in the environment can be moved by various means. The result is that these materials can become a hazard over a much wider area.

Carriers

The molecules of substances can be roughly separated into the following three classes: ionic, polar, and nonpolar. In Chapter 3, ionic bonding was described. Substances with ionic bonding are known as **ionic substances**. Ionic compounds are usually solids with fairly high melting points and are made up of a collection of positively and negatively charged ions. **Polar substances** contain molecules that have an uneven distribution of charge in the molecule but do not break up into positively and negatively charged particles. As is illustrated in Figure 6.7, these molecules have a positive end and a negative end because the electrons are not evenly distributed throughout the molecule, which can be seen for water. Nonpolar molecules have

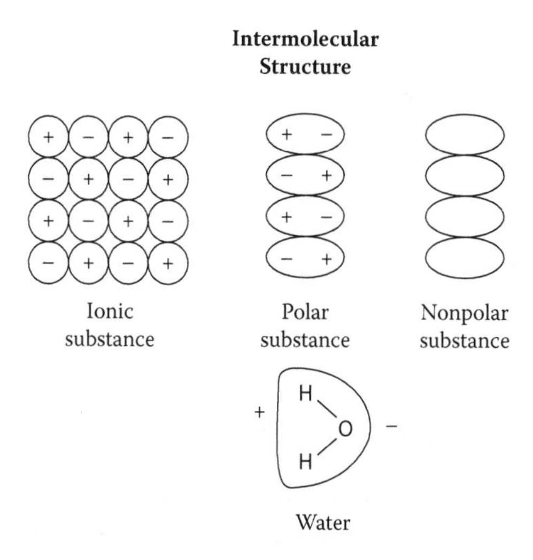

Figure 6.7 Illustration of the intermolecular structure of ionic, polar, and nonpolar substances.

even distributions of electrons and therefore no excess positive or negative charge at any point. Other than pure elements, truly **nonpolar substances** are rare, but many substances are reasonably close. Fats and oils are good examples of such substances.

One of the key features that controls what happens to toxins in the environment is the type of solvent in which the material dissolves. The old rule of thumb in chemistry is "like dissolves like." Polar substances tend to dissolve in polar solvents, and nonpolar substances tend to dissolve in nonpolar solvents. Because solvents are usually liquids or semisolids at room temperature, ionic substances are rarely solvents; however, ionic substances usually dissolve in the most polar solvents. One of the best highly polar solvents is water, and it dissolves many ionic compounds as noted earlier in Chapter 3.

Generally, environmental toxins can be classified as either water soluble or fat soluble. Each class presents its own unique problems. **Water-soluble toxins** spread very rapidly, as they dissolve in water and water is widely available in the environment. Water is referred to as the **carrier** for such substances. These materials usually find their way into groundwater and eventually contaminate wells. The solubility in water sometimes works to our advantage, in that the poison may be diluted to the point that it is no longer toxic; however, some poisons may be toxic even at very low levels, especially in the case of chronic exposure. In these cases, the water simply distributes the poison far and wide at a fairly rapid pace. Water-soluble poisons also have more rapid access to the cell because water is the solvent of life. Once dissolved in water, the poison can distribute itself in any water with which it comes into contact. This distribution can include the water inside the cells of an organism. The main protection the cell has is the cell membrane, which may or may not be easily penetrated by the poison.

The other class of poisons is **fat-soluble toxins**. These toxins do not move rapidly in the environment, as they do not dissolve in water, which is the main medium other than air capable of moving molecules in the environment. These molecules are fairly nonpolar, and as a result are not the type of molecule that usually dissolves in a polar substance such as water. These materials are referred to as fat soluble because they are usually found dissolved in the fatty or oily material of cells and organisms. These substances can be moved slowly by water because they usually dissolve slightly. Although water can be the carrier of these substances, they are generally spread by some other mechanism. Most commonly they are spread by organisms that have the toxins dissolved in their fatty tissue. Because fat-soluble toxins generally move around the environment much more slowly than the water-soluble variety, these toxins are a much more localized problem, but it also means that the toxins remain highly concentrated and therefore can be quite dangerous.

Bioaccumulation and Biomagnification

Based on the observations made earlier, one would not expect fat-soluble toxins to move around in the environment very quickly; however, it is a little more complicated than that. Normally it is said that fat-soluble toxins do not dissolve in water, but that is not true. Fat-soluble toxins do not dissolve to any great extent in the water, but

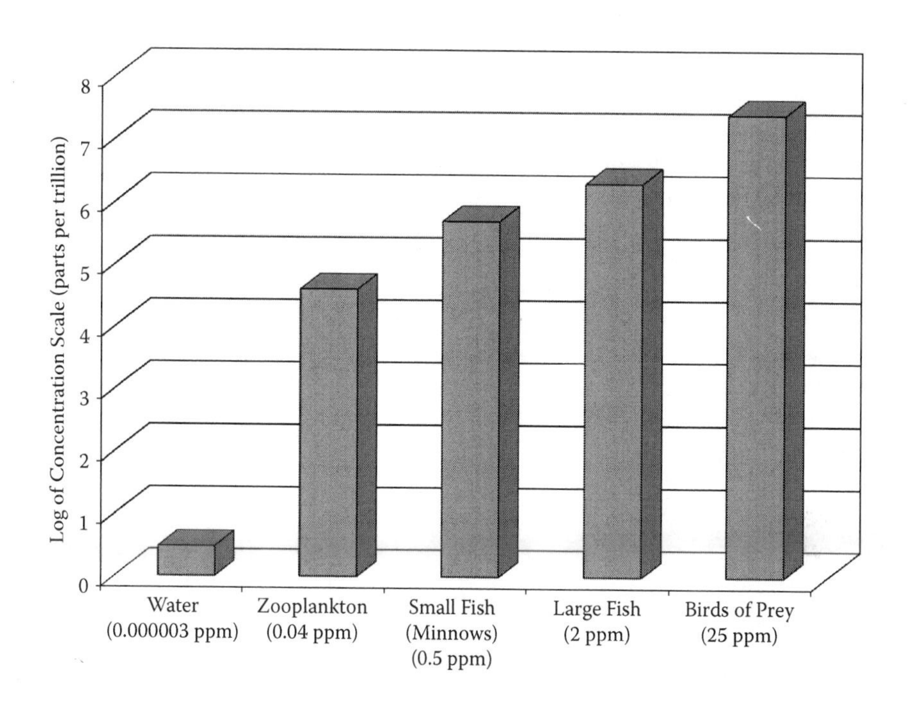

Figure 6.8 Concentration of DDT in water and various organisms in the marine food chain in Long Island Sound.

they do dissolve slightly. When these toxins come in contact with an aquatic organism either through the gills of a fish or in some other manner, these materials will immediately dissolve in the fats in the organism since they are much more fat soluble than water soluble. As the organism comes in contact with more and more water, the toxins accumulate in the organism. This process is known as **bioaccumulation**.

As organisms feed on one another and move up the food chain, the fats are digested in the predator organism, and the toxins find their way into the fat of the new organism. When this process is repeated multiple times, the effect is magnified and is referred to as **biomagnification**. The process of biomagnification can have rather dramatic effects. This was illustrated by a study on the levels of DDT in various marine organisms in Long Island Sound. The zooplankton present in this area are at the bottom of the food chain and are consumed by small fish such as minnows. The minnows serve as food for larger fish, which in turn are consumed by birds of prey such as peregrine falcons, brown pelicans, and ospreys. The concentrations of DDT found in the water and various organisms are shown in Figure 6.8. As can be seen, the very small amount of DDT in the water is taken up by the zooplankton and concentrated nearly 10,000-fold. (The change does not appear this great on the graph because a nonlinear log scale was used. The scale is similar to the pH scale in that each division is 10 times the previous division.) Small fish such as minnows eat the zooplankton and concentrate the DDT another 10-fold. The large fish and the birds of prey finally concentrate the DDT to its highest level. The net result is an increase in DDT concentration from water to the tissues of the birds of prey of nearly 10 million times. This tremendous concentration of an environmental toxin through several levels of the food chain is astonishing. The DDT in these birds of prey, as noted earlier, interferes with the ability to make strong eggshells. As a result, the eggs break during incubation and no offspring are produced.

DISCUSSION QUESTIONS

1. What are some of the key differences between the study of toxicology in the laboratory and the study of toxic substances in an environmental setting? Which endeavor seems more challenging and why?
2. Differentiate between acute and chronic toxicity.
3. Why are the LD_{50} and ED_{50} almost always applied to acute toxicity situations?
4. What are some of the advantages and disadvantages of using animal data to determine the toxicity of substances?
5. Pure malathion has minimal toxicity in rats, but the presence of very small amounts of impurities causes the toxicity to increase considerably. This is an example of what type of an effect? Name the effect and relate it to this situation.
6. Explain the relationship between normal oxygen and ozone. Of what type of phenomenon is this an example?
7. Why are very fine particulates more dangerous when inhaled than when larger particles are inhaled?
8. The compound $KClO_3$ is stable at room temperature. When heated it will break down to produce oxygen; however, the rate is fairly slow. When MnO_2 is added to the mixture, the production of oxygen is much faster. What principles concerning the rates of chemical reactions are illustrated here? Why?
9. Explain how carbon monoxide interacts with the blood leading to impairment and death.
10. How do enzymes differ from ordinary catalysts?
11. Discuss the toxic nature of each of the following relative to each other: liquid mercury, mercury vapor, and mercury compounds.
12. If a group of people are exposed to a low enough level of lead, it is likely that most of them will not develop lead poisoning. What might cause a few individuals in this situation to develop lead poisoning?
13. DDT, despite its environmental problems, is inexpensive. Should the poorer countries of the world be allowed to use it to combat malaria or should some other approach be taken? Why?
14. Explain how anticholinesterase neurotoxins stop the transmission of nerve impulses.
15. Given the lack of clarity concerning the long-term toxicity of BPA, what do you believe that the government should do? How would the precautionary principle relate to such a discussion?
16. Why is it impossible to avoid all contact with carcinogens?
17. Teratogens often produce dramatic and tragic effects in newborns, but mutagens may be the more common agent in miscarriages and birth defects. Why do you think that might be?

18. Persistent pesticides are generally seen as a problem when placed in the environment. Why is persistence as problem? Are there any ways in which persistence might be helpful?
19. Explain how a substance found in water at 30 parts per trillion might be found in an organism at the top of the food web at 30 parts per million.

BIBLIOGRAPHY

Ames, B.N. and Gold, L.S., Cancer prevention and the environmental chemical distraction, in *Politicizing Science: The Alchemy of Policymaking*, Gough, M., Ed., Hoover Institute Press, Stanford, CA, 2003, pp. 117–142.

Anon., *The Beaver Fur Hat—The Fashion of Europe*, White Oak Society, Deer River, MN, 2001, http://www.whiteoak.org/historical-library/fur-trade/the-beaver-fur-hat/.

Anon., Dioxin poisoning scars Ukrainian presidential candidate, *Environmental News Service*, December 13, 2004, http://www.ens-newswire.com/ens/dec2004/2004-12-13-03.asp.

Call2Recycle®, http://www.call2recycle.org.

Clark, P., *Tennessee Reclamation Project: More Lessons Learned for the New Millennium*, BlueRidgeHighlander.com, http://www.theblueridgehighlander.com/polk_county_tennessee/reclamation_project/index.html.

Clayton, R., *Endocrine Disruptors in the Environment: A Review of Current Knowledge*, 2nd ed., Foundation for Water Research, Buckinghamshire, UK, 2002.

Corrosion Doctors, *Mad as a Hatter*, http://www.corrosion-doctors.org/Elements-Toxic/Mercury-mad-hatter.htm.

Crummett, W.B., *Decades of Dioxin: Limelight on a Molecule*, Xlibris Corporation, Bloomington, IN, 2002.

Damstra, T., Barlow, S., Bergman, A., Lavlock, R., and Van Der Kraak, G., Eds., *Global Assessment of the State-of-the-Science of Endocrine Disruptors*, prepared on behalf of World Health Organization, International Labour Organisation, and United Nations Environment Programmme, World Health Organization, Geneva, Switzerland, 2002.

DES Cancer Network, *Timeline: A Brief History of DES (Diethylstilbestrol)*, DES Cancer Network, http://www.descancer.org/timeline.html.

Hileman, B., Yushchenko poisoning, *Chemical and Engineering News*, 82(51), 13, 2004.

Hill, J.W., *Chemistry for Changing Times*, Macmillan, New York, 1992, p. 735.

Hodgson, E., *A Textbook of Modern Toxicology*, Wiley, Hoboken, NJ, 2004.

Kungolos, A.G., Brebbia, C.A., Samaras, C.P., and Popov, V., Eds., *Environmental Toxicology*, WIT Press, Southampton, UK, 2006.

Landis, W.G. and Yu, M.-H., *Introduction to Environmental Toxicology: Impacts of Chemicals upon Ecological Systems*, Lewis Publishers, Boca Raton, FL, 2004.

Paasivirta, J., *Chemical Ecotoxicology*, Lewis Publishers, Chelsea, MI, 1991.

Quinion, M., *Mad as a Hatter*, World Wide Words, March 3, 2001, http://www.worldwidewords.org/qa/qa-mad2.htm.

Ritter, L., Solomon, K.R., Forget, J., Stemeroff, M., and O'Leary, C., *Persistent Organic Pollutants, An Assessment Report on: DDT-Aldrin-Dieldrin-Chlordane-Heptachlor-Hexachlorobenzene-Mirex-Toxaphene, Polychlorinated Biphenyls, Dioxins and Furans*, The International Programme on Chemical Safety, World Health Organization, Geneva, Switzerland, 1995.

Ritter, S., Bisphenol A, *Chemical and Engineering News*, 89(23), 13–22, 2011.
Van Patten, P., The Mad Hatter mercury mystery, *Wrack Lines*, 2(1), 13–16, 2002.
Wright, P., *Nitric Oxide: From Menace to Marvel of the Decade*, briefing document prepared for the Royal Society and Association of British Science Writers, May 1996.

CHAPTER 7

Energy and Modern Society

The two things that are critical to maintain any civilization are materials and energy. It is essential that materials should be available to build tools on which the functioning of any culture is based, and that these tools have some source of power (energy) to operate them. As noted in Chapter 1, before the Industrial Revolution the sources of energy were fairly simple. Mechanical jobs (moving things, grinding things, mashing things, and stirring things) generally used human power, animal power, water power, or perhaps wind power. When heat was required to warm a building, to cook, to fire pottery, or to extract metals from ores (smelting), the usual practice was to burn wood.

The move toward an alternative source of energy on a major scale probably occurred first in England. By the end of the 16th century, England was nearly deforested and the switch to another fuel was almost inevitable. Because England has been described as an island built on **coal**, switching over to coal as a fuel was not surprising. As the Industrial Revolution gained strength and spread to other countries, so did the use of coal as a fuel. This phenomenon lasted until the early years of the 20th century when coal was overtaken by **petroleum**.

The petroleum industry had its origin in the need for improved lamp oils in the mid-19th century. In the early part of that century, common people generally burned tallow candles for lighting; however, persons of higher economic status used oil lamps. The cost consideration was not the lamp, but the oil. The oil used in these lamps was usually whale oil or sperm oil, both of which were derived from whales. Whales became more and more difficult to hunt, and the price went higher and higher. As a result, the search was on for a lamp oil that nearly anyone could afford. At first, certain liquids were distilled out of coal and sold under the name kerosene. It was found that a "better kerosene" could be obtained from petroleum, a byproduct of some salt wells and natural springs. In 1859, Colonel Drake drilled the world's first oil well at Titusville, Pennsylvania, to provide a more reliable source of kerosene for oil lamps. Later, however, petroleum was found to be a good fuel for other purposes, particularly for the internal combustion engine. Although petroleum never completely replaced coal as an energy source, it certainly took over as the predominant energy provider. Coal continued to be in use in certain industries such as the electric industry. In 2011, 42% of the electricity generated in the United States came from coal.

Petroleum along with natural gas dominated many industries, but with the disappearance of the steam locomotive **oil** became the fuel of choice for the transportation industry. Trains, cars, buses, boats, and planes all used various petroleum products to move people and goods from point A to point B. In the United States, in the first half of the 20th century nearly all of this petroleum was produced domestically. In recent times, however, creeping affluence has led to the ever-increasing consumption of oil products at a rate higher than the American oil fields are producing. A little noticed fact is that U.S. peak oil production occurred around 1970 and has continued to fall.

The United States was partly dependent on imported oil when the Yom Kippur War broke out between Israel and some of its neighboring states in 1973. After hostilities ended, there were many after-effects, including the determination by some Muslim nations to apply their economic muscle in a unified effort against Israel and its allies. This took the form, in early 1974, of a withdrawal of oil supplies from Israel, the United States, and many other Western nations. The United States was drastically affected. Gas lines became common at service stations. People had to wait for hours in their cars for gasoline. Suddenly the plentiful supply of oil that everyone assumed was available was not available. The price skyrocketed. Gasoline that had sold for $0.33/gal in 1970 was being sold for $1.25/gal in 1980. The price of a barrel of crude oil went from $3 to $30 in the same time period. Americans for the first time started realizing the importance of energy, and that without it everything comes to a halt.

The oil embargo did not last very long, but prices remained high. The Organization of Petroleum Exporting Countries (OPEC)* realized that they could keep prices high by controlling the oil supply. Fortunately or unfortunately (depending on one's perspective), OPEC's plan to control and limit the oil supply never worked well, and it fell apart around 1980. The supply went up and the prices came down. With cheaper energy, Americans gradually forgot about the oil crisis and energy conservation. The problem was not solved, though; it had simply taken a temporary holiday.

In 1998, the price of crude oil in the world market began to increase and after a brief decline from 2000 to 2002 it increased rapidly, topping out in 2008 at around $125/bbl. The price per gallon at the pump began to rise along with the price of crude oil, with the price above $3/gal after the disruption of fuel supplies in the aftermath of Hurricane Katrina. Since then on two occasions the price has topped $4/gal. How high will fuel prices go? Will they go to $5/gal or perhaps $6/gal? Can we afford the high prices? Will the supply of oil last? What are the alternatives? These questions and many others about energy are dealt with in this chapter.

ENERGY SOURCES

As we know from the first law of thermodynamics, all energy comes from somewhere but we cannot create it. Because energy can be transformed from one form to another, many energy sources can be traced to other energy sources. There are some

* OPEC includes Algeria, Gabon, Indonesia, Iran, Iraq, Kuwait, Libya, Nigeria, Qatar, Saudi Arabia, United Arab Emirates, and Venezuela.

energy sources for which the ultimate source of the energy is not known. This type of source is known as a **primary energy source**. Many energy sources, however, clearly derive their energy from some other energy source. An energy source that gets its energy from another clearly identifiable energy source is known as a **secondary energy source**.

The common primary energy sources include solar, nuclear, geothermal, and tidal. In each of these cases, we cannot trace the ultimate source of the energy. Conventional **nuclear energy** releases energy which is locked away in the nucleus of certain atoms, but we do not know precisely from where this energy originates. In the same vein, **solar energy**, the energy production process in the sun, is a different type of nuclear process. Still, this energy comes from the creation of more stable nuclei, and the source of such energy cannot be clearly defined.

Geothermal energy is the release of energy trapped under the Earth's crust. We know that heat is present in the Earth's interior and has been there essentially since the Earth was formed. Tracing the origin of this heat is difficult. The heat appears to be produced by the radioactive decay of unstable nuclei deep within the Earth. Once again, the source appears to be nuclear in origin. **Tidal energy** is another source rooted deeply within the fundamental processes of the universe, but in this case it is not a nuclear process. The tides are created by gravitational force of the moon and the sun acting on the oceans of the Earth. The energy comes from the orbital motion of the moon and the rotation of the Earth. These are forces that have been present since the origin of the solar system and thus represent a primary energy source.

Some of the secondary energy sources that are used on a large scale around the world include **fossil fuels** (coal, oil, and natural gas), **water power**, **wind power**, biomass, and electricity. All of these energy sources, with the exception of some electric power, derive their energy from the same source—the sun. **Biomass energy** is the direct use of living matter as a fuel, and **fossil energy** is the use of ancient biological material as a fuel source. Because all life depends on photosynthesis to provide energy, any use of living matter as a fuel has solar energy as its primary energy source. The winds are driven by the sun, and the sun's heat evaporates water to produce clouds from which the rains come which feed the streams that flow over the great hydroelectric dams of the world. Only **electric power** is different from these types of energy.

The majority of electric power is in fact derived indirectly from solar energy. Around 40% of the electric power is generated by coal-fired generating plants, but important contributions are made by **natural gas** and **hydroelectric power**, as shown in Figure 7.1. The only important player that cannot be traced to solar energy is nuclear power. Wind power is a minor but increasingly important energy source for electric power. Other sources, such as geothermal, tidal, direct solar, biomass, and petroleum, do generate a small amount of electricity.

Figure 7.2 shows the relationships among the various energy sources. The common primary energy sources are shown on the top, with the secondary energy sources shown below them. As noted earlier, all of the secondary sources arise exclusively from solar energy with the single exception of electricity.

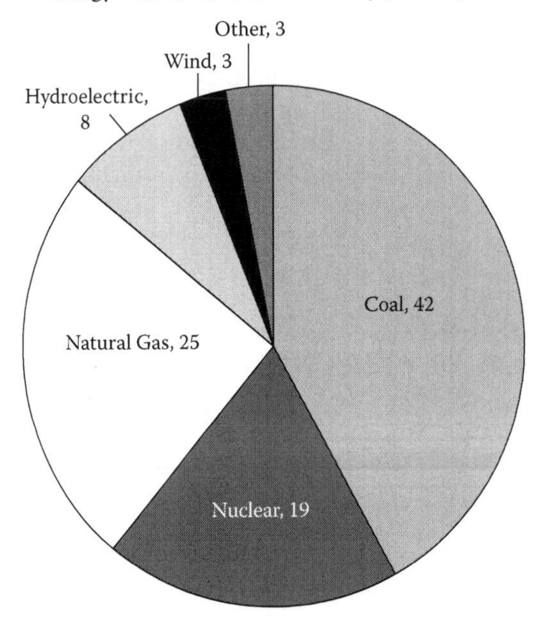

Figure 7.1 Percentage of various energy sources used to generate electricity in the United States in 2011.

ELECTRICITY

We are going to discuss electricity before discussing the other sources because this source is unique and pivotal in modern society. All of the other relationships between sources are dictated by the practical, natural relationships between each of the sources. We cannot change photosynthesis, we cannot change the weather, we cannot change the nature of the atomic nucleus, and we cannot change which fossil fuels happen to be buried in the Earth. Humans may tinker with these energy sources, but their fundamental relationships are fixed. Electricity, however, is a source we choose to generate because it is convenient to do so. One could, for instance, use compressed natural gas to operate a hair dryer, but most people find electricity more convenient.

Certain energy sources can be and are used directly without conversion to electricity. Wood (biomass) has been used to provide heat since before the beginning of recorded history and also was used as a fuel for the steam engine. All fossil fuels can be used to generate heat, and petroleum produces gasoline, diesel fuel, and jet fuel that power the modern transportation industry. Solar energy can also be used directly as more and more homes in certain regions of the world are installing solar heating devices in their homes. Geothermal energy can be used for heating purposes because it is usually brought to the surface as hot water or steam. Iceland, for example, has plentiful geothermal energy and uses that energy extensively for heating purposes.

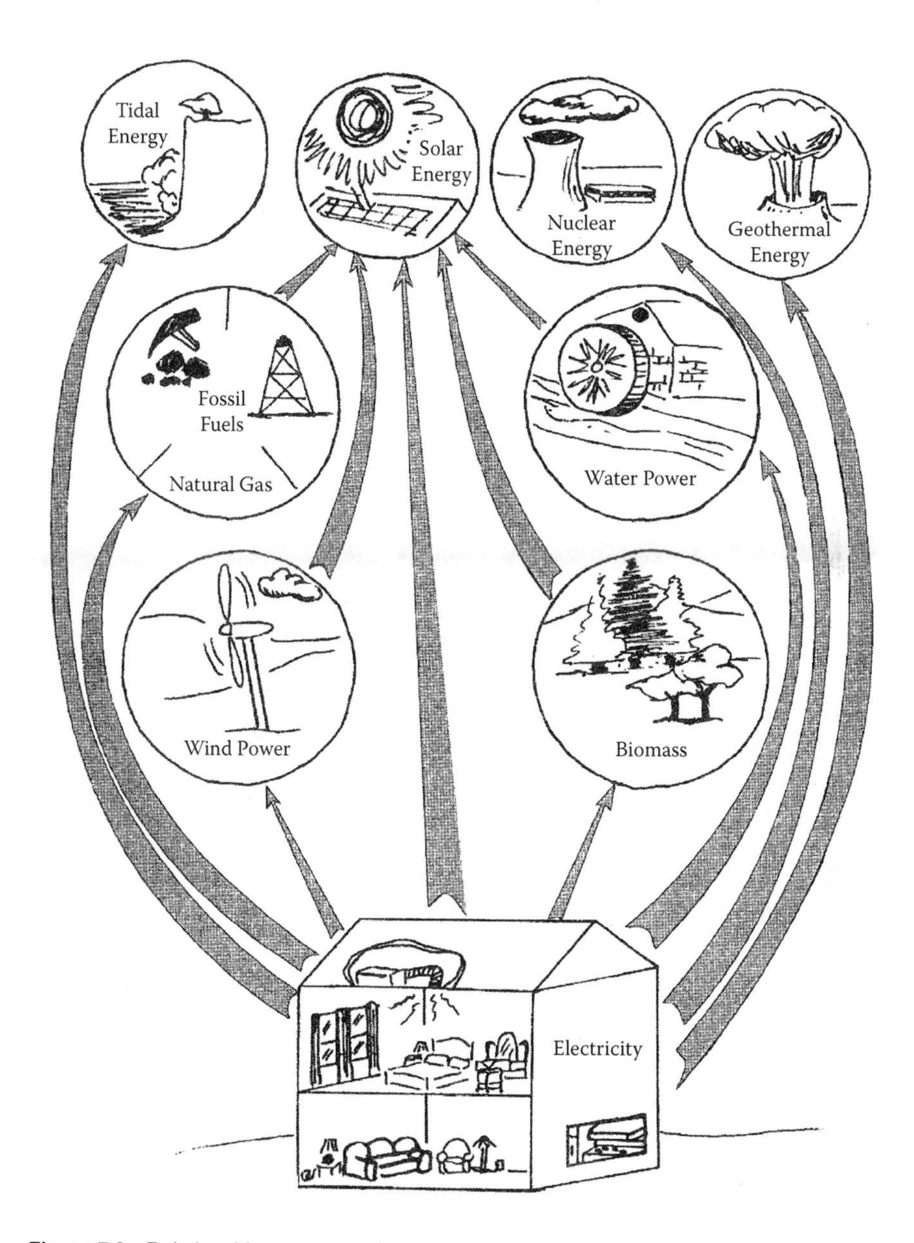

Figure 7.2 Relationships among various energy sources.

For other energy sources, it is simply impractical to use them directly to any significant extent. These energy sources are nearly always converted into electricity. The reasons vary from source to source. For water and tidal power, the problem is principally location. In the 18th and 19th centuries water power was used for many purposes, including powering textile mills and grinding grain. Many tidal mills for grinding grain could be found in 18th-century England, France, and New England. Some of the European mills date back to the 11th century. To use energy of this type directly, one had to be located in the right place and mechanical energy was required as well. Water or tidal power could not be used for heating. Because many of us live away from the coast or streams, tidal power and water power are not useful to us, but these power sources can be used to generate electricity that is versatile and can be transported to us.

Wind power has some of the same restrictions as water and tidal power. The windmill is certainly not well suited for producing heat energy and is similar to water power in this respect. The wind, however, unlike flowing water, is found almost everywhere on the Earth, hence it is not completely restricted by location. Wind power, however, is usually converted into electricity because it is a more versatile energy source in this form.

Nuclear energy produces heat and certainly could be used directly, but it is not practical to do so. The complexities and expense of a nuclear reactor make it impossible to build one for every location and application. The only convenient way to distribute nuclear power is to convert the nuclear power into electricity.

Modern society has developed a considerable affinity for electricity as the final form of energy to be supplied to the consumer. We have done so because in most cases electricity is a more convenient form of energy. We can cook; run a refrigerator, radio, and television; and heat and air-condition a home with it. We pay a price for this convenience, though. The second law of thermodynamics tells us that any energy conversion process will increase the entropy or disorder of the universe. Electricity is a fairly low-entropy or ordered form of energy; as a consequence, the conversion of nearly any form of energy into electricity will require the concurrent production of some disordered or high-entropy energy, usually heat energy. The result is that, although electricity is more convenient, it is usually the least efficient form of energy. For example, the typical coal-fired electric-generating plant converts about 35% of the coal energy into electricity. The most efficient way to use energy is to use it directly as it comes to us, but direct use of energy is not always practical.

WIDELY USED ENERGY SOURCES

This section discusses some of the energy sources that have been widely used in the world over the past two or three decades. Nuclear power and biomass could be included, but will be discussed separately. Nuclear power is discussed in a separate section because it is a complex and special subject requiring a more detailed discussion. Biomass is also treated separately because it has aspects of being both a widely used and an emerging energy source.

Fossil Fuels

The fossil fuels include *coal*, *oil*, and *natural gas*. All three of these fuels are the remains of ancient organisms that existed between a million and perhaps 600 million years ago. These organisms were gradually buried under sediment that was later compressed into sedimentary rock. It is between these layers of rock that the fuels are found today. Coal was the first of the fossil fuels to be used on a large scale and may still be used long after the others are depleted. It is oil, however, that powers much of modern society and on which whole economies rise and fall; therefore, given its pivotal role in our modern culture, let us start by considering oil as an energy source.

Oil

Oil seems to have been formed from single-celled plants and animals that were found in the oceans about 570 million years ago. When these organisms died they found their way to the bottom of the ocean, where they mixed with the sediment on the ocean floor. With the passage of time and the collection of even more sediment, these dead organisms were subjected to heat and pressure that gradually converted them into the liquid petroleum we use today.

Oil, because it is a liquid, is not always easy to extract from the ground and is often found at inconvenient locations. Such locations include offshore locations and locations far north of the Arctic Circle. As discussed in Chapter 13, the production of oil from offshore oilrigs can produce devastating environmental results in the case of a blowout, as was seen in the BP disaster in the Gulf of Mexico. The removal of oil from the North Slope of Alaska is a major concern, given the fragile ecology of the region.

Not every country has the same amount of petroleum under its land. This untapped oil supply is known as the oil **reserve**. The amount of any resource that is yet to be removed from the ground is referred to as the reserve of that resource. Reserves are often described in two different ways, total and proven. Proven reserves are those that can clearly be demonstrated as existing based on geological studies. Total reserves cannot be known but rather have to be based on estimates because these reserves include both identified and unidentified material.

Although the United States is by far the world's greatest consumer of oil, it has only 2.2% of the world's known proven oil reserves. The United States has burned about 80% of the oil it has ever discovered. Huge amounts of money, totaling about $250 million, were spent between 1980 and 1984 in a search for more oil, but during this period the United States used another 5% of the oil it had and found almost no new oil.

As seen from Figure 7.3, over half of the world's proven oil reserves are found in the Middle East. Saudi Arabia alone accounts for around one-fifth of all the world's oil reserves. The United States, along with many other Western industrialized societies, is economically dependent on the petroleum-exporting nations of the world. Net petroleum importing regions include the United States, Europe, and the major Asian countries of Japan, China, and India. These are three of the major industrial regions of the Earth. How long and at what rate can the world's industrialized countries demand and keep receiving ever increasing amounts of oil?

In 1956, the American geophysicist **Marion King Hubbert** predicted that American oil production from domestic sources would peak sometime between 1965 and 1970. Hubbert did not miss the date by far, as domestic oil production peaked in 1971. Because his prediction was made well in advance and was reasonably accurate, the peak in the U.S. domestic oil production became known as **Hubbert's Peak**. As

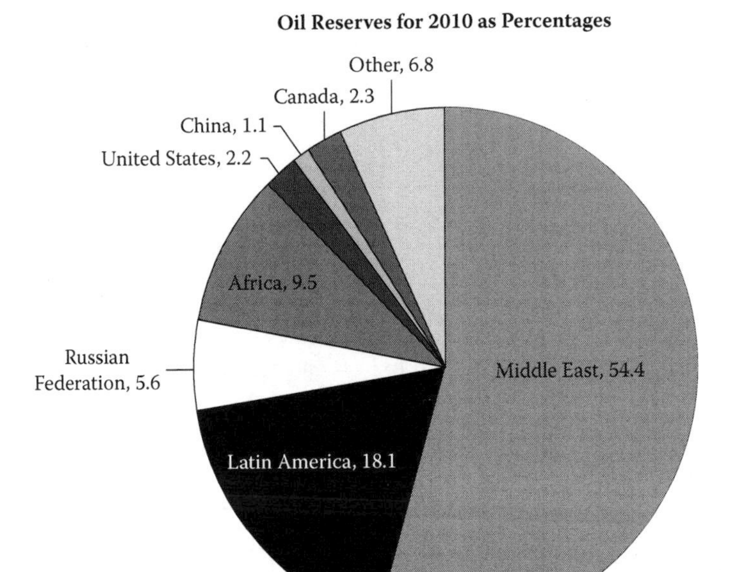

Figure 7.3 Oil reserves by region, 2010. (Based on BP, *Statistical Review of World Energy 2011*, http://www.bp.com/statisticalreview/.)

domestic production fell, the United States became more and more dependent on imported oil. In the early 1990s, the amount of oil imported began to exceed the domestic production.

Importing oil is only a short-term solution because petroleum is a nonrenewable resource. Logically, petroleum will run out some time in the future; however, predicting when this will happen is more complex than might be expected on first impression. At the end of 2010, the oil industry indicated that there were 1380 billion bbl of oil in global proved reserves. If one divides this by the current annual production rate of 30 billion bbl, the result would suggest that the supply should last for about 46 years, but this estimate is misleading for a number of reasons.

First, this analysis assumes that production will proceed at a constant rate. This is unlikely to be true, as the developing countries of the world will likely increase their demand for oil year after year. One of the fastest growing economies in the world is China. The Chinese nearly doubled their oil consumption from 2000 to 2010 (see Figure 7.4). The increase in oil consumption in China was greater than the decline in U.S. oil consumption, but both countries are competing for a share of total global oil production that is only increasing slowly. As the consumption rate goes up, the remaining oil will run out sooner. Most areas of the world show an increase in oil consumption, even if oil consumption is decreasing in some places, as can be seen for the European Union in Figure 7.4.

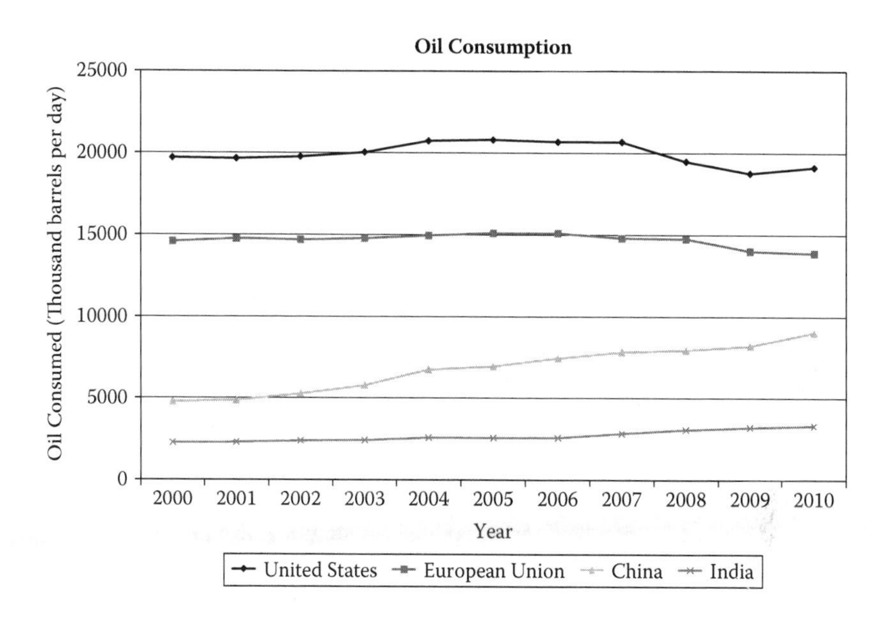

Figure 7.4 Oil consumption for selected countries in thousands of barrels per day. (Based on BP, *Statistical Review of World Energy 2011*, http://www.bp.com/statisticalreview/.)

Our simple analysis also further assumes that all of the oil in the ground can be extracted with the same ease. Unfortunately, the longer that an oil field has been used, the more difficult it becomes to pump out the oil from the ground; hence, the rate of flow of oil from a field always increases to a maximum and then decreases toward zero as the field ages. Furthermore, the last oil in a field is more expensive to extract than the rest of it. Therefore, the question is not just "How much oil is left?" but also "How much oil is left at what price?"

Another issue is the use of proved reserve quantities provided by national governments and corporations in these calculations. First, proved reserves are very conservative numbers and certainly do not represent all of the oil present. Additionally, many governments and corporations will for very logical, political, and business reasons manipulate these numbers. Usually, the desire is to hide reserves to drive up prices and keep control of the market. The result is that oil reserves keep growing over time, which appears to be a good thing; however, the problem is that the increase in oil reserves is not based on new oil discovery. This increase gives the impression that new oil is being found when in fact it is just hidden. There is a fixed amount of oil out there, and although we are probably underestimating the total amount it will still run out one day.

With increasing demand and oil production staying roughly steady, oil will eventually become more and more expensive. As noted earlier, oil started a fairly steady climb in price in the year 2002. The price of crude oil and gasoline has continued to increase since that time with some fluctuation due other market forces.

There are at least two unconventional sources of oil that may be able to increase the supply of oil. One of these sources is the **heavy oil** of Venezuela and the other is the **tar sand** and **oil shale** deposits of Canada and the former Soviet Union. In fact, Canada has such a large quantity of tar sands that, if these sands are included, the Canadians are third in oil reserves behind Saudi Arabia and Venezuela. These oil sources are not easy to extract and have serious problems. The heavy oil of Venezuela contains heavy metals and sulfur that must be removed. The use of tar sand and shale comes at a high environmental price. These materials must first be strip mined and then the oil must be extracted by a process that may cause air pollution.

It is possible that the use of energy conservation will allow the oil supply to last longer, but it is difficult to get a precise idea of the exact time left before the world's supply of oil runs out. The supply could last anywhere from 40 to 100 years depending on the factors we have been discussing; however, sooner or later we will use it up. It is not a matter of "if we will run out of oil…," but rather "*when* we run out of oil…."

In addition to its short supply, there are other problems related to oil. Chapter 9 discusses the key role that the internal combustion engine plays in the formation of certain types of air pollution. Furthermore, because petroleum is made up almost exclusively of carbon and hydrogen, its use as a fuel produces carbon dioxide when burned, which we will learn contributes to global warming in a major way.

Oil must be pumped out from the ground much as water is pumped out from a well. In some cases, the oil is pumped out easily, but in other locations the material must be forced from the ground using steam, water, various gases, and other materials. These substances are forced into the well at one location and that pressure forces the petroleum to exit at another, as illustrated in Figure 7.5. The injection of many of these materials into oil wells increases their yield but can lead to groundwater pollution. When water is injected deep underground, it will dissolve salt and other materials, some of which are toxic. These materials, along with detergents that may be added to increase effectiveness, may find their way into aquifers from which water is removed for commercial, industrial, or domestic use.

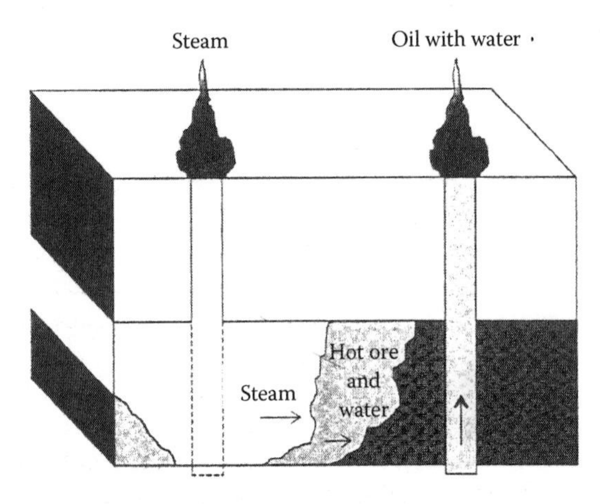

Figure 7.5 Diagram showing the use of steam to force oil from an oil well.

Table 7.1 Advantages and Disadvantages of Oil

Advantages	Disadvantages
Inexpensive	Release of CO_2, CO, NO, and volatile organic compounds
Easily transported	Potential oil spills
High net energy yield	Groundwater contamination
Versatile in use	Limited supply (30 to 90 years)

One alternative for injection wells is to use carbon dioxide. Carbon dioxide can be trapped when fossil fuels are burned to produce energy and can then be injected underground to force out oil. This gas will not produce toxic problems with the groundwater and will then be trapped underground. This topic will be discussed in greater detail in the later discussion on coal.

If oil can cause such severe problems and is in such short supply, why do we use so much of it? For one thing, historically it was cheap and, as a result, Western civilization was built around it. It is now difficult suddenly to change the way we supply energy for our wants and needs. Despite oil embargoes and attempted price fixing by OPEC, the price still remains low enough to make it fairly competitive with other forms of energy. Coal may be the least expensive of the fossil fuels, but it is less convenient. Oil is easy to transport, particularly over land. Oil pipelines can move oil quickly from its source to refineries, and oil can also be easily pumped onto very large tankers for transport over the ocean. See Table 7.1 for a review of the advantages and disadvantages of oil.

In the final analysis, one of the key reasons why we use so much oil is the transportation industry. Nearly all engines used today for transportation use liquid fuels. Vehicles must be able to carry fuel and to be refueled at various locations. Liquid fuels are easier to handle and to dispense than solid or gaseous fuels, so petroleum will probably continue to be used for transportation for the foreseeable future. Although oil does not dominate in other industries the way it does in transportation, it is a very versatile fuel that can be used for space heating and the generation of electricity. As a result, petroleum is very profitable to refine and distribute.

Natural Gas

Natural gas is chemically very similar to petroleum. Both are collections of hydrocarbons and are often found together. The prevailing theory is that both are often produced together as a result of some of the same processes. Generally, as the biological material is slowly converted into hydrocarbons, the higher the temperature the more likely it is that the product will be gaseous. As a result, the composition will vary from place to place, with some deposits containing a higher percentage of gas than others. Major deposits of natural gas are trapped in shale and sandstone.

People became aware of natural gas as a result of some of the early oil gushers, which were due to underground pressurized gas. In the early years of oil exploration, this gas was considered to be a nuisance. The oilmen of the day knew the gas was a fuel, but they had no clear idea of how to use it. Furthermore, they had no way to

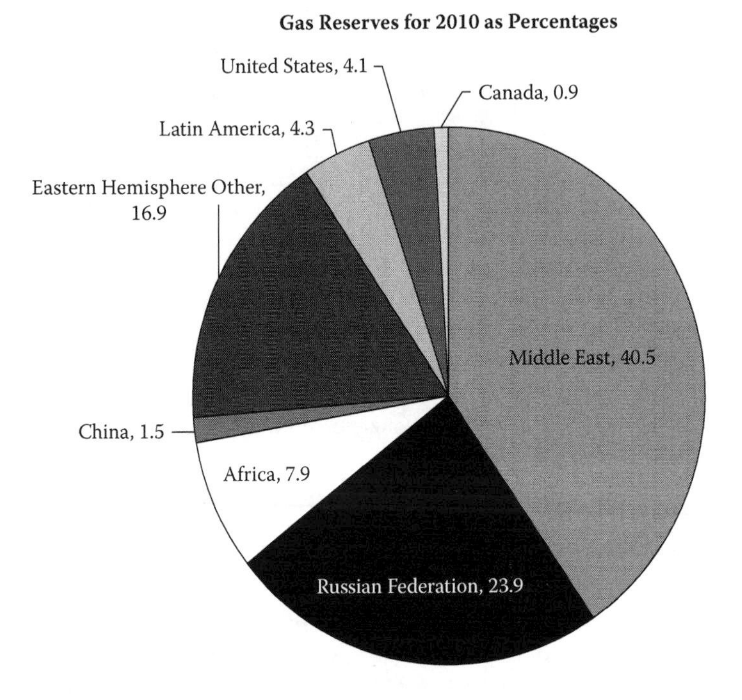

Figure 7.6 Natural gas reserves by region for 2010. (Based on BP, *Statistical Review of World Energy 2011*, http://www.bp.com/statisticalreview/.)

transport the gas. It was not until the 1920s that the advantages of natural gas as a fuel became apparent, and the high-pressure pipelines needed to move the gas over long distances became available.

Although it was in the United States that the fuel potential of natural gas was first noted, most of the easily accessible natural gas is found elsewhere on the planet. Figure 7.6 shows where the worldwide reserves of natural gas are found. The transport of natural gas is somewhat more problematic than oil. For transportation over land, there is little difference from oil other than the fact that gas pipelines must be gastight; however, shipping natural gas over the ocean is a much more expensive proposition. Gases have very low density; therefore, a little bit of gas takes up a very large amount of space. To put natural gas on a tanker as a gas to transport it is simply impractical. The result is that most natural gas is used on the same landmass of its production. This is a serious problem for most of the industrialized world except for Europe. As seen from Figure 7.6, most of the world's natural gas is in the Eastern Hemisphere. The entire Western Hemisphere (Latin America, United States, and Canada) accounts for only a little more than 9% of the natural gas on the planet.

At first glance it would appear that having so much of the world's natural gas in the Eastern Hemisphere would be a major problem somewhere down the road; however, the numbers may not be a clear cut as one might think. The numbers in the figure represent only the natural gas that is recoverable by conventional means, but a considerably larger quantity of natural gas is trapped in shale and sandstone which until recently was considered inaccessible. Tapping this large unconventional reserve of natural gas in now considered to be technically and economically feasible.

The change that made this new gas supply accessible is **horizontal drilling** combined with **hydraulic fracturing (fracking)**. Hydraulic fracturing in its simplest terms is designed to release gas that is trapped in various kinds of rock (shale, sandstone, etc.). A well designed for this purpose has two concentric pipes. When the well reaches gas-bearing rock, water containing various added chemicals is blasted down one of the pipes at high pressure. This high pressure forces the rock to fracture and release the gas. The spent fracking liquid followed by the natural gas moves to the surface through the other pipe. This process releases otherwise trapped gas from shale and sandstone layers. The process, which was developed in the late 1940s, was never used extensively because a given well could fracture only a relatively small amount of rock. Horizontal or directional drilling changed drastically the amount of rock that could be fractured.

Fracking combined with horizontal drilling is illustrated in Figure 7.7. Recent technology has allowed the direction of drilling in a well to change as much as 90 degrees to make it possible to drill horizontally through a specific layer of rock. So, instead of drilling fairly quickly through layers of rock, the drilling can be turned to the horizontal and can drill within one layer for as much as 5000 feet. This allows a given well to fracture a huge amount of shale or sandstone.

Almost from the time that it was developed, the process of fracking has been controversial. The major concern is for the safety of groundwater, because fracking can contaminate water intended for drinking and irrigation purposes. Methane from the natural gas can get into the groundwater, or the groundwater could be contaminated by the fracking water mixture.

Originally it was thought that contamination of groundwater by methane was very unlikely because the shale being fractured is so deep in the ground (5000 feet or more) and unlikely to connect with groundwater near the surface. However, some evidence exists of water wells being contaminated with methane. The question remains as to how the methane got there. It is possible for fractures from fracking to line up with other existing fissures in the ground or for abandoned oil or gas wells to provide a path for methane to the surface water. Another possibility is that cracks in the concrete well casing could allow gas to leak from the well into the groundwater.

Probably the greater concern for groundwater contamination comes from the fracking flowback water. Once the fracking fluid is injected into the well under high pressure, up to 75% of that liquid will return to the surface. The amounts of material involved here are huge, with two to four million gallons being injected along with thousands of gallons of fracking chemicals. As is true with any deep injection well, the material forced up from deep within the ground is likely to contain salt, toxic materials, and radioactive elements. In addition, the fracking fluid itself contains a cocktail of chemicals, many of which may be toxic. Many of these chemicals remain unknown because drilling companies refuse to divulge them, as they claim that the formulation is a proprietary trade secret. The government may eventually force the publishing of these chemicals, but some of them are currently known and some are toxic.

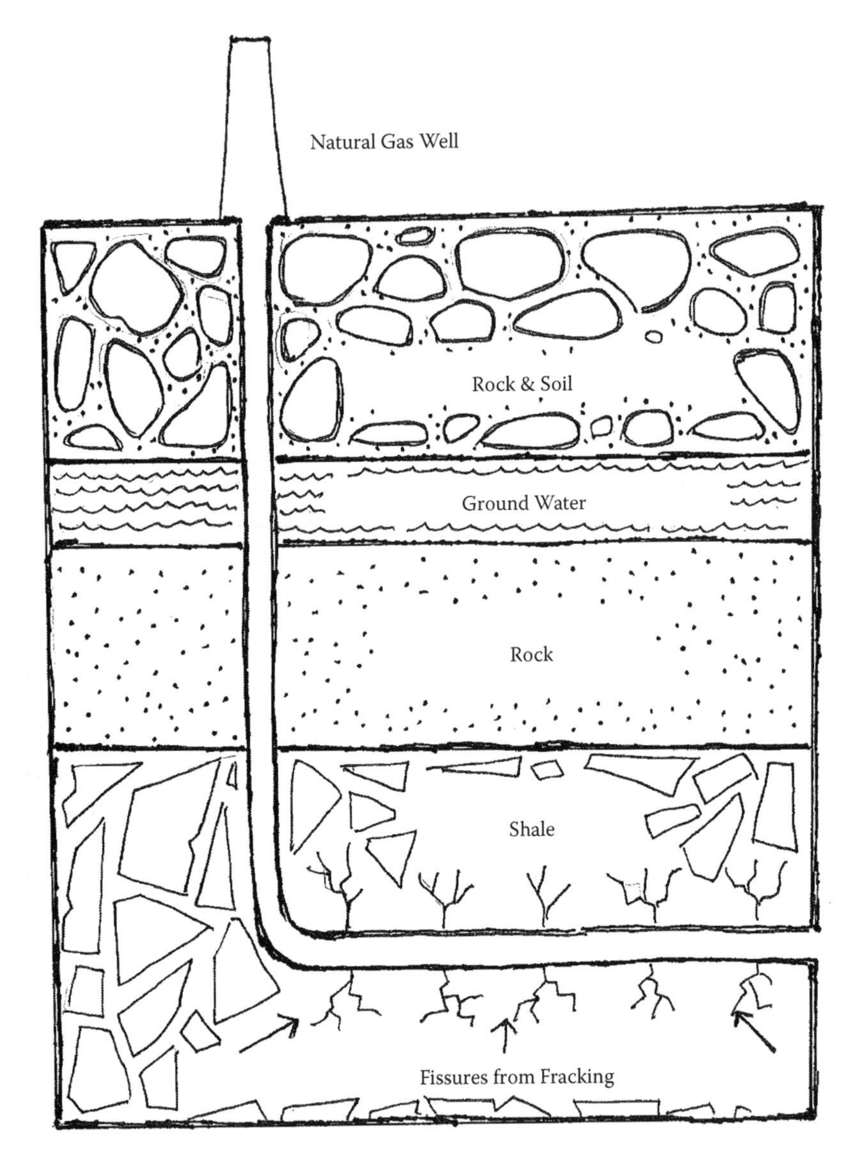

Figure 7.7 Diagram showing the use of hydraulic fracturing (fracking) combined with horizontal drilling to release natural gas from shale. (Illustration by Mark Brincefield.)

So the question is, "What to do with several million gallons of spent fracking fluid so it will not contaminate the groundwater?" Clearly, such material is not something that you can allow to flow into the nearest stream, so holding ponds are usually constructed; however, a simple holding pond would allow the liquid to seep into the groundwater. Many states require that holding ponds be lined with material that will keep the fracking liquid out of the soil and, therefore, the groundwater. Of course, liners can leak and heavy rain and wind can cause the ponds to overflow. New York is considering requiring watertight tanks for the storage of flowback water.

Fracking is becoming a major player in the natural gas industry in the United States because it opens up a very large reserve of natural gas that was previously inaccessible. This source of gas could increase U.S. reserves by anywhere from 275% to 775%, which could make the need to import gas from the Eastern Hemisphere unnecessary for the foreseeable future.

As attractive as this large supply of gas is, it is not without environmental consequences. Concerns about groundwater contamination have been demonstrated in a number of cases and give cause for concern. What remains to be determined is whether fracking and related processes are inherently environmentally unsafe or whether with sufficient governmental control the processes can be rendered sufficiently safe for use as a method of extracting natural gas.

If it necessary to import natural gas from Eastern Hemisphere countries, there are ways of doing it, but they are not without difficulties. One solution involves cooling the natural gas below its boiling point, which is –259°F (–162°C), and converting it into a liquid (**liquefied natural gas, LNG**). Once the natural gas is converted into a liquid, it can be transported by sea because as a liquid it has 1/600th the volume it had as a gas. Shipping natural gas in this manner is an expensive proposition. Each of the LNG tankers costs about $150 million, and a fleet of them is required. For each export site, it is necessary to have a liquefaction facility at about $2 billion each, and for each import site there must be a regasification plant at $1 billion each. Some of this infrastructure already exists, as LNG accounts for about 12% of natural gas imports.

There are safety concerns with LNG because by its very nature it is a concentrated, very volatile fuel. It is important to make clear that LNG cannot explode in the usual sense of the word. A container full of LNG would tend to exclude air and without air there could be no rapid burning, which would lead to explosion. However, if there is a leak or rupture that allows either liquid or gaseous material to escape, then the material could mix rapidly with air and burn. Such fires would be very hot and would quickly burn back toward the source of the leak. This type of fire is exactly the worrisome element of LNG and, although technically it is not an explosion, can become a very dangerous, raging inferno. The history of LNG goes back to 1912 in West Virginia and over that history has been relatively safe; however, the potential size and seriousness of an accident are still worrisome. The most serious accident to date occurred in Cleveland, Ohio, in October 1944 when an LNG tank ruptured and the resulting fire incinerated about a 1 square mile area of the city; 128 people were killed and 225 to 400 injured. The only other major accident in the United States occurred at Cove Point, Maryland, when natural gas got into the electrical generating system and one worker was killed. Outside the United States, in early 2004 a fire at a liquefaction facility in Skikda, Algeria, killed 27 people and injured 72.

The other option for the overseas transport of natural gas would be the use of undersea pipelines in areas where the undersea distances are not large. This would be a convenient way to move natural gas to England or Japan, for example. It would even be possible to deliver Asian natural gas to North America via a pipeline through the Bering Strait. It is not clear that such pipelines are practical, though, given that building a gas pipeline from the North Slope of Alaska to the lower 48 states has turned out to be extremely problematic and expensive.

Much of the interest in natural gas is driven by the fact that it is easy to use and is a clean-burning fuel. Because it is a gas and flows easily through a pipeline, it can be mixed with air and burned to produce a very hot flame, producing few pollutants, if any. About the only gases produced at all are water vapor, carbon dioxide, and nitric oxide. Only the nitric oxide is normally classified as a pollutant and is released purely due to the high temperature of the flame. The carbon dioxide and water are the inevitable end products of the combustion of a hydrocarbon.

Like oil, natural gas is a versatile fuel; however, natural gas has found its principal application in space heating, drying, water heating, and cooking. Another major use of natural gas is the generation of electricity, which represents about 31% of natural gas use in the United States. Natural gas electricity-generating plants are less expensive to build than coal-fired plants, cost less to operate, and produce electricity very efficiently.

Natural gas is beginning to find applications in the transportation industry. Natural gas is an ideal fuel for the internal combustion engine. This is partly because liquid fuels must be vaporized and mixed with air to be burned in the engine, whereas natural gas need only be mixed with air because it is already a gas. This fuel burns cleaner, yielding considerably fewer pollutants than ordinary gasoline. The obvious problem, of course, relates to containing a gaseous fuel in a fuel tank. This problem can be handled by compressing the gas, but unfortunately a compressed tank the size of a conventional gas tank will take a vehicle only about one-fourth as far as current similar gasoline-powered cars. Due to the lack of range and refueling locations, natural gas-powered vehicles have for the most part been restricted to commercial fleets.

The ultimate supply of natural gas is a subject of much debate. Using very conservative numbers and considering only the domestic supply, the United States has only about a 12- to 13-year supply of natural gas remaining. On the surface, such a calculation would suggest a serious problem, but fracking has appeared to change all of that. Based on various estimates of the amount of gas that could be released by fracking, one can get supply estimates ranging from 35 years to almost 100 years. These numbers, along with the difficulties associated with importing natural gas other than from Canada, are driving the tremendous interest in fracking.

The world supply, based on current data, should last about another 60 years. The problem, as noted earlier with petroleum, is that the size of the reserves is often far from the real production potential. Vigorous exploration and drilling of more difficult gas fields along with known fields that are simply underreported could stretch the supply considerably. One factor that could increase the rate of consumption of natural gas and help in solving the world oil supply problem is the conversion of natural gas into

Table 7.2 Advantages and Disadvantages of Natural Gas

Advantages	Disadvantages
Burns hotter than other fossil fuels	Difficult to transport by sea
Causes little air pollution	Releases CO_2 and NO
Easily transported over land by pipeline	Limited supply (60 years or more; 12 to 100 years for the United States)
Versatile fuel	
Produces electricity more efficiently than oil or coal	

liquid fuels. A number of forces are driving this idea. First, despite the limited supply of natural gas, the supply of oil is even more limited; however, liquid fuels are generally more convenient to use than gaseous fuels. Second, if we convert natural gas into a liquid fuel then it is much easier to ship it across water via tanker.

A number of approaches are available to convert natural gas into liquid fuel. These processes are an area of fervent research, and new or improved methods are appearing all the time. Generally, two types of liquid fuels are created. One is a liquid nearly identical to gasoline that can be used with little modification, if any. Alternatively, the natural gas can be converted into methanol, which can be used in the internal combustion engine with some modification. Methanol is believed by some to produce less air pollution and is currently used as a fuel in Formula One race cars.

If these conversion processes become economically viable, then the lifespan of the natural gas supply may be considerably shorter than indicated earlier. When the supplies of oil and natural gas begin to fade out then the world will need to turn elsewhere to power the societies in which we all live. At this point, we would need to either abandon fossils fuels or turn to coal. See Table 7.2 for a review of the advantages and disadvantages of natural gas.

Coal

Of all the fossil fuels, coal probably has the worst reputation. Modern society could hardly wait to get rid of coal because it has been implicated in some of the worst smog incidents in history; however, as the reserves of oil and perhaps natural gas begin to play out, coal may begin to look more attractive again. The supply of coal is simply enormous. Based on proved reserves, the supply should last about another 120 years at the current rate of use. If usage would increase then the length of supply could decrease, but if new reserves are found then the number of years' supply could increase. Length of supply estimates vary from the conservative value based on proved reserves to 600 years based on other estimates. The United States alone probably has a 240-year supply based on its current production rate.

Just what is coal? Is it a good fuel to burn just because there is a lot of it? The formation of coal seems to be more of a natural ongoing process than the circumstances that led to the formation of oil and natural gas. However, it is worth noting that the conditions in some geological periods may be more conducive to coal formation than in other periods. Coal is formed from the remains of terrestrial plants that we believe were growing in swamps or bogs. These plant remains, which were subjected to heat and pressure as part of the sedimentation process, were gradually converted into coal.

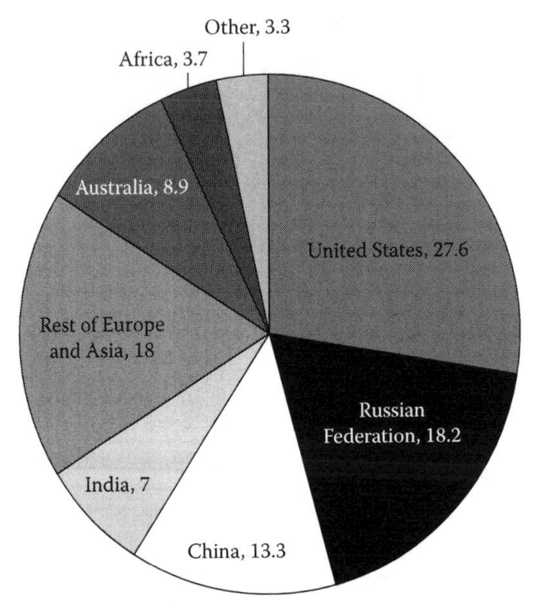

Figure 7.8 Coal reserves by region for 2010. (Based on BP, *Statistical Review of World Energy 2011*, http://www.bp.com/statisticalreview/.)

The process of coal formation begins with the formation of peat, a dark brown spongy material that looks like decayed wood. As moisture is removed and the carbon content increases, the material passes through various forms of coal: **lignite**, **subbituminous**, and **bituminous**. The last stage of coal development is the production of anthracite. Each type is drier and higher in energy content than the type from which it was formed. It takes about 440 million years to complete the process, but only about 1 million years to get to the lignite stage. Although **anthracite coal** has some nice properties such as burning without smoke and having the highest energy content, it is a relatively rare form. The highest fuel-content coal that is available in large quantities is bituminous. This type of coal contains about 10% water and 83% carbon.

The distribution of coal reserves around the world is much different than for oil and natural gas, which is related to the fact that the material from which coal is formed is different from oil and natural gas. As can be seen from Figure 7.8, over half of the world's coal reserves are found in the United States, Russia, and China. The Middle East, which has extensive oil and natural gas reserves, has little coal, if any. Only Russia seems to have good supply of all types of fossil fuels.

One of the biggest problems with coal is taking it out of the ground. Both oil and natural gas, since they are fluids (gases or liquids), can be pumped out from the ground. A solid such as coal needs to be mined. The choices are either digging a tunnel down to the coal or removing the earth above the coal seam. Both approaches have been used and each presents its own unique problems.

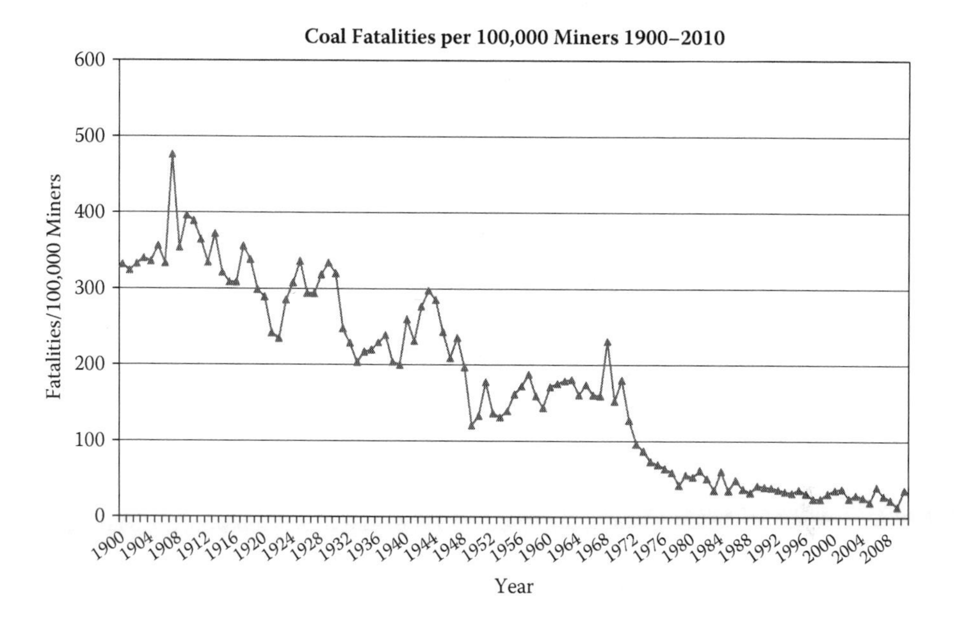

Figure 7.9 Coal mining fatalities in the United States per 100,000 miners, 1900–2010. (Based on USDOL, *Coal Fatalities for 1900 through 2011*, Mine Safety and Health Administration, U.S. Department of Labor, Arlington, VA, 2012, http://www.msha.gov/stats/centurystats/coalstats.asp.)

The tunnel approach is the oldest coal mining method because it can be done with simpler tools. The history of **tunnel mining**, whether for coal or anything else, is not a pleasant story. Miners have generally worked under the worst of conditions, either as slave labor or for extremely low wages. Since 1900, over 100,000 men have been killed in coal mine accidents in the United States and more than 1,000,000 permanently disabled. In Figure 7.9, one can see that the situation has improved over the years. Coal mining fatalities hit a peak of 3242 deaths in 1907 but declined to a low of 18 in 2009.

Much of the history of the coal mining industry in the first part of the 20th century was punctuated by violence and confrontation. Appalachian coal was the material that fueled much of the American Industrial Revolution. As the industrialists of the Northeast moved deeper and deeper into the mountains to feed their insatiable need for coal, they bought up the mineral rights to a large percentage of the mountain areas of West Virginia, eastern Kentucky, and western Virginia. Most of these rights were purchased for a very small amount from the isolated residents who understood neither the value of the coal rights they were selling nor the disruption that mining for coal would cause. When the operators began to mine for the coal, the local residents were hired to work in the mines, but they were too few in number. The coal mining companies began to import workers and housed them in towns that the companies built and owned. By the 1920s, the operators were bringing in immigrants directly from Eastern Europe and dropping them into various isolated Appalachian valleys. Many of these immigrants barely spoke English. In 1923, the number of coal miners in the United States peaked at about 862,000.

Many of these workers were abused, as all of the power was with the coal companies. The companies owned and controlled everything, including the mines, the miners' homes, the stores, and the schools. Over the next 50 years, the number of coal miners steadily dropped in response to two factors. First, the mines began to mechanize, which reduced the demand for workers, and, second, the demand for coal gradually declined as society began to use other energy sources. Because a decrease in mine workers would be expected to result in a decrease in mine fatalities, Figure 7.9 shows fatalities per 100,000 miners to compensate for fluctuations in the size of the workforce. The fatality rate did steadily decline after 1907. The increase in safety was the result of both company and union activity. Although the companies seemed little concerned for the welfare of the workers, it was a major problem for the mine operators if a mine had to be temporarily shutdown because of a major mine accident. The unions began to become a major force in the coalfields during the Great Depression. The unions wanted higher wages for their workers, more freedom for their workers in the company towns, and improved safety for their workers in the mines. Despite these efforts, in the 1960s there were still over 150 deaths per 100,000 workers. To make certain safety features of mines were uniform across the industry, the Coal Mine Health and Safety Act was passed by Congress in 1969. Shortly after the act was passed, the fatality figures fell into the range of 20 to 50 deaths per 100,000 workers and have remained in this range to date. In a similar fashion, mine-related injuries have declined as well.

Tunneling underground is always an inherently dangerous process. Even with the best technical safeguards, the possibility of a roof collapse is always present. This danger is increased by the various gases that seep out from underground coal seams. In the 19th century, canaries were taken underground often to detect odorless poisonous gases. The canaries were generally more sensitive to these gases than humans and would succumb more quickly. The death of the canary would signal the miners to get out. One gas that could cause death by another route was methane. Because methane is flammable, if enough of it seeped out and a spark was produced to ignite it, an explosion would occur. Such explosions trapped or killed many miners. Today, underground mines in the United States are required to have alarm systems to detect such gases and ventilation systems to remove the gases.

For those miners who evaded death or injury by trauma, the years of working underground and breathing coal dust led to a degenerative lung disease known as *black lung disease*. In 1969, 50,000 miners and ex-miners suffered from this condition; the current number is estimated at about 100,000.

For obvious reasons, the process of removing the ground over the coal seam (**overburden**) is safer for mine employees. Unfortunately, this process, known as **strip mining**, has significant drawbacks. The strip mining approach is generally much more environmentally destructive, as huge amounts of materials have to be moved to expose the coal seam, and when the coal mining is completed this mess often is left as is. In flat areas, strip mining results in an unsightly, unproductive, partially filled hole in the ground; however, in mountainous areas, strip mines can present even more significant problems. Extensive soil erosion due to the exposed bare dirt may clog rivers with sediment. In the past, large mudslides have occurred when heavy rains have saturated strip mine overburden and caused it to slide down a mountainside. The extensive environmental problems caused by strip mining led the U.S. Congress to pass the Federal Surface Mine Reclamation Act in 1977. Because strip mining itself is a relatively inexpensive mining method, the reclamation process becomes the most expensive part of the process. Companies often try to avoid doing an adequate job of restoration; therefore, enforcement is difficult and important.

An additional issue for both surface and underground mines is acid drainage. The exposure of underground materials to air and water leads to leaching of various toxic substances and to the bacteriological conversion of underground sulfur deposits into sulfuric acid. This acid mine drainage has been known to destroy all life in some Appalachian Mountain streams for several miles.

Coal does not transport well. Being a solid, coal cannot be fed into a pipeline and made to flow from one place to another. For small loads, coal is usually transported on trucks, whereas for larger loads train cars or river barges are the method of choice. Whichever way the coal is moved, considerable physical handling is necessary. One alternative is the slurry pipeline. In these pipelines, the coal is mixed into a slurry with water and the mixture is moved by pipeline, like oil. In the United States, many of these pipelines would need to be in the West, but the tremendous water requirements (2700 gal/min) may make them impractical in an area where water is scarce.

The burning of coal for energy presents significant air pollution problems that will be discussed further in Chapter 9; in addition, coal produces more carbon dioxide per unit of energy produced than other fossil fuels which has direct implications on global warming (see Chapter 11). The higher production of carbon dioxide comes directly from the fact that coal contains a much higher percentage of carbon than the other fossil fuels.

Unless there is a fairly quick move away from fossil fuels in the near future, coal will become more important in the near term. There is simply more coal than other fossil fuels. Coal can fill the gap until alternative energy sources are completely developed. Considerable research is being directed at converting coal into gases or liquids that are more convenient to burn. Such fuels are collectively known as *synfuels*. Currently, these fuels are not economically competitive, but as oil and natural gas get more expensive they may become competitive. See Table 7.3 for a review of the advantages and disadvantages of coal.

Water Power

The fossil fuels are not a sustainable resource. Given enough time the supply will be exhausted; however, there are other energy sources that are sustainable. These energy sources can be used indefinitely without concern for depletion of the energy resource. Of the sustainable energy sources, water power has been used to the greatest extent worldwide for commercial power generation. Roughly 8% of all electric power generation in the United States is from water power.

Table 7.3 Advantages and Disadvantages of Coal

Advantages	Disadvantages
Supply could last 120 to 600 years	Dangerous to mine (tunnel mining)
Inexpensive	Environmentally destructive (strip mining)
High net energy yield	Difficult to transport
	Releases more CO_2 per unit of energy than other fossil fuels
	Dirty to burn (releases CO, SO_2, NO, and particulate matter)

Water power is a secondary energy source that in most cases derives its power from the sun. The solar input of energy is through the water cycle, which will be discussed in Chapter 8 (see Figure 8.1). The sun provides the energy for evaporation, which sends the water vapor into the atmosphere. Eventually, this moisture returns to the ground as rain or snow. Some of the precipitation falls on high ground and enters rivers and streams. This water has potential energy due to its location. As the water flows back to the ocean, the potential energy is released. By forcing some of the water to flow over a waterwheel or through a turbine, some of this energy can be captured.

Waterwheels have been used to generate mechanical power as far back as ancient times. These devices were usually used to run gristmills and sawmills. The large-scale need for water power did not occur until it began to be used to generate electricity. The first hydroelectric plant, which generated electricity to power 16 brush-arc lamps, was put into operation in 1880 at the Wolverine Chair Factory in Grand Rapids, Michigan. In 1882, a gentleman built a dam to generate electricity for his own lighting needs. A neighbor wanted to use some of the electricity as well; hence, the first electric utility was born.

The use of hydroelectric power varies from place to place around the world. Norway gets 95% of its electric power from water generation and Canada gets 58%, whereas the United States gets only 6% of its power from water. The United States generates so much hydroelectric power in absolute terms that it is the fourth largest generator of hydroelectric power after China, Brazil, and Canada.

It is estimated that only about 10% of the worldwide potential for hydroelectric generation has been realized. Many of the untapped sites are in less-developed countries, and because many of these countries have limited fossil fuel reserves the development of hydroelectric power appears to be very useful. China probably has the most aggressive plans for developing hydroelectric power. As seen from Figure 7.10, China's consumption of hydroelectric power has been increasing steadily, particularly since 2000. The most controversial of the Chinese hydroelectric projects is the Three Gorges Dam across the Yangtze River, which was begun in 1994 and was completed in 2006. This is the largest hydroelectric dam in the world. It is higher than the Washington Monument and more than 100 times that height in width, and its reservoir has displaced 1.4 million people. Three Gorges Dam has all of the advantages and disadvantages of other dams, which will be discussed later, but they are magnified because of the sheer size of the dam. It generates

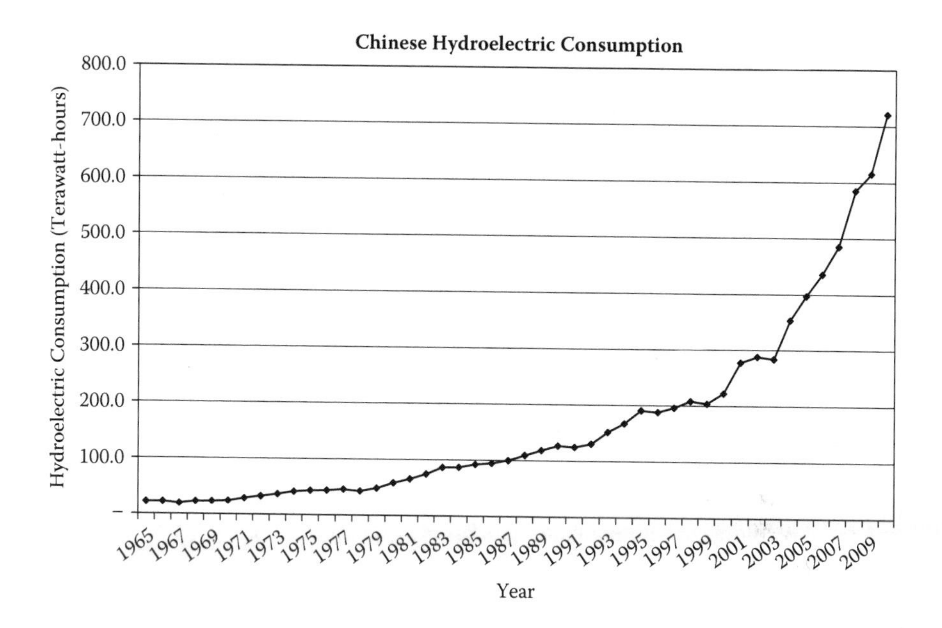

Figure 7.10 Chinese hydroelectric consumption since 1965. (Based on BP, *Statistical Review of World Energy 2011*, http://www.bp.com/statisticalreview/.)

roughly 80 billion kilowatt-hours of electricity per year, but Chinese officials at the highest levels in government have admitted to a large number of serious issues that need to be addressed.

The generation of hydroelectric power usually involves building a dam that in addition to providing electric power can be an effective means of flood control. The tendency in hydroelectric power generation is to build large dams to generate massive amounts of electricity. Such dams are more problematic in less-developed countries because of the high cost of construction. Once constructed, however, they will provide electricity with little maintenance and a low cost of operation. In addition, most large dams have a long operating life.

A major issue with the construction of a dam is the reservoir that develops behind the dam. To be sure, these reservoirs are great sources of recreation and irrigation water, but they come at the expense of flooding hundreds of square miles of land. When the land is not valuable and the population is sparse, this process may not be a significant problem. In many cases, however, large populations of people are displaced or thousands of acres of valuable farmland flooded, or both.

In tropical areas, the problem is even worse. The slow-moving, nutrient-rich waters of the tropics are home to certain snails that are an important part of the life cycle of a parasitic flatworm. This flatworm causes a debilitating disease known as schistosomiasis. Large dams built in tropical regions often take fast-flowing rivers and convert them into very slow-moving reservoirs that become perfect homes for the snails. Outbreaks of schistosomiasis usually occur whenever large reservoirs are developed in tropical regions.

Dams also interrupt the natural process of sediment flow through a river. The rich farmland found at the mouth of many rivers is a result of the natural process of deposition of sediment washed down the river. When this process is interrupted, the sediment fills in behind the dam. With time, the reservoir will fill up with sediment. Small dams on rivers with high sediment loads can become useless in a short period of time; however, at the mouth of the river, farmers will be buying fertilizer to replenish their fields. A good example of this is found on the Nile River in Egypt. The Nile Delta was once one of the richest farming areas in the Middle East; however, with the construction of the Aswan High Dam the enriching sediment of the Nile River no longer reached the delta. The Egyptian farmers were forced to turn to fertilizer to keep their fields fertile.

In Europe and the United States, there is very little development of new hydroelectric sites. Many of the potential sites are highly populated areas, valuable farmland, or scenic river valleys. Most people consider the price of building dams in these places to be simply too high. For example, large amounts of electric power could be generated by damming up the Colorado River in the Grand Canyon, but that is not going to happen.

Finally, dams often interfere with delicate ecosystems. The dam produces a large slow-moving lake, as noted earlier. The ecology of a lake is different from the ecology of a river; hence, there is a drastic disruption in the biological interactions. Such projects may cause disruption of economically important species such as salmon, or other species that might be found on the endangered species list. In environmental circles hydroelectric dams are not well accepted. The reasons for this are the ecological considerations, despite the fact that hydroelectric power is a nonpolluting renewable source of energy.

From a purely chemical and energy perspective hydroelectric power is nearly ideal. The energy is derived in a renewable manner from a constant supply of falling water that is available on a continuous basis. Most other forms of renewable energy are intermittent; for example, wind power only works when the wind blows and solar power does not work at night. Water power does not produce air pollution or contribute to global warming (see Chapter 11).

Given the positive energy considerations of dams and the negative ecological implications of these same dams, the use of dams to generate electric power forces some difficult choices even for the most environmentally concerned individuals. Until very recently, it was assumed that dams would remain in place as long as they usefully served the purpose for which they were built. In recent years, however, several dams have been removed by their owners, including the Quaker Neck Dam on the Neuse River in North Carolina, the Waterworks Dam on the Baraboo River in Wisconsin, and a dam on the Clyde River in Vermont. A significant event took place in the summer of 1999 that led to removal of the Edwards Dam on the Kennebec River in Maine against the owner's wishes. Although this was an old (1837) and small dam by industry standards, it was a functional dam capable of generating 3.5 MW of electricity. The decision by the U.S. Federal Energy Regulatory Commission to remove it has had tremendous implications for many larger dams that are considered to interfere with the migratory patterns of various species of fish. Many such

Table 7.4 Advantages and Disadvantages of Water Power

Advantages	Disadvantages
Renewable energy source	High costs of construction
Continuous energy source	Few new sites available in Europe and United States
Releases no CO_2, particulates, CO, SO_2, or NO	Severe disruption of people and ecosystems when reservoirs are created
Provides flood control	Sediment flow affected
Availability of potential sites	Disruptive to many ecosystems
High net energy yield	Encourages disease in tropical areas
Low operating and maintenance costs	
Long operating life	

dams are found in the American Northwest, where the conflict between the ecology of fish and electric power generation could become intense, as this area receives about 70% of its electricity from hydroelectric generation. Court cases involving the Federal Energy Regulatory Commission, power companies, environmentalists, and American Indian Tribes are almost continuous.

Those existing dams that are not causing serious problems should probably be used to their fullest. The site location and building of new dams should be done only with the utmost care. The ideal would be to extract electric power from a river without damming it up. There are devices that can do this, and they may become more important in the future. See Table 7.4 for a review of the advantages and disadvantages of water power.

NUCLEAR POWER

Early in the evening of April 27, 1986, a worker at a radiological monitoring station in northeastern Poland went out to check the radiation level. He was surprised to find a reading of approximately 700 times the normal level and made a remark that can roughly be translated as, "Oh, hell, the damn needle's stuck again." After cleaning the instrument, he took readings again early the next morning and found about the same results. It was through this observation and other similar findings that the world became aware that a serious nuclear episode had occurred in the USSR. The event was the catastrophic nuclear accident at the **Chernobyl** nuclear power station in Ukraine.

It is very difficult to present nuclear power objectively without sensationalizing the issue. Nuclear power is probably not the shining star of energy salvation it was once thought to be, nor is it likely to be the harbinger of doom it became after the Chernobyl disaster. Nuclear power is of sufficient complexity and importance to warrant devoting a separate section to this method of power production.

History of Nuclear Fission and Energy

In 1940, with the involvement of the United States in World War II looming on the horizon, the main interest in nuclear fission was in getting the reaction to go out of control—that is, to make a bomb. By the mid-1940s, the United States had made the nuclear bomb and used it, thereby changing the nature of warfare forever. Of more interest to us in discussing nuclear energy is some of the work done along the way to creating the atomic bomb, the Manhattan Project. At the University of Chicago under the stadium seats of the football field, the first sustained nuclear reaction was carried out in 1942. This work made it clear that a reactor could be designed to use nuclear fission to produce useable energy. In 1946, the **Atomic Energy Commission**, which later became the **Nuclear Regulatory Commission** (**NRC**), was formed, but its main purpose was the production of more and better nuclear weapons. By the 1950s, the interest in using nuclear technology to produce energy grew rapidly. In 1958, the first civilian nuclear reactor was put into operation in Shippingport, Pennsylvania. Optimism for nuclear power ran high in the 1960s. It was projected that by the end of the century nuclear power should produce 21% of the world's commercial power and 25% of that in the United States. By the 1970s, much of that optimism faded as activists began to point out many problems with nuclear power plants. In 1974, the orders for new nuclear power plants had peaked and have continued to decline since then. After this point, events, including the **Three Mile Island** incident in 1979 and the Chernobyl accident in 1986, soured public opinion on the nuclear industry.

As the world entered the 21st century, attitudes toward nuclear power began to change. Three Mile Island and Chernobyl faded in our collective memories, and new concerns over global climate change (Chapter 11) and air pollution (Chapter 9) began to take their place. The nuclear industry and their advocates pointed out that nuclear power releases no carbon dioxide to increase global warming and under proper operation emits no pollutants into the air or water. As the first decade of this century passed, new nuclear reactors were proposed and planned. Then, on March 11, 2011, the northeast shore of Japan's main island was struck by an off-shore magnitude 9.0 earthquake followed by a roughly 50-foot tsunami. These events produced a major disaster at the **Fukushima Daiichi** nuclear power plant located on the coast of Japan.

The disaster had an immediate impact on American public opinion such that support for new nuclear power plants in the United States dropped from 49% to 41%. It is not clear if this shift is permanent, but the government is proceeding with plans to approve new nuclear plants. In February of 2012 the construction of a new American nuclear power plant was approved by the NRC. This represents the first new construction approved since 1979. It would appear that the future of nuclear power is somewhat uncertain but appears to be on the upswing.

Nuclear Power Plant Design

Before considering the design of a **nuclear reactor**, it is necessary to consider the factors necessary to sustain a nuclear reaction. As discussed in Chapter 4, a sustained, controlled nuclear reaction could be obtained if on average one neutron from each fission produced a second fission and so on. A normal sample of uranium does not normally undergo fission, so what is required to cause nuclear fission to occur and generate nuclear power?

The first problem is that uranium is only 0.7% U-235, which is the only easily fissionable uranium isotope. Most uranium is U-238, which is not easily fissionable. To produce nuclear fuel the uranium has to be enriched in U-235. A key problem for the

Manhattan Project during World War II was the enrichment of uranium. The amount of enrichment necessary to produce sustainable nuclear fission depends on the reactor design, but most American reactors require a level of about 3%.

Even after the uranium is enriched to the proper concentration of U-235, sustainable fission will not necessarily occur. The likelihood that emitted neutrons will strike and cause fission of other nuclei depends on many factors such as total mass, density, and shape. A quantity of material with the requisite mass, density, and shape to support nuclear fission is known as a **critical mass**.

The material in a nuclear bomb must be forced to critical mass suddenly so that a rapid uncontrolled chain reaction can occur. The fuel in a nuclear reactor must be brought to critical mass gradually and brought away from critical mass in the same way. The fuel is usually made into rods that are about the diameter of a pencil and about 15 to 20 feet long. Each **fuel rod** contains the energy equivalent of about three railroad carloads of coal. The usual reactor contains about 40,000 such rods. A set of rods will last about 3 to 4 years. Shaping the fuel into rods intentionally makes it so that the fuel cannot be at critical mass without other materials being present. With the rod shape of the fuel, most of the neutrons escape at high velocity, having encountered very little that would slow them down. If a bundle of these rods is placed together in the reactor, nuclear fission will not occur because the neutrons escaping from one rod hit the next rod at too high a velocity and either pass on through or are captured by U-238 nuclei. It is necessary to place a moderator between the rods to slow the neutrons down. Luckily, a very common, plentiful material is useful as **moderator**: ordinary water. Although water has some disadvantages, it is used in most nuclear reactors worldwide. Other moderators are graphite (a form of carbon) and heavy water, which is water in which all the hydrogen atoms have a mass of two. This hydrogen isotope is known as *deuterium*.

If we place a bundle of fuel rods into water, we can get sustainable nuclear fission to occur, but if it takes off too fast we have no control over it. To maintain control over the reaction **control rods** are inserted to absorb neutrons. The reaction is controlled by selecting how many control rods are inserted and how far they are inserted. By adding control rods more neutrons are absorbed and the reaction slows down, whereas by removing control rods fewer neutrons are absorbed and the reaction speeds up. Control rods are usually made of materials such as boron or cadmium.

Because the purpose of a nuclear reactor is to produce heat, it is necessary to remove this heat. In reactors that use heavy or ordinary water as the moderator, the moderator is the **coolant**. The water has a high heat capacity and, by circulating the water over the fuel rod/control rod assemblies (known as the reactor core), the heat can be carried away. In one design, this water can be allowed to boil and turn a steam turbine before being condensed and returned to the reactor core. Some individuals have referred to this as an exotic way to boil water. Alternatively, the water can be kept under high pressure; therefore, it cannot boil. This very hot water is passed through coils over which water from another circulating system passes. This water boils and turns a steam turbine. This design, known as a pressurized water reactor, is shown in Figure 7.11. The advantage of such a design is that the water turning the turbine is not contaminated by radioactive material from the core.

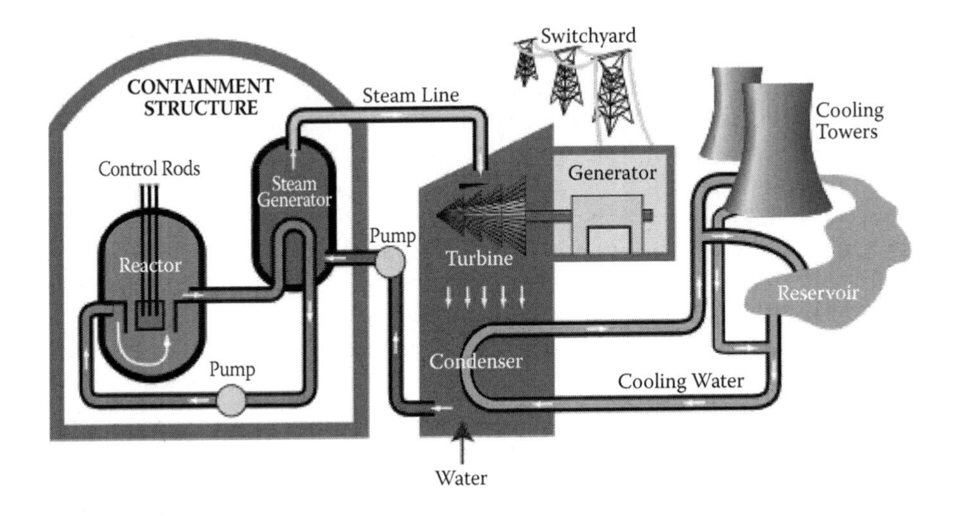

Figure 7.11 Pressurized water nuclear reactor. (Courtesy of Tennessee Valley Authority.)

In graphite-moderated reactors the choice of coolant is less obvious. Because graphite is a solid, it cannot serve as the coolant; therefore, any fluid with a reasonably high heat capacity can be chosen. Water is still often the choice because it is inexpensive and has a very high heat capacity; however, carbon dioxide has been used in Great Britain.

Nuclear Fuel Cycle

Nuclear fuel is mined from the ground much as coal, iron, or copper might be. The uranium is radioactive when it is found. It is naturally radioactive and is not created by the humans who dig it up. Humans, however, concentrate the natural radioactivity. Uranium ore contains too little uranium for use in the nuclear industry; therefore, it is processed to increase the concentration of uranium until it is nearly pure U_3O_8. The uranium is converted into UF_6, which is a gas. The material is then placed in gas centrifuges, where the lighter U-235 hexafluoride is partially separated from the heavier U-238 hexafluoride. Ultimately, the percentage of U-235 is increased from 0.7 to 3. Finally, the UF_6 is converted into UO_2 and made into fuel rods. When these rods have been used for 3 or 4 years to produce nuclear power they are removed and replaced. Spent fuel rods may not be able to support nuclear fission, but they are still highly radioactive. Such rods cannot be thrown in the nearest dump. These rods are initially stored in special water tanks designed for the storage of spent fuel rods and moved after a few years into giant above-ground concrete casks. Currently, both the water tank storage and the cask storage are located onsite at the nuclear plant where the spent fuel was produced.

This is intended to be temporary storage, but is it? As of now, there is no approved long-term storage site for spent fuel rods from American nuclear power plants. As a result, it is not clear that we have a fuel cycle. To have a cycle one needs to return the material back to its source. Nobody is willing to do this because the material has been changed, and there are now more highly radioactive isotopes compared to uranium present. The general belief is that it will take 10,000 to 240,000 years before the rods will return to the simple radioactive level of uranium. If the radioactive material were only uranium, one could argue that if it were diluted enough with rocks and soil it could be returned to the ground from which it came.

Because the spent fuel rods cannot simply be returned to the ground, then perhaps a permanent storage facility could be developed. A repository site in Nevada called **Yucca Mountain** has been under study since 1983. In 1987, Congress narrowed down potential sites and chose Yucca Mountain as the only site to be studied; in 2002, it designated this site as the single official U.S. nuclear waste repository. The entire process, however, was always embroiled in politics, with the various states where potential sites were located fighting hard to make sure such a repository was not placed there. In 2006, Harry Reid of Nevada was the majority leader of the Senate, and early in the Obama administration it became clear that Yucca Mountain was unlikely to become the radioactive waste site for the nuclear power industry. This decision came despite the expenditure of around $12 billion on developing the site. In 2012, the Blue Ribbon Commission on America's Nuclear Future, which was created by Secretary of Energy Steven Chu, recommended that the process of finding a waste site start over from the beginning. As the debate drags on and the fuel rods pile up at nuclear reactors, the nuclear industry continues to wait for a decision by the government on where and how these wastes will be stored for the next several hundred thousand years.

Rather than storing or burying the spent fuel, another option would be to reprocess the fuel. The material still contains some unused U-235 and plutonium-239 (Pu-239), which is another fissionable isotope. The Pu-239 is created from the more plentiful U-238 in the fuel rods through the following reactions:

$$^{238}_{92}\mathrm{U} + ^{1}_{0}n \rightarrow ^{239}_{92}\mathrm{U}$$

$$^{239}_{92}\mathrm{U} \rightarrow ^{239}_{93}\mathrm{Np} + ^{0}_{-1}e$$

$$^{239}_{93}\mathrm{Np} \rightarrow ^{239}_{94}\mathrm{Pu} + ^{0}_{-1}e$$

The Pu-239 is a very good fuel for nuclear reactors. This material undergoes fission in a manner similar to U-235 and could be mixed with U-238 to make useable fuel rods. It is also a fact that Pu-239 is the ingredient of choice for nuclear bombs. It was concern about nuclear weapons proliferation that moved President Gerald Ford in 1976 to stop the reprocessing of nuclear fuel in the United States. France, Russia, and the United Kingdom have continued reprocessing and have become the main reprocessors of nuclear fuel rods in the world. These countries reprocess their own fuel rods and contract out to reprocess rods from other countries such as Japan and Germany.

By reprocessing one can get more useable energy per unit of uranium. Of all the estimated fission energy available from all of the uranium (U-235 and U-238), we get about 5% and 6% without and with reprocessing, respectively. By being more efficient we generate slightly less waste.

Another approach to reprocessing nuclear fuel is use of a **breeder reactor** (also known as a *fast-neutron reactor*, or simply a *fast reactor*). These reactors operate in such a way as to maximize the reactions that produce Pu-239 shown earlier. It turns out that these reactions can be encouraged using much faster neutrons. It turns out, of course, that U-235 does not undergo fission well with fast neutrons. Pu-239, however, does not have this problem. Pu-239 will undergo fission with fast neutrons, and the fast neutrons very efficiently convert U-238 to Pu-239. The breeder reactors produce Pu-239 from U-238 and produce energy by Pu-239 fission at the same time. These reactors are called breeders because they produce their own Pu-239 fuel faster than they use it. Using breeder reactors and new fuel reprocessing technologies, one can theoretically extract 99% of the fission-based energy content of the uranium fuel. There is still nuclear waste for this procedure, but much less waste per unit of energy produced.

Currently, no commercial-scale breeder reactors are in operation, although there are some large prototype reactors operating. The United Kingdom built a prototype reactor at Dounray, Scotland, but shut it down in 1994. The United States has built several breeder reactors but none has ever produced commercial electric power. In 1986, France put the only commercial breeder reactor, the SuperPhénix, into full-scale operation but had to shut it down twice, once in 1987 and again in 1990 for repairs. This reactor is the only breeder reactor to have actually produced commercially available electricity. In 1997, the reactor was shut down due to cost and political opposition. Other countries that have built breeder reactors of some type include the former USSR, India, Germany, and Japan.

Given some of the apparent advantages of breeder reactors, why have they not become more commercially successful? The reasons probably fall into two categories, technical and economic. Most of these reactors have had serious technical difficulties that require a plant to shut down, often for extensive periods of time. One common problem comes from the coolant. These plants cannot use water as the coolant because the reactions use fast neutrons and water is a moderator that will slow them down. As a result, molten sodium metal is most commonly used as the coolant. Sodium is a very reactive, dangerous metal when exposed to air or water.

Economically, current breeder reactors are simply too expensive to operate. Although the breeder reactors use nuclear fuel much more efficiently, uranium fuel from mining sources is simply too inexpensive and too plentiful. Given this state of affairs, it is simply more economical to operate a standard light-water reactor even with the poor fuel efficiency. Based on current and projected use of standard nuclear reactors, the supply of uranium is placed anywhere from 80 to 230 years.

It appears that the nuclear fuel cycle problem is going to persist for some time. Recycling the fuel seems to have technical problems, and disposal is a technical and political "hot potato" that defies solution at the moment.

Nuclear Safety

No other nuclear power issue gets the public's attention faster than the safety issue. Because of our fear of nuclear weapons, we sense something ominous and fearful about nuclear projects of any kind. Nuclear energy is complicated and technical, and most of us do not understand it; we tend to fear what we do not understand. Add to this the nuclear disasters at Chernobyl and Fukushima, and it is no wonder that the public is scared. But, fear aside, what is the truth about nuclear safety?

First, it needs to be made clear that a nuclear explosion such as that from a nuclear bomb is not possible in a nuclear reactor. Weapons-grade uranium requires a U-235 content of about 90%, which is a long way from the nuclear fuel rod concentration of about 3%. If a nuclear explosion is not a danger from nuclear power, then what is?

1. Meltdown of the reactor core
2. Release of radioactive steam
3. Chemical explosion

Meltdown of the reactor core can occur if the coolant is removed from the core. There are any number of circumstances that might lead to a loss of coolant, including malfunctioning valves, faulty pumps, or a loss of electrical power. When the core of a nuclear reactor becomes uncovered, then there is no coolant to carry away the heat, which raises the possibility that the temperature will climb up to a point where the reactor core will simply melt.

Loss of coolant can lead to a meltdown even if the control rods have been fully inserted. Unless the fuel rods are new or slightly used, the fission reactions that have already occurred in the fuel rods will have produced highly radioactive fission products. The decay of these products produces a tremendous amount of heat, called **decay heat**. The heat output falls rapidly to roughly 1% of the initial decay heat level after the first year, but for the first few weeks after loss of coolant the heat will be sufficient to melt the reactor core. Clearly, once the core has been converted into a molten mass, one loses any control over it. The only options are to throw coolant (water, liquid nitrogen, etc.) at the core in very large volumes as quickly as possible. It is considered possible that a meltdown might melt through to the water table and contaminate groundwater, although we do not believe that has occurred in any case to date.

The release of radioactive steam is probably the least serious of the likely accidents. There are any number of ways in which release could occur, including the accidental opening of a steam valve, release of steam through a pressure relief valve because pressure has built up too high, or from an event that cracks a pipe in the coolant system. Most modern reactors are built with a containment structure around all radioactive parts of the plant. The containment structure, which is basically a building around a building, should contain any radioactive steam that is released accidentally.

Although a nuclear explosion is impossible, a chemical explosion is a distinct possibility. The explosion most likely would come from hydrogen gas generated during a partial or complete meltdown. At high temperatures (2700°F or 1500°C), steam will react with zirconium, which is found in the fuel casing material, to produce hydrogen gas:

$$Zr + 2H_2O \rightarrow 2H_2 + ZrO_2$$

In the case of graphite reactors, the carbon of the graphite can also react with the steam:

$$C + H_2O \rightarrow H_2 + CO$$

If the hydrogen concentration builds up to high enough levels, the red-hot core will set off an explosive reaction with the oxygen of the air:

$$2H_2 + O_2 \rightarrow 2H_2O$$

How likely are any of these events? Given the primary and the secondary backup systems in nuclear power plants, it is difficult to see how anything could go wrong, but of course nothing is foolproof. Therefore, as is true of most things in life, we are left with probability and risk. The amount of risk we are willing to take varies with the perceived consequences. Most of us are willing to try ice skating because the consequences of falling down are not that great; however, fewer of us are interested in trying bungee jumping because the perceived consequences are much greater. We also are more willing to take a risk if the probabilities are small. What does one do when the probabilities are very small, but the perceived consequences are potentially enormous? Such is the dilemma posed by nuclear power.

In 1975, Professor Norman Rasmussen of the Massachusetts Institute of Technology carried out a study for the Atomic Energy Commission. He concluded that the probability of a simple meltdown with all radiation contained was *one chance in 100 years for 100 reactors*. There are a little over 400 reactors worldwide. The accident at Three Mile Island in Middletown, Pennsylvania, on March 28, 1979, was similar to this type of a situation. The incident arose from routine maintenance when a filter was being changed on the reactor cooling system. Some air was accidentally sucked into the cooling system, interrupting the flow of water. The backup cooling system was switched on, but no water flowed because the valves on that system were closed. The warning lights that were supposed to tell the operators that no water was flowing were covered by the maintenance tags hanging from the valves that were being repaired. With the operators unaware that water was not flowing through the reactor, no action was taken. The temperature began to rise rapidly. The pressure rose until a steam-release valve opened to release some of the pressure, but when the pressure went back down the valve failed to close. About a million curies of radionuclides were released into the atmosphere. The pressure dropped drastically and finally the emergency core cooling system came online. When the water level came up to the appropriate level the emergency system was turned down. What the operators did not know was that a 1000-ft^3 (28,000-L) bubble of steam and hydrogen gas was in the system. It rose to the top and uncovered the core. The core overheated. By the time the problem was discovered and corrected it was too late to do anything but wait and pray. The uncertainty went on for 13-1/2 hours before it was learned that core meltdown had been avoided. The damages from this one accident cost over $1 billion and of that $187 million came from federal taxpayers. This reactor was shut down and sealed.

Was Three Mile Island one of the four accidents expected by Professor Rasmussen for the century? Rasmussen's study also addressed the likelihood of a "worst-case" accident. Such an accident would be one where the containment structure was breached, there was a large population nearby, and the weather conditions took the radiation toward high-population areas. Rasmussen set the probability at *one chance in 10,000,000 for 100 reactors.* Have we already had that accident at Chernobyl?

The incident that occurred in the wee hours of the morning of April 26, 1986, at the Chernobyl nuclear power station had a major irony associated with it. The exercise that led to the destructive accident was an experiment designed to test safety procedures and equipment. Without going into the details of the test, a few key points about the test are worth noting. First, this test of safety equipment needed to be done at very low power and the best time to do this would be when the reactor was in the process of being shut down for maintenance. In April 1986, Reactor 4 was scheduled to be shut down for routine maintenance. The engineers had waited for 2 years to redo this test, as a previous test in 1984 had failed to give satisfactory results. This was a one-shot test, as it would be about a year before another reactor was scheduled for maintenance. To check this safety system, all of the other safety or backup systems were turned off or disabled.

For some important technical reasons, the nuclear power reactor needed to be powered down slowly. The scheduled shutdown of the Chernobyl reactor 4 began at 1:00 a.m. on Friday, April 25. The timing of this shutdown is of interest because it coincided with the beginning of the May Day holiday break on Saturday. (In Communist USSR, the May Day holiday is one that probably rivals Christmas elsewhere.) The power was carefully lowered until about midday, when the shutdown was suspended to provide for an unexpected power demand. The shutdown process resumed at 11:10 p.m.

It was still a couple of hours before the experiment could be run. It is unclear as to what was going through the minds of the engineers in those fateful few hours before the accident, as many of them were killed in the accident. Were they concerned that the experiment could not be repeated soon if not done then? Were they preoccupied with the May Day holiday? Were they tired and not thinking clearly? Perhaps all of these things were factors. Whatever the case, the engineers pushed on with their experiment. The operators took manual control of the reactor and continued to decrease power to the level necessary to carry out their experiment. Computer-controlled power-down would have been safer, but they might not have been able to level out at the power they wanted for their experiment. When the power dropped too far, they pulled nearly all of the control rods to increase power. Finally, at 1:00 a.m. on April 26 they were ready to carry out the experiment. The power range at which they were operating was too low for safe operation of the reactor; they had achieved this low level of output only by tinkering with the reactor controls manually and overriding the computer-driven operating systems. Shortly after 1:00 a.m., the reactor became unstable and tried to shut itself down, but it could not because all of the safety systems had been turned off or compromised. At 1:22 a.m., a warning was received from the computer that the reactor was in a dangerous state. At 1:23 a.m., the engineers proceeded with their experiment. At 1:23:40 a.m., the reactor began to go out of control. The power output surged from 5% of normal to 100 times normal within 3 or 4 seconds, as all of the safety and emergency systems had been disabled or compromised. The core broke up and water was converted rapidly into steam. The steam pressure blew open the reactor. Graphite reacted with the steam to produce hydrogen gas that mixed with the air and exploded. The explosion ripped open the building and threw radioactive debris several thousand feet into the air. The exposed graphite in the core of the reactor produced a fire that took 10 days to be put out.

The results of this accident were devastating—36 people were killed outright in the incident, another 237 people developed acute radiation sickness, and it was necessary to evacuate 135,000 people from the nearby area. Most of these people will never be allowed to return to their homes. The estimates of the final death toll from Chernobyl for both direct and indirect deaths vary from 5000 to 100,000 people. Many of these deaths may occur years afterward as a result of cancer caused by this event. It may never be totally clear which deaths associated with Chernobyl have been caused by radiation from Chernobyl and which have not.

Three other reactors at Chernobyl continue to operate despite the radiation danger to the workers. The design of the Chernobyl reactors has been criticized in many quarters, but Ukraine needed the electric power and the hard currency that could be obtained by selling this power. In 1995, however, the Ukrainian government signed a memorandum of understanding with several other nations in which they pledged to close the Chernobyl plant by the year 2000. The Chernobyl reactors were shut down permanently on December 15, 2000.

Is Chernobyl Rasmussen's the *one in 10,000,000 years per 100 reactors* accident, or is it an event that is likely to be repeated in 5 or 10 years as some people have suggested? The answer to this question depends on whom one asks. It can be pointed out that this accident was the result of human error, and that the reactor designs in the West are generally superior to the Chernobyl type. Therefore, the likelihood of a repeat of such an accident is still controversial.

The more recent nuclear accident at Fukushima, Japan, was a serious, major accident but not as catastrophic as the Chernobyl accident. This accident was not caused by human error but rather by a confluence of events that were unforeseen in either magnitude or timing. The plant was built with standard earthquake precautions and designed for a tsunami of up to 33 feet.

When the earthquake struck off of the coast of Japan, the Fukushima Daiichi power plant came though the quake as planned. The major earthquake triggered an automatic shutdown of the three operating reactors at Fukushima. Because the earthquake disrupted external electric power to the plant, the diesel emergency generators started up to provide power to the cooling system. The reactors responded properly to the quake and all was well until the tsunami hit roughly an hour after the earthquake. The tsunami produced 49-foot waves, which were much higher than what was envisioned when the plant was designed. The waves knocked out the emergency generators, and the battery backup for the generators took over. The batteries are designed to last about 8 hours, normally long enough to bring the generators back online. Because of the devastating nature of the quake and tsunami, the generators could not be brought back up quickly; eventually, the batteries were completely discharged and all of the cooling pumps stopped. The decay heat from the fuel rods boiled the water away, which exposed the core. The core continued to heat and eventually melted.

As the temperatures continued to climb, eventually the steam became hot enough to react with the zirconium fuel rod casings to produce hydrogen gas. The hydrogen gas mixed with the air and the steam and was vented out of the primary containment structure to reduce the high pressure. The mixture was vented into the secondary concrete reactor building in which the primary structure was housed. Once in the reactor building the mixture was sparked and exploded in at least two of the reactors. These explosions tore the roof from the buildings housing these two reactors.

At this point, the only option for the workers at the Fukushima plant was to use huge volumes of water to remove heat from these reactors as quickly as possible. Much of this water was initially pumped from the ocean and became contaminated by contact with the radioactive core and various leaked radioactive materials. A serious attempt was made to keep this water from returning to the ocean, but still some lower level contaminated water was discharged into the sea. It took until October for temperatures in the reactors to be low enough to be considered stable with a normal rate of water circulation. In December, the reactors were deemed shut down.

The message from this disaster may simply be that it is difficult to consider all possible compromising events in the design and operation of a nuclear power plant. If the plant had been sited at 65 feet above sea level instead of 32 feet above sea level, then perhaps this story would have had a much different ending. All in all, nuclear power is probably as safe or safer than most other forms of power, but when things go wrong with nuclear, they go very wrong.

Much research in the past 10 to 15 years has gone into totally redesigning reactors. Design changes have included a move toward plants with a large number of smaller reactors and fuel fabricated in such a way that meltdown will not occur even if there is a prolonged coolant loss. Of particular interest are gas-cooled, pebble-bed reactors, which are different from light-water reactors in a couple of ways. The fuel is in the form of pebbles rather than the typical fuel rod. The pebbles are made principally of graphite with small kernels of uranium dioxide embedded in them. Each kernel is encased in a special carbon layer, then in a silicon carbide ceramic layer, and finally in another special carbon layer. Each pebble has about 15,000 of these kernels. The reactor utilizes about 330,000 pebbles containing kernels of uranium dioxide and about 100,000 "dummy" pebbles containing no kernels. These reactors contain no water; therefore, the graphite is the moderator. This type

Table 7.5 Advantages and Disadvantages of Nuclear Fission Power

Advantages	Disadvantages
Does not release CO_2, particulate matter, SO_2, or NO	Disposal of fuel waste difficult and currently an unresolved problem
Low environmental impact at the plant site, assuming normal operation	Nuclear plant safety issues
Fuel supply of 80 to 230 years	High plant construction and operation costs
Fuel and fuel wastes very small in mass and volume compared to other fuels	Cannot be used in transportation

of system virtually removes the possibility of a meltdown. The ceramic cover on the kernels will not melt, even at very high temperatures. So, should there be a loss of coolant, the fuel will not melt. These systems operate at very high temperatures: 900°C (1652°F) rather than the 300°C (572°F) in pressurized light-water reactors. The higher temperatures produce more efficient energy transfer. A noble gas such as helium is usually used as the coolant. The helium is stable at any temperature and will not react with any of the reactor components.

Public opinion was clearly poisoned against nuclear power by the Three Mile Island and Chernobyl accidents. As time has passed, perceptions of the nuclear industry are gradually changing. Even so, it is difficult to be objective about nuclear power, and the Fukushima accident has only made the issue more confusing. Nuclear power clearly has potential, but its future is still clouded in controversy. See Table 7.5 for a review of the advantages and disadvantages of nuclear fission power.

Nuclear Fusion

Because there are huge quantities of hydrogen in ocean water, the idea of fusing nuclei for the production of energy is attractive. The biggest drawback of this approach is that several million degrees of temperature are required to get the process started; also, the reactions are difficult to carry out in a facility under controlled conditions. So far the only successful application of this technology has been the hydrogen bomb. A conventional nuclear fission bomb is set off in the midst of deuterium (H-2) and tritium (H-3) to give the high temperature necessary for the fusion reaction, shown as follows:

$$ {}^{2}_{1}\mathrm{H} + {}^{3}_{1}\mathrm{H} \rightarrow {}^{4}_{2}\mathrm{He} + {}^{1}_{0}\mathrm{n} $$

The most significant problem in carrying out such a reaction under controlled conditions is that all containers that currently exist vaporize at the high temperatures required for nuclear fusion. Current research is centered on using lasers to produce the high temperatures and using magnetic fields to isolate the material in a fixed location. These reactions have been carried out in the laboratory. At this point, it would take 25 to 30 years just to develop a design, not including scaling up to prototype or commercial plant. This technology may be viable sometime in the late 21st or 22nd century.

BIOMASS

Probably the oldest fuel on Earth is wood. In early days, humans usually lived in areas containing a considerable number of trees. These people used the wood from trees for warmth, cooking, and keeping animals at a distance. Wood has continued to be the fuel of choice for most people in the developing world because it can be harvested by nearly anyone and can be burned directly in fairly crude stoves or in open fires. Wood is an example of a biomass energy source. Generally, biomass energy refers to the combustion of materials derived from current biological resources to provide energy. Biomass fuels include wood, charcoal, field wastes, animal wastes, aquatic wastes, urban trash, wood gas, synthetic natural gas, ethanol, and methanol.

Biomass fuels derive their energy directly or indirectly from the sun through photosynthesis:

$$H_2O + CO_2 + \text{Sunlight} \rightarrow \text{Biomaterials} + O_2$$

This process is direct for plants, but animals obtain the energy indirectly by eating the plants or eating animals that ate the plants. Most biomass fuels are from plants, but things such as animal wastes can be used as well. The energy from photosynthesis is locked up in the biomaterial as chemical energy. This energy is released when the biomass material is burned:

$$\text{Biomaterials} + O_2 \rightarrow H_2O + CO_2 + \text{Heat}$$

Biomass has been considered in a separate section because it is very difficult to categorize. It is both a widely used energy source and an emerging energy source. Wood is the oldest fuel used by humans, but novel new fuel ideas are coming out from the research on biomaterials.

Biomass may or may not be a renewable resource. Much depends on the balance between the two reactions given earlier. Because inputs into the first reaction (carbon dioxide, water, and sunlight) are virtually inexhaustible, the process can go on continuously; however, if we carry out the combustion process too rapidly then eventually we will run short of the biomaterials.

Some of the less-developed countries of the world are currently experiencing a fuelwood crisis. In many of these countries, the people are simply too poor to afford any other type of fuel. In these situations, if the population is too high or the number of trees is too low, the supply will either run out or become too expensive. Another important reason to keep the rate of use of biomaterials in line with the production of biomaterials is the possibility of global warming, which is related to the increase in atmospheric carbon dioxide. Burning biomaterials will release carbon dioxide into the atmosphere, as one can see from the second reaction of this section. But, if we burn these biological products no faster than we form them, then the net amount of carbon dioxide released into the atmosphere will be zero.

Biomass fuels are very versatile. First, they can be produced in various forms. Because most biomass materials start out in the solid form, the simplest approach is to burn them directly. Wood is sometimes converted into another solid form, charcoal; however, if liquid or gaseous fuels are desired they can be created from many of the solid forms. Biofuels are also versatile in their uses. They can be used for heating, electricity generation, or transportation. Some types of biofuels are better for one purpose than for another.

Liquid biofuels are of key interest in transportation. These applications vary from using the fuel as the complete auto fuel to using it as a gasoline additive to stretch the petroleum supply. Most of the commercial interest has centered around two fuels, **ethanol** (ordinary grain alcohol) and **biodiesel**. Biodiesel can be used in the place of diesel fuel with no major engine modification. Biodiesel can be sold mixed with regular diesel or in the pure form. The fuel mixture is identified by a "B" number (B followed by the percentage of biodiesel); for example, pure biodiesel would be B100 and 20% biodiesel would be B20. Fuel above 20% biodiesel would require a slight modification in the engine, whereas fuels below 20% could be used simply by pumping the fuel into the fuel tank.

The use of ethanol in a standard gasoline engine is somewhat complicated. Most standard gasoline engines will not run properly on ethanol mixtures with high percentages of ethanol. The percentages of ethanol in fuel mixtures are denoted by "E" numbers (E followed by the percentage of ethanol), the most common blends being E10 (10% ethanol and 90% gasoline) and E85 (85% ethanol and 15% gasoline). The E10 mixture, also known as gasohol, can be used in any standard gasoline engine and has been sold in the United States since the 1970s. Cars that use higher percentage ethanol blends must either be manufactured to burn these fuels or be modified to do so. In the United States, high-ethanol blends such as E85 are not found at most fueling stations. As a result, most of the cars engineered to run on E85 are flexible fuel vehicles (FFVs). These cars are able to adjust themselves to accommodate a wide range of ethanol–gasoline mixtures. Currently, cars and light trucks of this type are available in the United States from most major automobile manufacturers.

Ethanol can be produced by the fermentation of many biomaterials, including sugarcane, sugar beets, corn, and other grains. The advantage of ethanol is that it can be produced from very ordinary crops that are plentiful in many countries with inadequate or limited oil reserves. The United States alone produced 52.8 billion liters of ethanol in 2011, which is about 5.8% of the U.S. automotive fuel by volume. Most of this ethanol was produced from corn. Brazil has been the most aggressive country in trying to use ethanol for its transportation system as it can be produced from sugarcane. At one time, the country had hoped to convert 90% of its cars to run on pure ethanol. By 1990, the program had fallen on hard times due to the presence of cheap oil and a poor sugarcane harvest; however, the program has since rebounded, and today 80% of new cars in Brazil are flexible-fuel vehicles (with the exception of diesel vehicles). Currently, the United States and Brazil are world leaders in ethanol production, producing over 85% of all the global ethanol fuel.

Biodiesel, although certainly of interest in the United States, has been principally produced in Europe and more recently in South America, mainly in Brazil and Argentina. Europe and South America together account for 84% of the world's production of biodiesel. Ethanol is made from sugars and other related materials found in plants, whereas biodiesel is made from oil found in the seeds of plants or occasionally from animal fats. It should be made clear that diesel engines do not run on the oils but rather on the fuel made from the oils using a process called *trans-esterification*. Some of the oil sources include soybeans, corn, sunflowers, peanuts, rapeseed, cottonseed, canola, and animal fat. Sometimes the oils from deep-frying are also used.

Gaseous biofuels are also available. Natural gas can be produced from the anaerobic decomposition of urban refuse and sewage sludge. One can recover this gas either by inserting pipes into a landfill or by placing sewage sludge into an anaerobic digester. In less-developed countries, animal waste and plant residues can be fed into special anaerobic digesters, called *biogas digesters*, to produce methane. The methane is a more versatile fuel than burning the materials directly. An additional advantage is that the digester residue can be used as a fertilizer.

Solid biomass fuels are generally used for heating and cooking in fairly low-tech applications; however, in several locations around the world electric generating plants have been built that use wood or some other solid biofuel as the energy source. Generally, the same observations are true for the gaseous biofuels. Most biogas electricity generation is from small generators that are run from biogas digesters.

The most controversial aspect of biomass fuel generation is that it competes with food production for the use of farmland. If biomass fuel is used as a minor supplementary fuel source, then the conflict is not a major issue; however, if the plan is to produce major quantities of fuel that would compete, for example, with the petroleum industry, then there could be a major problem. If the United States, for example, decided to use all of its corn crop for transportation fuel, still only about 12% of the fuel required in this country would be produced. One option would be to use non-crop sources such as grasses to obtain the raw materials for fermentation. The processes for carrying out this approach are not perfected, and considerable research and development work would be required to make these processes practical. In addition, although this approach would certainly expand the capacity for making ethanol, it probably would not solve our entire fuel shortage problem.

Another issue that arises with the production of fuel ethanol is the amount of energy required to produce it. The obvious energy input is the necessity to distill ethanol away from the fermentation mixture. Fermentation of sugars will produce only about 20% ethanol. To get a higher concentration of ethanol, heat must be used to vaporize the ethanol, which can later be collected by cooling. Other players in this process include the energy consumed in planting, harvesting, and transportation. In the end, one gets only about 25% more energy out than is put in. The situation is much better in Brazil, because after the juices containing the sugar are removed the residue can be used as fuel for the distillation. This arrangement makes the sugar-cane process much more efficient.

Another problem is related to the water content of many biofuels. Biofuels often contain water, as they are derived from living organisms and water is the solvent of life. The water does not burn and represents weight that one pays to transport without any addition to the fuel content. If one burns the fuel wet, some of the fuel energy goes toward vaporizing the water, but removing the water requires energy, as well. See Table 7.6 for a review of the advantages and disadvantages of biomass.

Table 7.6 Advantages and Disadvantages of Biomass

Advantages	Disadvantages
Versatile	May compete for cropland
Renewable (ideally)	High moisture content
No net CO_2 increase (ideally)	May cause soil depletion, soil erosion, or water pollution from fertilization and cultivation
Emits less SO_2 and NO than fossil fuels do	

EMERGING ENERGY SOURCES

Solar Energy

Of all the sources of energy, the most obvious choice to exploit is solar energy. It is there, it is free, and it does not belong to anyone. All one has to do is collect it. The main issues and variations in solar energy are the methods of collection. The source is always the same, but there is some variation in the availability of the sun's energy with climate. In a sunny dry climate, such as that found in the U.S. Southwest, the sun shines nearly all the time during the day, and the percentage of possible sunshine is 90 or more. Such areas are ideal for solar energy production. By contrast, regions such as the U.S. Pacific Northwest have lots of cloud cover throughout the year; however, one should not jump to the conclusion that solar energy production is not possible in this area. Although the percentage of possible sunlight is less than 60, it certainly is not zero or anything close to that. These lower percentages simply mean that energy production from sunlight will be less efficient than in the U.S. Southwest.

Passive Collection of Solar Energy

Passive collection is, as its name implies, a system that allows the solar energy to be collected without active intervention. Specifically, the **passive collection of solar energy** or **passive solar heating** is the process by which solar energy is captured directly within a structure without any energy storage or circulating devices. From the definition this is clearly a system that can only be used for heating.

The idea is simple. It is based on the principal of the greenhouse (also discussed in Chapter 11). Solar radiation is made up of many kinds of radiation, including visible light, which will pass through the glass in a greenhouse and find its way to the floor of the greenhouse (or to some structure in the greenhouse). This light will either be reflected off the floor or absorbed by it. The reflected light may go right back out through the glass, but it is not the reflected light we are interested in. The absorbed light will be converted to heat, which raises the temperature of the floor. Some of the energy will heat the air in the greenhouse and some will be converted into heat radiation. This heat radiation, unlike visible light, cannot pass through the glass but will be absorbed by it. The glass will pass some of the heat outside, but it will also send some of the heat back inside. The ease with which solar radiation can get in and the resistance to heat energy getting out create a warming effect inside the structure (see Figure 7.12).

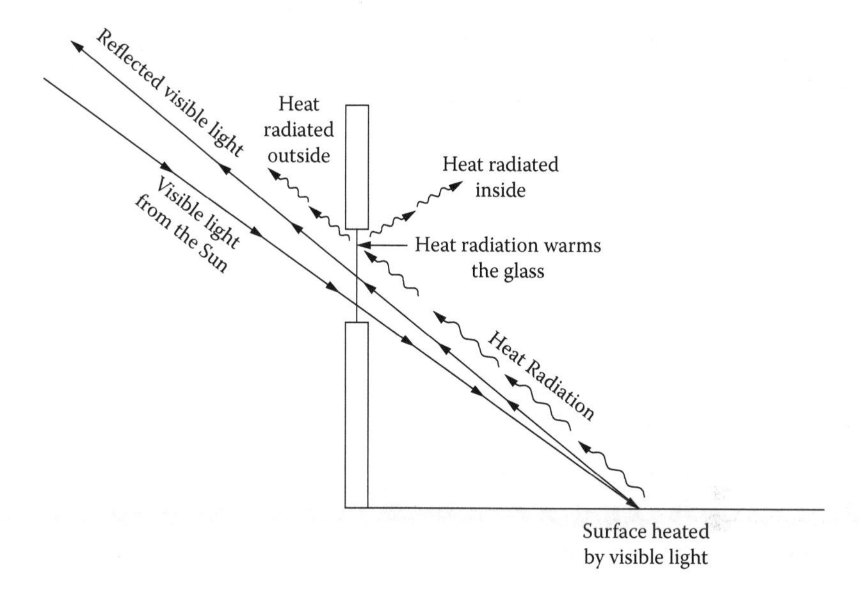

Figure 7.12 The process of passive solar heating through a window.

This same process can be used in any structure containing glass windows. In the Northern Hemisphere, where the sun passes across the southern sky, windows with a southern exposure are the most effective (the reverse would be true in the Southern Hemisphere). In temperate climates, which are only mildly cool in the wintertime, this type of solar heating can have a significant effect on one's heating requirements. In nearly any climate requiring heating, passive solar heating can help to lower heating bills. The general approach is to open drapes or blinds during the daytime as the sun passes by the windows. In the evening, these drapes or blinds are closed to help insulate the structure and keep the heat in. One advantage of the passive approach is that one does not have to buy any new equipment, but simply take advantage of the sunlight coming through the window.

Active Collection of Solar Energy as Heat

In this type of solar energy collection, we are going to construct a type of device to collect solar energy by heating some substance. The approach known as **active solar heating** will be discussed first. Active solar heating is the process by which solar energy is captured in specially designed collectors that concentrate the solar energy and store it as heat for space heating and possibly water heating. Generally, these systems have three components. The first part is a type of solar energy collector. These collectors usually consist of glass on the front side to allow the sun's rays through and some type of black material on the back to absorb a maximum fraction of the solar radiation. A fluid is circulated between the black material and the glass to carry the heat away. The fluid can be air, water, antifreeze, or any of a large number of other

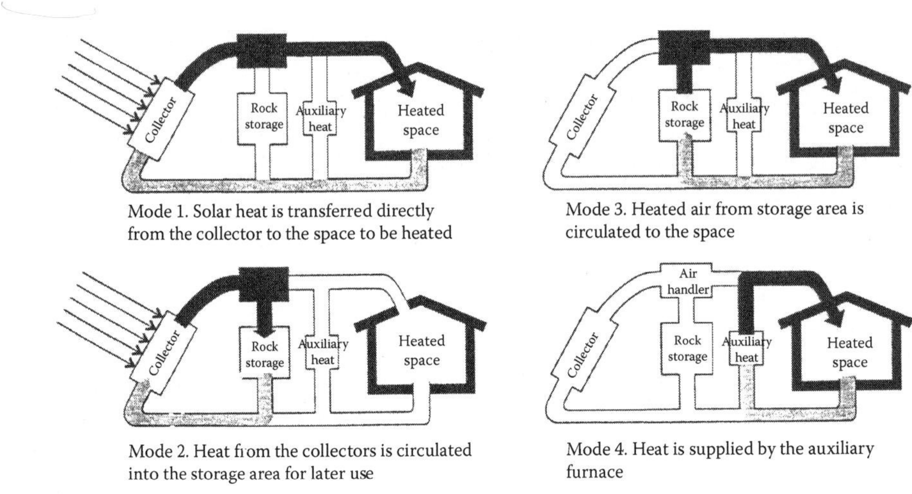

Figure 7.13 Hot-rock energy storage system.

fluids. The second stage is a type of heat storage device that can hold large amounts of heat for a long period of time. This part of the system is essential to take care of nights and cloudy days. Possible heat storage materials include water, antifreeze, or rocks. Diagrams of a hot-rock system are shown in Figure 7.13. The final part of the system is a method by which the heat is brought from the collector or the storage unit to the living space. The disadvantage of such active systems is the initial expense of installation. The cost would be more than the money spent to install a new furnace system; however, the investment can be recovered over time because the energy source is free.

A closely related application of solar heating is the heating of water for domestic use. The designs are very similar to those for heating air except that hot water is the ultimate goal. China has been the major user of solar hot water; that country has 70.5% of the world's solar hot water and solar heating capacity.

The active collection of solar heat can also be used for the production of electricity. This fairly traditional method to produce electric power involves the use of mirrors in various configurations to track the sun and focus sunlight on some fluid such as water or oil. In such facilities, the fluid on which the sunlight is focused can be heated to very high temperatures. These high-temperature liquids can be used to generate steam to drive a steam turbine and generate electricity.

Unfortunately, these types of solar plants face a dilemma. They require large amounts of water to condense the spent steam back into water so the condensed water can recirculate through the system; however, these plants are best located in areas with much sunshine and where land is inexpensive. This last requirement is necessary because concentrating large amounts of solar energy requires installing collection mirrors over a large amount of land. Areas with large amounts of inexpensive, sunny land are otherwise known as deserts, and they generally do not contain much water. Obviously, if there is a river nearby, then water can be drawn from the river. An innovation to cut down on the water requirements is the Stirling engine, which converts heat into motion without the use of an external fluid. This engine design is quite old, having been developed by Robert Stirling in 1816. These engines require some time to warm up and produce power and do not respond quickly to change; however, for continuous electric generation over long periods of time such an engine is ideal.

Collection of Solar Energy as Electricity

Currently, considerable research and development are going into the use of photovoltaic cells to convert sunlight directly into electricity. Various materials that would produce an electric current under the influence of light have been known since 1839, but it was not until 1954 that Bell Laboratories developed a practical photo cell. They were much too expensive to generate any quantity of electricity, though. These cells were used mainly as light meters for photographic applications. The American space program changed all of this. When Vanguard I went up in 1958, the only practical way to produce electricity in space was by photovoltaic cells. The radio in this satellite contained six solar cells producing electricity at a cost of about $2000/W of power capability. Because solar energy is free, the only cost reflected here is for manufacturing the device. By 1970, the manufacturing cost of a photovoltaic cell had dropped to $100/W of power capability. In 1990, the cost was only about $5/W of power capability. If one assumes the cell will last about 30 years and is 5% efficient, the cost in 1990 would have been about 38¢/kilowatt-hour (kWh). This cost fell to 15 to 30¢/kWh, and currently the cost is usually in the range of 6 to 12¢/kWh. The cost of electricity from photovoltaic cells is expected to continue to fall and should at some point be competitive with conventional electric generation. Currently, solar photovoltaic cells provide only 0.05% of the global electric generating capacity, but from 2004 to 2009 photovoltaic capacity increased at the rate of 60% per year, including a 7-gigawatt increase in 2009, which was the largest single-year increase in photovoltaic so far recorded.

Conventional solar cells are made by placing together layers of silicon, each containing different impurities. By carefully choosing the impurity for each side and using just the right amount, the device can generate an electric current when it is struck by light. A diagram of such a cell is shown in Figure 7.14. Each one of these cells produces only a small amount of electricity. To use solar cells to, for example, provide the electrical needs for a home, a large number of these cells would have to be wired together into a panel.

There are two problems that need to be overcome to use solar cells as a power source at home or at the office. First, normal commercial electricity that all of our electrical appliances are designed to use is alternating current (AC), whereas solar cells produce a direct current (DC). To use the current from the photovoltaic cells, the DC would have to be run into an inverter and converted into AC. Another problem common to all forms of solar power is night. To maintain power overnight and for very overcast days some type of electrical energy storage device is required.

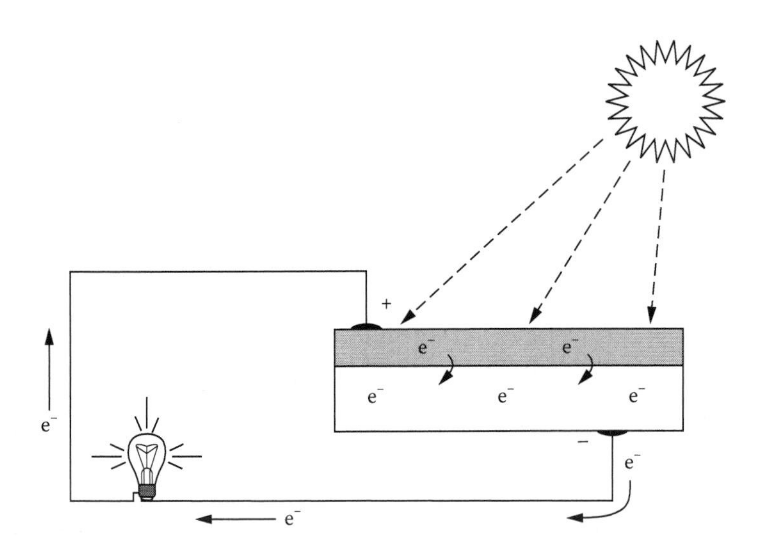

Figure 7.14 Diagram of the operation of a photovoltaic cell. (From Cunningham, W.P. and Saigo, B.W., *Environmental Science: A Global Concern*, William C. Brown, Dubuque, IA, 1992. With permission of The McGraw-Hill Companies.)

Solar power may soon become very attractive in making the individual homeowner energy independent. Solar power can be produced either on or off the grid. The grid refers to the interconnected set of electrical wires over a large area that allows electric power to be moved from place to place. If the solar electric is produced on the grid, then one can draw on the grid at night and at other times of low solar-generating capacity. In addition, any excess power that the homeowner generates can be sold back to the power company. See Table 7.7 for a review of the advantages and disadvantages of solar energy.

Table 7.7 Advantages and Disadvantages of Solar Energy

Advantages	Disadvantages
The energy is free.	Back up systems are necessary for nights and cloudy days.
Passive solar energy does not necessarily require a "system."	The initial costs may be high in some cases.
Photovoltaic systems and active heat collection systems are low to moderate cost in all but a few systems.	Passive solar heating may require constant adjustment.
Systems have very little environmental impact.	
Systems will pay for themselves in time.	

Wind Energy

The idea of harnessing the power of wind is not new. The Babylonians used wind power to pump water as early as 1700 BCE, and sailing ships have used the wind for transportation for thousands of years. Wind energy is a sustainable power source that ultimately derives its energy from the sun. The sun heats the Earth unevenly, giving rise to temperature variations. The temperature variations in turn produce pressure variations that cause the air to flow to equalize pressures (see Chapter 8).

As America was being settled in the 19th century, windmills figured to be important during the migration across the Great Plains. With little surface water, windmills were the key to pumping out underground water to the surface for human and animal use. This arrangement worked well because the Great Plains had fairly high-velocity sustained winds. In the late 20th century and up to the present, wind has mostly been used to generate electricity. Large numbers of wind machines are grouped together at a single site called a **wind farm**. California currently has several wind farms, and they generate enough electricity to meet the residential needs of the city of San Francisco. Currently, the leaders in wind power are the United States, China, and Germany, in this order. In the United States, Texas is the leader, as it produces about one-fourth of the wind power in the country. California was for many years the leader in the production of wind power but has been overtaken in the last several years by Texas and Iowa.

Clearly, wind machines generate electric power only when the wind blows. It is, therefore, important to conduct wind surveys before deciding where to locate a wind farm. Wind farms should generally be located in an area with strong and sustained winds. Wind, like many other resources, is not evenly distributed throughout the world. Some areas with strong and sustained winds include the Great Plains of the United States, mountain passes, and coastlines. Surveys have been conducted, and the areas suitable for wind power are very well known. In the United States, for example, the top five states for wind power potential are Texas, Kansas, Montana, Nebraska, and South Dakota. All of these states are situated either partly or wholly in the Great Plains. California is not a highly ranked state in wind power potential, but has for years produced large amounts of wind power because the state was the first to put considerable effort into producing this type of power.

The only disadvantage of wind power appears to be the wind machines themselves. When they dot the landscape they are very visible. The more scenic the landscape, the greater the likelihood that objections will be raised. Furthermore, the closer the wind farms are to urban areas, the greater the odds that objections will be raised. Despite the fact that many objections have been raised by local residents against wind farms, there are locations that work out well. Many parts of the western American Great Plains have been in slow economic decline for decades. These areas are rural and have very high wind energy potential; for example, the states of Texas, Kansas, and Montana have enough wind power potential to supply all of the

Table 7.8 Advantages and Disadvantages of Wind Power

Advantages	Disadvantages
High net energy yield	Only practical to build in windy areas
Free energy source	Visual pollution of scenic areas
Quick construction of wind farms that are relatively inexpensive	Backup or storage required for low wind periods
Low operating and maintenance costs	
No release of CO_2, particulates, CO, SO_2, or NO	

electrical power for the entire United States. The farmers in these areas no doubt would welcome the wind machines, as the lease income on their land might provide the income to keep them solvent.

Another option to avoid clashes over wind machines as eyesores is to place them offshore. These machines, although more expensive to build and maintain, are out of sight and away from civilization. Such an approach works well on shallow, wind-swept coastlines with low population density and has produced wind power generation facilities in the Irish Sea and the North Sea. Even this approach can have its pitfalls, though. Such was the case with the Cape Wind project in Nantucket Sound in Massachusetts. The offshore project developers misjudged the recreational boating nature of the area and ran into well-funded and well-organized major opposition from local residents, many of whom were quite wealthy. Although this project has passed on legal challenge, the project continues to face stiff opposition.

One can only suspect that wind power has a great future if it is managed carefully, so as to avoid unnecessary confrontations with local citizens concerned about the beauty of their area. After all, wind power is renewable and emits no harmful gases into the atmosphere. See Table 7.8 for a review of the advantages and disadvantages of wind power.

Geothermal Energy

Inside the Earth is a vast reservoir of heat energy. The **core** of the Earth, about 1800 miles (2900 km) below the surface, is a molten mass of iron and nickel maintaining a temperature of several thousand degrees. It is generally recognized that the core of the Earth has always been a very hot molten mass; however, the source and means of maintenance of this heat are somewhat less clear. Radioactive decay of elements such as uranium, thorium, and potassium certainly supply at least some of the heat.

A somewhat cooler layer called the **mantle** surrounds the core. This layer is composed of semi-molten rock known as *magma* and extends up to about 20 miles (32 km) beneath the surface. Above the mantle is the solid **crust** of the Earth. Although a tremendous amount of thermal energy is locked in the interior of the Earth, there is no practical way to get to it directly. Fortunately, the Earth's crust is not a uniform layer of solid rock and soil. The crust varies in thickness, has cracks and flaws in it, and is made up of a number of different plates. The magma heats the underside of the Earth's crust to a fairly high temperature. The heat at that depth

Table 7.9 Advantages and Disadvantages of Geothermal Energy

Advantages	Disadvantages
Sustainable energy source	Not available in all locations
Easy to exploit in some locations	Cannot be used directly for transportation
Produces relatively less CO_2	Can cause some air pollution from H_2S
Has a high net energy yield	Causes some water pollution

cannot generally be accessed; however, at many locations on the Earth's surface, special geological features, such as faults in the Earth's crust or irregularities in the structure of the crust, produce pockets of hot water or steam near the surface. These heated reservoirs can be tapped and used for energy production. In some cases, the hot water or steam comes to the surface naturally as geysers or hot springs, but often it is necessary to drill a well into the heat reservoir.

The oldest geothermal field exploited for energy production is at Larderello, Italy. The first electric generator was placed there in 1904 and is still in operation today. Most of the power for the train system of Italy comes from electricity generated at Larderello.

Because the occurrence of accessible geothermal energy depends on chance formations in the Earth's crust, the availability of this type of energy will depend strongly on location. The United States has the greatest availability of geothermal energy and also has, in absolute terms, the greatest existing geothermal energy production capacity. However, as a fraction of the total energy production capacity, the geothermal capacity is very small. The countries that have done the most to date with geothermal resources relative to their national energy situation are New Zealand, Iceland, Japan, Italy, and the Philippines. The Philippines generates about a quarter of its electric power from geothermal energy. In Iceland, 99% of heating in the capital city of Reykjavik is done using geothermal energy. In the United States, most of the geothermal resources are in the western third of the country. California has the greatest geothermal resources of any state and has done the most to develop them. Hawaii generates about 25% of its energy needs from geothermal.

Because the steam and hot water in these systems come from deep within the Earth's crust, these fluids can carry up a variety of pollutants that are found in the rocks. One pollutant is hydrogen sulfide (H_2S), which is a poisonous and corrosive gas with a horrible odor similar to rotten eggs. Other materials include carbon dioxide, arsenic, boron, mercury, and high levels of salt. The levels of these pollutants generally produce fewer problems than fossil fuels if the geothermal pollutants are handled properly. See Table 7.9 for a review of the advantages and disadvantages of geothermal energy.

HYDROGEN AS FUEL

Over the past several years, some discussions have taken place about using hydrogen as a fuel. There are many things about hydrogen that in many ways make it an ideal fuel. Hydrogen will combine with oxygen to give a large amount of energy:

$$2H_2 + O_2 \rightarrow 2H_2O$$

This energy-releasing reaction is nearly pollution free. The only product is water vapor, which can hardly be considered as a pollutant. However, if the energy is released by burning the hydrogen, then the heat coming from the hydrogen flame will produce nitric oxide in the air.

The current focus is on the **hydrogen fuel cell**, which combines hydrogen and oxygen to give an electric current without producing excess heat. Prototype cars powered by fuel cells have already been produced. Without the excess heat there is no danger of forming nitric oxide in the air. The fuel cell is very similar to a battery, except that instead of having the active materials on the electrodes, these materials are in the gaseous form flowing over the electrodes. The electrodes are separated by a proton exchange membrane that only allows protons (H^+ ions, hydrogen atoms without an electron) to pass through it. To complete the reaction, electrons are forced to pass through a wire.

The reaction at the electrode in contact with hydrogen gas is

$$H_2 \rightarrow 2H^+ + 2e^-$$

or

$$2H_2 \rightarrow 4H^+ + 4e^-$$

At the electrode in contact with oxygen, the $4H^+$ and the $4e^-$ are accepted:

$$4H^+ + O_2 + 4e^- \rightarrow 2H_2O$$

The net effect of these two reactions is

$$2H_2 + O_2 \rightarrow 2H_2O$$

which is the same as the reaction for combining hydrogen and oxygen, shown at the beginning of this section.

Although there is a tremendous amount of uncombined oxygen in the world, finding hydrogen as the free, elemental gas is a difficult problem. Most of the hydrogen in the world is combined in the very familiar compound water; however, it takes a lot of energy to release the hydrogen. In fact, breaking down water into hydrogen and oxygen will take exactly the same amount of energy as one gets out of combining hydrogen and oxygen to give water. The splitting reaction is just the opposite of the combination reaction:

$$2H_2O \rightarrow 2H_2 + O_2$$

Basically, hydrogen gas acts as an energy carrier in this process. We get the energy from somewhere else to produce the hydrogen; hence, the hydrogen is not an energy source in and of itself. Currently, most hydrogen is produced from natural gas and steam in a two-step process:

$$CH_4 + H_2O \rightarrow CO + 3H_2$$

$$CO + H_2O \rightarrow CO_2 + H_2$$

Currently, this process works well to produce hydrogen, but the method has some obvious limitations. Once again, one is relying on fossil fuels that have limited supply and will produce carbon dioxide, which is responsible for global warming (see Chapter 11). Another proven method involves the use of coal to release the hydrogen from steam, but this approach also has the same problems as natural gas.

There are probably a couple of options for the production of hydrogen in the long run. The simplest is to break down water using electricity. We already know how to build facilities to do this procedure; however, the big question is from where to get the electricity. To avoid the ultimate reliance on fossil fuels, the electricity has to be produced from renewable sources of energy. We are a long way from being able to meet this goal. Currently, renewable energy, including hydroelectric, accounts for about 13% of our electricity production; in addition, it would take more electricity than is currently produced in the United States to produce the hydrogen needed if every U.S. vehicle were to run on fuel cells. The other option for producing hydrogen from water is the thermal decomposition of water at a very high temperature. Nuclear power plants can produce the necessary heat to carry out this reaction from the intense heat of the reactor; however, this technology is yet to be proven.

As a fuel, hydrogen is most similar to natural gas except that it is considerably less dense and therefore contains about one-fourth the amount of energy per unit volume when compared to methane. As a result, hydrogen has most of the same transportation and storage issues that natural gas has, but is somewhat more difficult to handle.

If one envisions a hydrogen fuel cell car, one would need to store some hydrogen onboard to keep the car running and provide a reasonable running distance (e.g., 300 miles). One option would be liquefaction of the hydrogen; however, hydrogen is even more difficult to liquefy than natural gas, as it has to be cooled to –423°F (–253°C). The most reasonable option would be to compress the gas, but this would still take up a lot of space and the tanks would be heavy. Of the hydrogen cars built to date the most common option is gas compression.

Other options involve absorbing large amounts of hydrogen onto some material, that would release the hydrogen on demand. Several metals will work to store hydrogen, but the main drawback with metals would be their rather high weight. Various forms of carbon can be used that would have a much lower weight. The main drawback so far with the carbon types of fuel tanks is the cost, which may run as high as $30,000 per vehicle.

Much is made of hydrogen as the answer to our fuel woes. Experts generally agree that cars powered by hydrogen fuels cells have a future but disagree on how soon such an approach would be viable. Many policymakers expect hydrogen-fueled cars to make significant inroads into the market in the next 20 years, but others suspect that this day may be several decades away.

ENERGY CONSERVATION

All of the energy sources we have discussed so far have one thing in common. All energy sources, with the possible exception of passive solar heating, cost money. In energy, as in many other things in life, there is no "free lunch." The more one uses, the more it costs. Why do we use more? Many of the explanations have to do with advancing technology, increased population, and increases in the gross domestic product. All of these factors will drive up energy consumption, but the fact is that much of the energy we use is wasted. In the United States, for example, over 40% of all the energy used is unnecessarily wasted. The percentage of wasted energy is not as high in other countries, but is most likely a long way from zero.

The best energy source is **energy conservation.** This approach has several advantages. First, it costs the least to build. Even if the energy source is renewable and environmentally friendly, it is always less expensive and less intrusive not to build the power plant at all. Second, energy conservation costs the least to operate, as fewer energy facilities mean lower operating costs. And, finally, it reduces the impact on the environment. In contrast to producing energy, energy conservation can actually save an individual or corporation money. By reducing our energy consumption to a minimum level, we can reduce pollution, cut emissions of gases related to global warming, and save ourselves money. Energy conservation is the one place where an individual can have a direct impact based on individual choices. Some things are easy and can be done quickly, whereas other strategies take more time and effort.

Home, Apartment, or Room

There are a number of fairly simple things that many of us can do in our personal lives to save energy. Many of these ideas can be found in a brochure from the U.S. Department of Energy entitled, "Energy Savers, Tips on Saving Energy & Money at Home" (http://www.energysavers.gov/tips/). Instead of turning the heat up in the winter when one is feeling cold, put on a sweater. Generally, set thermostats relatively low in the wintertime (65 to 68°F, 18 to 20°C) and relatively high in the summertime (72 to 75°F, 22 to 24°C). In addition, at times when no one will be at home, set the heat even lower and the air-conditioning even higher. Also, turning the heat down at night is not a problem for most people. This process can be automated by installing a programmable thermostat.

One of the most inefficient devices ever invented was the **incandescent light bulb**. Although Thomas Edison did us all a great service by inventing this type of light bulb, it is an old, outdated technology. The incandescent light bulb converts only 5% of the energy into light; the rest is wasted as heat. Currently, most fixtures for light bulbs can be switched to **compact fluorescent light (CFL) bulbs**. These bulbs are about four times more efficient than incandescent ones and will last 10 times longer. The CFL bulbs are somewhat more expensive than the incandescent ones, but they will easily pay for themselves over time with longer life and energy savings. Incandescent bulbs for most common applications are being phased out in the United States.

Table 7.10 Relationship of Brightness in Lumens to Wattage of Traditional Incandescent Light Bulbs

Traditional Incandescent Bulb Power (Watts)	Brightness (Lumens)
40	~450
60	~800
75	~1100
100	~1600

Source: USDOE, *Lumens: The New Way to Shop for Light*, U.S. Department of Energy, Washington, DC, 2011 (energy.gov/sites/prod/files/lumens_placard-green.pdf).

One complaint that has been launched against CFL bulbs is that they contain mercury, which may be released when they are discarded. On the other hand, CFL bulbs save energy compared to incandescent bulbs. Also, in areas where much of the electrical energy is generated through coal combustion, their use may in fact reduce the total load of mercury in the atmosphere, as coal contains a small amount of mercury that will be released when it is burned. In areas with little coal-generated power, the mercury load may increase, but CFL bulb recycling programs can help keep the mercury out of the environment.

Another option is to use **light-emitting diode (LED) bulbs**. LED bulbs have been around for a long time but only recently have been increased in brightness to be useable for most household lighting applications. These bulbs last 25 times longer than incandescent bulbs and contain no mercury but are more expensive at this time than CFL bulbs. LED bulbs may well be the bulb of the future.

Most of us are used to buying light bulbs based on the wattage of power used in incandescent bulbs; however, for the same brightness, CFL bulbs and LED bulbs use much less power. Bulbs are now rated in **lumens** of brightness. A comparison of brightness in lumens to the wattage of traditional incandescent bulbs is shown in Table 7.10. By checking the number of lumens of brightness one can pick the correct bulb for any application. All bulbs are now labeled in lumens.

Home electronics is another area of potential energy waste. We are talking about such things as TVs, VCRs, DVD players and recorders, audio systems, cellphone chargers, and computers. Most of these items will continue to draw power even if they are turned off or not in use. To save energy, either unplug the electronic equipment when not in use or plug the equipment into a power strip and switch the power strip off when the item is not in use. Some people believe that computers work better if they are never turned off, but this is not the case. Computers need to reboot from time to time; powering down and powering back up will accomplish this task and will save power while one is not using the computer. When buying a new computer, one should consider a laptop, as laptops generally consume a lot less power than desktop models.

One can cut down on hot water use by taking short showers rather than baths. Generally, unless one tends to take a very long shower, showers will nearly always consume less water than baths. When using automatic dishwashers and automatic clothes washers, wait until a full load is accumulated before using them. The same energy is used whether the load is full, half full, or nearly empty. When buying an appliance, one should look for the ENERGY STAR® label. These appliances meet strict guidelines from the U.S. Environmental Protection Agency and U.S. Department of Energy.

Let us add a word about fireplaces. Nearly everyone loves to curl up in front of a roaring fire in the fireplace. There is good likelihood that one will waste more energy than one gets from the fireplace. A good fire will draw a large amount of air up the chimney. This air must be replaced and will be drawn in from the outside. This outside cold air will be heated by the heating system. Hence, the air will get heated and then run out of the chimney. One can cut down on this problem by turning down the central heating system, but the rest of the house, outside the room where the fireplace is, will cool down.

Vehicles

Energy is required whenever one drives a car. There are a number of simple things one can do to save energy and money. We often use our cars even for short distances, to get just one thing. Consider the alternatives. Could I walk or ride a bike to get the item and get some exercise while I am at it? Could I combine several trips into one trip? One longer trip will generally take less gas than several smaller trips when added together.

Most of the energy consumed by a car is used to accelerate it to a given speed. Maintaining a certain speed does require energy but a lot less than acceleration. "Jackrabbit starts" (rapid accelerations) waste gas, and sudden stops will waste energy as well because a lot of energy is used to heat up the brakes. Continually changing speeds will waste gas because it is necessary to keep accelerating to get back up to higher speeds. When appropriate, one should consider using the cruise control to maintain a constant speed. Avoid high speeds, as they are often above the optimal fuel economy level for the vehicle. On the other hand, do not sit in a car with the engine idling, as the car is burning fuel and emitting pollutants and carbon dioxide, and the car's engine is accomplishing nothing. Except for extremely short stops, it does not cost more fuel to start the car than is used. Furthermore, in the wintertime the best way to warm up a car is to drive it.

Cars should be well maintained. The better they are maintained, the better they run. Poorly running vehicles will waste gas and cost money in the long run. Keep tires properly inflated. Tires with low air pressure will flex as they turn and will produce resistance to the movement of the car, which will require more power from the engine.

Cars are inherently inefficient devices. Only 13% of the energy from burning the fuel in most cars actually reaches the wheels. When buying a car, be aware of the fuel efficiency. If you buy a car with low fuel efficiency, you will be paying for it as long as you own it. Cars vary widely on their fuel efficiency; therefore, a little research on the front end can pay off in the long run.

The innovation that takes improvement of gas efficiency in automobiles to another level is the **hybrid electric vehicle** (hybrid). These vehicles have both a gasoline engine and an electric motor, but all of the energy comes from gasoline. Hybrids are not, therefore, an electric car in the usual sense of the word, but rather a car that uses an electric motor and a battery to recover some of the waste energy that is lost in a conventional car. The engines in the hybrids are usually a little smaller than those in the usual car because the electric motor can give the engine a boost to make up for the difference.

The hybrid works by first using the battery and electric motor to start the gasoline engine in the usual manner. When the accelerator is pressed, the gasoline engine powers the car and uses the electric motor as needed to help. Once cruising speed is attained, the gasoline engine charges the battery that is the power source for the electric motor. At times more power will be needed; for example, in passing, the electric motor kicks in and assists the gasoline engine. On braking, in some models the electric motor reverses to help in slowing down the car in addition to the usual braking mechanisms. The reversed motor acts as a generator and charges the battery. The effect of this approach is to capture some of the kinetic energy of the car as it slows down. Once the car comes to a complete stop the gasoline engine shuts off. When the driver is ready to move again, simply pressing the accelerator will restart the engine and the car will move. This engine shutdown eliminates the waste of fuel while idling. The battery keeps all the auxiliary functions of the car running, such as dashboard displays, radio, CD player, and other electrical devices. Not all hybrids are created equal, and they vary considerably in gas mileage. Be sure to check the fuel efficiency before buying one.

The most recent hybrid vehicle to enter the market is the **plug-in hybrid electric vehicle** (plug-in). These vehicles are very similar to the hybrids except that the batteries can be charged by plugging them into an electrical outlet and the car can run exclusively from the electric motor without using gasoline at all. When driven short distances between charges, the car might not use the gasoline engine at all and, therefore, use no gasoline at all. When driven for longer distances between charges it operates as a standard hybrid vehicle.

The plug-ins have exceedingly high gas mileage and can be seen to decrease the dependence on fossil fuels; however, this observation is misleading. Much of the energy for these cars comes from electricity, and the amount of dependence on fossil fuels depends on the power source used to create the electricity. If the electricity is produced by coal, which we know is common, then little has been gained. Alternatively, if the electric power source is wind, then we are decreasing our dependence on fossil fuels. In the future, ideally more and more of our electricity will be "green."

DISCUSSION QUESTIONS

1. What is the difference between a primary and secondary energy source? What is the advantage of primary energy sources over secondary energy sources?
2. Name some common primary and secondary energy sources.
3. Why is electricity said to be more convenient and less efficient as an energy source?
4. Why are some energy sources in nearly all cases converted into electricity, whereas many others are often used directly? Give some examples of energy sources that are always converted into electricity and some that are used directly at times.

5. Given the distribution of the world's oil reserves and the fact that the United States consumes over 20% of the world's oil, discuss the likelihood that the United States can become oil independent (i.e., cease to import oil)? Is it possible to just import oil from friendly nations? How do you define friendly?
6. Why is petroleum used to such a large extent, given its short supply and many pollution problems related to its production and use?
7. Why is natural gas so difficult to move between continents?
8. What are some of the advantages of natural gas over the other fossil fuels?
9. Discuss some of the problems associated with hydraulic fracturing (fracking) and consider whether it might be worth doing in spite of the problems, considering that the U.S. natural gas supply could be extended from about 11 years to perhaps as long as almost 90 years.
10. Given that the United States has over a quarter of the world's coal reserves, should the country rely more on coal, in spite of its well-known pollution problems? Is there such a thing as clean coal?
11. What are some of the problems associated with deep tunnel coal mining and strip mining for coal? Which do you think is the best method for mining coal?
12. Why is hydroelectric power said to be renewable? What is the ultimate source of energy for hydroelectric power?
13. Why is hydroelectric power often viewed negatively in spite of the fact that it is a renewable source of energy?
14. Could a nuclear power plant produce a nuclear explosion? Why?
15. How is the power output of a nuclear power plant controlled?
16. Why do nuclear power plants, after they have been in operation for some time, continue to produce heat even after they have been shut down?
17. What are the current options for handling used fuel rods?
18. What are the most likely types of accidents associated with nuclear power plants and what are some of the likely consequences of each?
19. Do you feel nuclear power should be used? Why or why not?
20. What are some of the problems associated with the commercial production of electricity by nuclear fusion?
21. Biomass fuel can be considered as a renewable or nonrenewable energy source. Explain.
22. Solar heating can either be active or passive. Which type of solar heating might you prefer if you only plan to live in a house for one more year? Why? Which type of solar heating might you prefer if you plan to live in a house for more than 20 years? Why?
23. Given that solar energy is free, why is there a cost for photovoltaic solar energy?
24. Explain why wind power is considered a secondary energy source that receives its energy from the sun.
25. What are some of the considerations to be evaluated when locating a wind farm, other than wind potential?
26. Explain how geothermal energy reaches near the surface and how it is tapped as a commercial energy source.
27. If you were interested in investing a large sum of money in renewable energy, which type would you invest in and why? (Choose just one.)
28. Hydrogen fuel is an energy carrier and therefore a secondary energy source. In what ways could you generate hydrogen that would allow it to be a renewable energy source?

29. If one were to design a hydrogen car using a metal to absorb hydrogen, and assuming that magnesium and titanium are equally efficient on a per-mole of metal basis at absorbing hydrogen, which would be better and why?
30. Why is energy conservation considered the "best" energy source?
31. What are five fairly simple things that an individual can do to save energy?
32. Consider the following light bulbs, each of which would be used on average 2000 hours per year. Explain how they would compare to each other economically over a period of 10 years. Other than economics, are there any other factors that might have to be considered? (Some of the data below are from the Lowe's website at http://www.lowes.com.)

Incandescent bulb	*Compact fluorescent light bulb*	*LED bulb*
$0.35 each	$7.23 each	$40.16 each
60 W	13 W	12 W
$8/year electric cost	$1.73/year electric cost	$1.60/year electric cost
1000-hour lifetime	10,000-hour lifetime	25,000-hour lifetime
855 lumens	900 lumens	810 lumens

33. One of the newer entries onto the consumer car market is the plug-in electric hybrid vehicle. What are some of the pros and cons of this type of vehicle compared to hybrid electric vehicle and the conventional gasoline-powered vehicle?

BIBLIOGRAPHY

Anon., Matters of scale—hubris on the Yangtze, *World Watch Magazine*, 16(6), 2003, http://www.worldwatch.org/node/795.

Anon., *Potential for US Wind Energy Is 10.5 GW*, RenewableEnergyFocus.com, 2010, http://www.renewableenergyfocus.com/view/7446/potential-for-us-wind-energy-is-10-5-gw/.

Anon., *Hydrogen Fuel Tanks*, HydrogenFuelCarsNow.com, 2013, http://www.hydrogencars-now.com/hydrogen-fuel-tanks.htm.

Barta, P. and Spencer, J., Crude awakening as alternative energy heats up, environmental concerns grow, *The Wall Street Journal*, December 5, 2006, p. A1.

Bentley, R.W., Global oil and gas depletion: an overview, *Energy Policy*, 30, 189–205, 2002.

BP, *Statistical Review*, BP, 2011, http://www.bp.com/statisticalreview.

Brown, L.R., *Exploding U.S. Grain Demand for Automotive Fuel Threatens World Food Security and Political Stability*, Earth Policy Institute, Washington, DC, 2006 (http://www.earth-policy.org/Updates/2006/Update60.htm).

Brown, L.R., *Supermarkets and Service Stations Now Competing for Grain*, Earth Policy Institute, Washington, DC, 2006 (http://www.earth-policy.org/Updates/2006/Update55.htm).

Bui, Q., *History of Gasoline Prices in the United States*, Stanford University, Stanford, CA, 2010 (http://large.stanford.edu/courses/2010/ph240/bui1/).

Campbell, C.J., *The Imminent Peak of World Oil Production*, presentation to a House of Commons All-Party Committee, London, July 7, 1999 (http://www.greatexchange.org/ov-campbell,presentation_to_parliament.html).

Campbell, C.J. and Laherrère, J.H., The end of cheap oil, *Scientific American*, 278(3), 78–83, 1998.

Canine, C., How to clean coal, *OnEarth*, Fall, 21–29, 2005.

China Three Gorges Project Corporation, *History*, China Three Gorges Project Corporation, Yichang, China, 2002 (http://www.ctgpc.com/history/history_a.php).

Cunningham, W.P. and Saigo, B.W., *Environmental Science: A Global Concern*, 2nd ed., Wm. C. Brown, Dubuque, IA, 1992, pp. 375, 380.

Deffeyes, K.S., *When Oil Peaked: Hubbert's Peak in the 21st Century*, 2011, http://www.princeton.edu/hubbert/the-peak.html.

Deutch, J., Ansolabehere, S., Driscoll, M., Gray, P.E., Holdren, J.P., Joshow, P.L., Lester, R.K., Moniz, E.J., and Todreas, N.E., Eds., *The Future of Nuclear Power: An Interdisciplinary MIT Study*, Massachusetts Institute of Technology, Cambridge, MA, 2003.

Deutch, J., Forsberg, C., Kadak, A., Kazimi, M., Moniz, E., and Parsons, J., Eds., *Update of the MIT 2003 Future of Nuclear Power*, Massachusetts Institute of Technology, Cambridge, MA, 2009.

Deutch, J., Kanter, A., Moniz, E., and Poneman, D., Making the world safe for nuclear energy, *Survival*, 46(4), 65–80, 2004–2005.

Dickson, M.H. and Fanelli, M., *What Is Geothermal Energy?*, International Geothermal Association, Bochum, Germany, 2004, http://iga.igg.cnr.it/geo/geoenergy.php.

Farrell, A.E., Plevin, R.J., Turner, B.T., Jones, A.D., O'Hare, M., and Kammen, D.M., Ethanol can contribute to energy and environmental goals, *Science*, 311(5760), 506–508, 2006.

FERC, *A Guide to LNG: What All Citizens Should Know*, Office of Energy Projects, Federal Energy Regulatory Commission, Washington, DC, 2005.

Fickett, J.W., *US Natural Gas Reserves*, ClearOnMoney.com, June 21, 2012, http://www.clearonmoney.com/dw/doku.php?id=public:us_natural_gas_reserves.

Geller, H.S., *Energy Revolution: Policies for a Sustainable Future*, Island Press, Washington, DC, 2003.

Goodell, J., *Big Coal: The Dirty Secret Behind America's Energy Future*, Houghton Mifflin, Boston, MA, 2006.

Griffiths, D., Three Gorges Dam reaches for the sky, *BBC News*, May 19, 2006, http://news.bbc.co.uk/2/hi/asia-pacific/4998740.stm.

Gupta, R.B., Ed., *Hydrogen Fuel: Production, Transport, and Storage*, CRC Press, Boca Raton, FL, 2009.

Hebert, H.J., Study: Ethanol won't solve energy problems, *USA Today*, July 10, 2006, http://www.usatoday.com/tech/news/2006-07-10-ethanol-study_x.htm.

Hess, G., EPA probes fracking impact on water, *Chemical and Engineering News*, 89(46), 29, 2011.

Johnson, J., EPA issues fracking rules, *Chemical and Engineering News,* 90(17), 7, 2012.

Johnson, J., Federal rules for fracking, *Chemical and Engineering News*, 90(20), 8, 2012.

Johnson, J., Methane: a new 'fracking' fiasco, *Chemical and Engineering News*, 90(16), 34–37, 2012.

Johnson, J., No more Yucca Mountain, *Chemical and Engineering News*, 90(7), 33–34, 2012.

Johnson, J., Nuclear returns, *Chemical and Engineering News*, 90(8), 10, 2012.

Kemsley, J.N., Nuclear efficiency, *Chemical and Engineering News*, 88(37), 29–31, 2010.

Kennedy, J.H., Cape Wind clears hurdle to Nantucket Sound wind project, *Vineyard Gazette*, December 30, 2011 (http://www.mvgazette.com/article.php?33411).

Liu, Y., *Chinese Biofuels Expansion Threatens Ecological Disaster*, Worldwatch Institute, Washington, DC, 2007, http://www.worldwatch.org/node/4959.

Maine SPO, *A Brief History of the Edwards Dam*, Maine State Planning Office, Maine.gov, 2013, http://www.state.me.us/spo/sp/edwards/history.php.

McElroy, M.B., The ethanol illusion, *Harvard Magazine*, November–December, 33–35, 107, 2006.

McGivering, J., Three Gorges Dam's social impact, *BBC News*, May 20, 2006, http://news.bbc.co.uk/2/hi/asia-pacific/5000198.stm.

McMahon, T., Ed., *Historical Oil Prices Chart*, InflationData.com, June 14, 2012, http://inflationdata.com/inflation/inflation_rate/Historical_Oil_Prices_Chart.asp.

Miller, G.T., *Living in the Environment: An Introduction to Environmental Science*, 6th ed., Wadsworth, Belmont, CA, 1990, p. 426.

Mooney, C., The truth about fracking, *Scientific American*, 305(5), 80–85, 2011.

National Biodiesel Board, *What Is Biodiesel?*, Biodiesel.org, 2013, http://www.biodiesel.org/what-is-biodiesel.

Natural Gas Supply Organization, *Natural Gas Supply*, NaturalGas.org, 2011, http://www.naturalgas.org/business/supply.asp.

Natural Gas Supply Organization, *You've Got Shale: The "Where" and "What" of Shale Gas Formations*, NaturalGas.org, 2011, http://naturalgas.org/shale/gotshale.asp.

NEI, *Nuclear Energy: Just the Facts*, Nuclear Energy Institute, Washington, DC, 2012, http://www.nei.org/resourcesandstats/documentlibrary/reliableandaffordableenergy/brochures/justthefacts.

NEI, *World Statistics: Nuclear Energy Around the World*, Nuclear Energy Institute, Washington, DC, 2012, http://www.nei.org/resourcesandstats/nuclear_statistics/worldstatistics/.

NEI, *How It Works: Electric Power Generation*, Nuclear Energy Institute, Washington, DC, 2013, http://www.nei.org/howitworks/electricpowergeneration/.

NHA, *Hydro Facts*, National Hydropower Association, Washington, DC, http://www.hydro.org/hydrofacts/factsheets.php.

Nice, K., *How Stirling Engines Work*, Howstuffworks, http://travel.howstuffworks.com/stir-ling-engine.htm.

NIOSH, *Coal Workers' Pneumoconiosis: Black Lung Benefits*, National Institute for Occupational Safety and Health, Centers for Disease Control and Prevention, Atlanta, GA, 2008, http://www2a.cdc.gov/drds/WorldReportData/FigureTableDetails.asp?FigureTableID=527&GroupRefNumber=T02-13.

Palha, S., *The Emerging Natural Gas Crisis*, SafeHaven.com, 2006, http://www.safehaven.com/showarticle.cfm?id=6252.

Petkewich, R.A., Fluorescent bulbs trade off energy savings for mercury, *Chemical and Engineering News*, 86(41), 29, 2008.

Pimentel, D. and Patzek, T., Green plants, fossil fuels, and now biofuels, *BioScience*, 56(11), 875, 2006.

Piore, A., Planning for the black swan, *Scientific American*, 304(6), 48–53, 2011.

REN21, *Global Status Report*, Renewable Energy Policy Network for the 21st Century, Paris, 2010.

Romm, J.J., *The Hype About Hydrogen: Fact and Fiction in the Race to Save the Climate*, Island Press, Washington, DC, 2004.

Sawin, J.L., Mastny, L., Aeck, M.H., Hunt, S., MacEvitt, A., Stair, P., Cohen, A.U., Hendricks, B., and Mohin, T., *American Energy: The Renewable Path to Energy Security*, Worldwatch Institute and Center for American Progress, Washington, DC, 2006.

Seward, G.P., How long will global uranium deposits fuel the world's nuclear reactors at present consumption rates?, *Scientific American*, 300(3), 84, 2009.

Sperling, D. and Cannon, J.S., *Driving Climate Change: Cutting Carbon from Transportation*, Academic Press, Amsterdam, 2007.

Swenson, R., *The Coming Global Oil Crisis: Natural Gas*, http://www.hubbertpeak.com/gas/.

Tracy, D., Has solar push lost its luster? *Orlando Sentinel*, December 24, 2006.

Union of Concerned Scientists, *Clean Energy: How Hydrokinetic Works*, Union of Concerned Scientists, Cambridge, MA, 2008 (http://www.ucsusa.org/clean_energy/renewable_energy_basics/how-hydroelectric-energy-works.html).

USDOE, *Biodiesel Handling and Use Guidelines*, 3rd ed., DOE/GO-102006-2358, National Renewable Energy Laboratory, U.S. Department of Energy, Washington, DC, 2006, http://www.nrel.gov/docs/fy06osti/40555.pdf.

USDOE, *Proceedings of the Hydrokinetic and Wave Energy Technologies*, Technical and Environmental Issues Workshop, October 26–28, 2005, Washington, DC, U.S. Department of Energy, Washington, DC, 2006.

USDOE, *Alternative Fueling Station Total Counts by State and Fuel Type*, Alternative Fuels Data Center, U.S. Department of Energy, Washington, DC, 2012, http://www.afdc.energy.gov/fuels/stations_counts.html.

USDOE, *Clean Cities 2012 Vehicle Buyer's Guide*, DOE/GO-102012-3314, National Renewable Energy Laboratory, U.S. Department of Energy, Washington, DC, 2012.

USDOE, *Hybrid Electric Vehicles*, Alternative Fuels Data Center, U.S. Department of Energy, Washington, DC, 2012, http://www.afdc.energy.gov/afdc/vehicles/electric_basics_hev.html.

USDOE, *Installed Wind Capacity*, U.S. Department of Energy, Washington, DC, 2012, http://www.windpoweringamerica.gov/wind_installed_capacity.asp.

USDOE, *Maps and Data*, Alternative Fuels and Data Center, U.S. Department of Energy, Washington, DC, 2012, http://www.afdc.energy.gov/data/#tab/all.

USDOE, *Plug-In Hybrid Electric Vehicles*, Alternative Fuels and Data Center, U.S. Department of Energy, Washington, DC, 2012, http://www.afdc.energy.gov/afdc/vehicles/electric_basics_phev.html.

USDOE, *Energy Savers: Tips on Saving Energy & Money at Home*, Energy.gov, U.S. Department of Energy, Washington, DC, 2013, http://www.energysavers.gov/tips/index.cfm/.

USDOE, *Sunshot*, U.S. Department of Energy, Washington, DC, 2013, http://www1.eere.energy.gov/solar/sh_basics_water.html.

USDOL, *Coal Fatalities for 1900 through 2011*, Mine Safety and Health Administration, U.S. Department of Labor, Arlington, VA, 2012, http://www.msha.gov/stats/centurystats/coalstats.asp.

USEIA, *Natural Gas Imports, Exports, and Net Imports, Selected Years, 1949–2011*, U.S. Energy Information Administration, Washington, DC, 2011, http://www.eia.gov/totalenergy/data/annual/pdf/sec6_9.pdf.

USEIA, *Natural Gas Overview, Selected Years, 1949–2010*, U.S. Energy Information Administration, Washington, DC, 2011, http://www.eia.gov/totalenergy/data/annual/pdf/sec6_5.pdf.

USEIA, *Retail Motor Gasoline and On-Highway Diesel Fuel Prices, 1949–2010*, U.S. Energy Information Administration, Washington, DC, 2011, http://www.eia.gov/totalenergy/data/annual/pdf/sec5_59.pdf.

USEIA, *Electricity Explained: Electricity in the United States—Basics*, U.S. Energy Information Administration, Washington, DC, 2012, http://www.eia.gov/energyexplained/index.cfm?page=electricity_in_the_united_states.

USEIA, *Natural Gas Explained: Use of Natural Gas*, U.S. Energy Information Administration, Washington, DC, 2012, http://www.eia.gov/energyexplained/index.cfm?page=natural_gas_use.

McGivering, J., Three Gorges Dam's social impact, *BBC News*, May 20, 2006, http://news.bbc.co.uk/2/hi/asia-pacific/5000198.stm.

McMahon, T., Ed., *Historical Oil Prices Chart*, InflationData.com, June 14, 2012, http://inflationdata.com/inflation/inflation_rate/Historical_Oil_Prices_Chart.asp.

Miller, G.T., *Living in the Environment: An Introduction to Environmental Science*, 6th ed., Wadsworth, Belmont, CA, 1990, p. 426.

Mooney, C., The truth about fracking, *Scientific American*, 305(5), 80–85, 2011.

National Biodiesel Board, *What Is Biodiesel?*, Biodiesel.org, 2013, http://www.biodiesel.org/what-is-biodiesel.

Natural Gas Supply Organization, *Natural Gas Supply*, NaturalGas.org, 2011, http://www.naturalgas.org/business/supply.asp.

Natural Gas Supply Organization, *You've Got Shale: The "Where" and "What" of Shale Gas Formations*, NaturalGas.org, 2011, http://naturalgas.org/shale/gotshale.asp.

NEI, *Nuclear Energy: Just the Facts*, Nuclear Energy Institute, Washington, DC, 2012, http://www.nei.org/resourcesandstats/documentlibrary/reliableandaffordableenergy/brochures/justthefacts.

NEI, *World Statistics: Nuclear Energy Around the World*, Nuclear Energy Institute, Washington, DC, 2012, http://www.nei.org/resourcesandstats/nuclear_statistics/worldstatistics/.

NEI, *How It Works: Electric Power Generation*, Nuclear Energy Institute, Washington, DC, 2013, http://www.nei.org/howitworks/electricpowergeneration/.

NHA, *Hydro Facts*, National Hydropower Association, Washington, DC, http://www.hydro.org/hydrofacts/factsheets.php.

Nice, K., *How Stirling Engines Work*, Howstuffworks, http://travel.howstuffworks.com/stirling-engine.htm.

NIOSH, *Coal Workers' Pneumoconiosis: Black Lung Benefits*, National Institute for Occupational Safety and Health, Centers for Disease Control and Prevention, Atlanta, GA, 2008, http://www2a.cdc.gov/drds/WorldReportData/FigureTableDetails.asp?FigureTableID=527&GroupRefNumber=T02-13.

Palha, S., *The Emerging Natural Gas Crisis*, SafeHaven.com, 2006, http://www.safehaven.com/showarticle.cfm?id=6252.

Petkewich, R.A., Fluorescent bulbs trade off energy savings for mercury, *Chemical and Engineering News*, 86(41), 29, 2008.

Pimentel, D. and Patzek, T., Green plants, fossil fuels, and now biofuels, *BioScience*, 56(11), 875, 2006.

Piore, A., Planning for the black swan, *Scientific American*, 304(6), 48–53, 2011.

REN21, *Global Status Report*, Renewable Energy Policy Network for the 21st Century, Paris, 2010.

Romm, J.J., *The Hype About Hydrogen: Fact and Fiction in the Race to Save the Climate*, Island Press, Washington, DC, 2004.

Sawin, J.L., Mastny, L., Aeck, M.H., Hunt, S., MacEvitt, A., Stair, P., Cohen, A.U., Hendricks, B., and Mohin, T., *American Energy: The Renewable Path to Energy Security*, Worldwatch Institute and Center for American Progress, Washington, DC, 2006.

Seward, G.P., How long will global uranium deposits fuel the world's nuclear reactors at present consumption rates?, *Scientific American*, 300(3), 84, 2009.

Sperling, D. and Cannon, J.S., *Driving Climate Change: Cutting Carbon from Transportation*, Academic Press, Amsterdam, 2007.

Swenson, R., *The Coming Global Oil Crisis: Natural Gas*, http://www.hubbertpeak.com/gas/.

Tracy, D., Has solar push lost its luster? *Orlando Sentinel*, December 24, 2006.

Union of Concerned Scientists, *Clean Energy: How Hydrokinetic Works*, Union of Concerned Scientists, Cambridge, MA, 2008 (http://www.ucsusa.org/clean_energy/renewable_energy_basics/how-hydroelectric-energy-works.html).

USDOE, *Biodiesel Handling and Use Guidelines*, 3rd ed., DOE/GO-102006-2358, National Renewable Energy Laboratory, U.S. Department of Energy, Washington, DC, 2006, http://www.nrel.gov/docs/fy06osti/40555.pdf.

USDOE, *Proceedings of the Hydrokinetic and Wave Energy Technologies*, Technical and Environmental Issues Workshop, October 26–28, 2005, Washington, DC, U.S. Department of Energy, Washington, DC, 2006.

USDOE, *Alternative Fueling Station Total Counts by State and Fuel Type*, Alternative Fuels Data Center, U.S. Department of Energy, Washington, DC, 2012, http://www.afdc.energy.gov/fuels/stations_counts.html.

USDOE, *Clean Cities 2012 Vehicle Buyer's Guide*, DOE/GO-102012-3314, National Renewable Energy Laboratory, U.S. Department of Energy, Washington, DC, 2012.

USDOE, *Hybrid Electric Vehicles*, Alternative Fuels Data Center, U.S. Department of Energy, Washington, DC, 2012, http://www.afdc.energy.gov/afdc/vehicles/electric_basics_hev.html.

USDOE, *Installed Wind Capacity*, U.S. Department of Energy, Washington, DC, 2012, http://www.windpoweringamerica.gov/wind_installed_capacity.asp.

USDOE, *Maps and Data*, Alternative Fuels and Data Center, U.S. Department of Energy, Washington, DC, 2012, http://www.afdc.energy.gov/data/#tab/all.

USDOE, *Plug-In Hybrid Electric Vehicles*, Alternative Fuels and Data Center, U.S. Department of Energy, Washington, DC, 2012, http://www.afdc.energy.gov/afdc/vehicles/electric_basics_phev.html.

USDOE, *Energy Savers: Tips on Saving Energy & Money at Home*, Energy.gov, U.S. Department of Energy, Washington, DC, 2013, http://www.energysavers.gov/tips/index.cfm/.

USDOE, *Sunshot*, U.S. Department of Energy, Washington, DC, 2013, http://www1.eere.energy.gov/solar/sh_basics_water.html.

USDOL, *Coal Fatalities for 1900 through 2011*, Mine Safety and Health Administration, U.S. Department of Labor, Arlington, VA, 2012, http://www.msha.gov/stats/centurystats/coalstats.asp.

USEIA, *Natural Gas Imports, Exports, and Net Imports, Selected Years, 1949–2011*, U.S. Energy Information Administration, Washington, DC, 2011, http://www.eia.gov/totalenergy/data/annual/pdf/sec6_9.pdf.

USEIA, *Natural Gas Overview, Selected Years, 1949–2010*, U.S. Energy Information Administration, Washington, DC, 2011, http://www.eia.gov/totalenergy/data/annual/pdf/sec6_5.pdf.

USEIA, *Retail Motor Gasoline and On-Highway Diesel Fuel Prices, 1949–2010*, U.S. Energy Information Administration, Washington, DC, 2011, http://www.eia.gov/totalenergy/data/annual/pdf/sec5_59.pdf.

USEIA, *Electricity Explained: Electricity in the United States—Basics*, U.S. Energy Information Administration, Washington, DC, 2012, http://www.eia.gov/energyexplained/index.cfm?page=electricity_in_the_united_states.

USEIA, *Natural Gas Explained: Use of Natural Gas*, U.S. Energy Information Administration, Washington, DC, 2012, http://www.eia.gov/energyexplained/index.cfm?page=natural_gas_use.

McGivering, J., Three Gorges Dam's social impact, *BBC News*, May 20, 2006, http://news.bbc.co.uk/2/hi/asia-pacific/5000198.stm.

McMahon, T., Ed., *Historical Oil Prices Chart*, InflationData.com, June 14, 2012, http://inflationdata.com/inflation/inflation_rate/Historical_Oil_Prices_Chart.asp.

Miller, G.T., *Living in the Environment: An Introduction to Environmental Science*, 6th ed., Wadsworth, Belmont, CA, 1990, p. 426.

Mooney, C., The truth about fracking, *Scientific American*, 305(5), 80–85, 2011.

National Biodiesel Board, *What Is Biodiesel?*, Biodiesel.org, 2013, http://www.biodiesel.org/what-is-biodiesel.

Natural Gas Supply Organization, *Natural Gas Supply*, NaturalGas.org, 2011, http://www.naturalgas.org/business/supply.asp.

Natural Gas Supply Organization, *You've Got Shale: The "Where" and "What" of Shale Gas Formations*, NaturalGas.org, 2011, http://naturalgas.org/shale/gotshale.asp.

NEI, *Nuclear Energy: Just the Facts*, Nuclear Energy Institute, Washington, DC, 2012, http://www.nei.org/resourcesandstats/documentlibrary/reliableandaffordableenergy/brochures/justthefacts.

NEI, *World Statistics: Nuclear Energy Around the World*, Nuclear Energy Institute, Washington, DC, 2012, http://www.nei.org/resourcesandstats/nuclear_statistics/worldstatistics/.

NEI, *How It Works: Electric Power Generation*, Nuclear Energy Institute, Washington, DC, 2013, http://www.nei.org/howitworks/electricpowergeneration/.

NHA, *Hydro Facts*, National Hydropower Association, Washington, DC, http://www.hydro.org/hydrofacts/factsheets.php.

Nice, K., *How Stirling Engines Work*, Howstuffworks, http://travel.howstuffworks.com/stirling-engine.htm.

NIOSH, *Coal Workers' Pneumoconiosis: Black Lung Benefits*, National Institute for Occupational Safety and Health, Centers for Disease Control and Prevention, Atlanta, GA, 2008, http://www2a.cdc.gov/drds/WorldReportData/FigureTableDetails.asp?FigureTableID=527&GroupRefNumber=T02-13.

Palha, S., *The Emerging Natural Gas Crisis*, SafeHaven.com, 2006, http://www.safehaven.com/showarticle.cfm?id=6252.

Petkewich, R.A., Fluorescent bulbs trade off energy savings for mercury, *Chemical and Engineering News*, 86(41), 29, 2008.

Pimentel, D. and Patzek, T., Green plants, fossil fuels, and now biofuels, *BioScience*, 56(11), 875, 2006.

Piore, A., Planning for the black swan, *Scientific American*, 304(6), 48–53, 2011.

REN21, *Global Status Report*, Renewable Energy Policy Network for the 21st Century, Paris, 2010.

Romm, J.J., *The Hype About Hydrogen: Fact and Fiction in the Race to Save the Climate*, Island Press, Washington, DC, 2004.

Sawin, J.L., Mastny, L., Aeck, M.H., Hunt, S., MacEvitt, A., Stair, P., Cohen, A.U., Hendricks, B., and Mohin, T., *American Energy: The Renewable Path to Energy Security*, Worldwatch Institute and Center for American Progress, Washington, DC, 2006.

Seward, G.P., How long will global uranium deposits fuel the world's nuclear reactors at present consumption rates?, *Scientific American*, 300(3), 84, 2009.

Sperling, D. and Cannon, J.S., *Driving Climate Change: Cutting Carbon from Transportation*, Academic Press, Amsterdam, 2007.

Swenson, R., *The Coming Global Oil Crisis: Natural Gas*, http://www.hubbertpeak.com/gas/.

Tracy, D., Has solar push lost its luster? *Orlando Sentinel*, December 24, 2006.

Union of Concerned Scientists, *Clean Energy: How Hydrokinetic Works*, Union of Concerned Scientists, Cambridge, MA, 2008 (http://www.ucsusa.org/clean_energy/renewable_energy_basics/how-hydroelectric-energy-works.html).

USDOE, *Biodiesel Handling and Use Guidelines*, 3rd ed., DOE/GO-102006-2358, National Renewable Energy Laboratory, U.S. Department of Energy, Washington, DC, 2006, http://www.nrel.gov/docs/fy06osti/40555.pdf.

USDOE, *Proceedings of the Hydrokinetic and Wave Energy Technologies*, Technical and Environmental Issues Workshop, October 26–28, 2005, Washington, DC, U.S. Department of Energy, Washington, DC, 2006.

USDOE, *Alternative Fueling Station Total Counts by State and Fuel Type*, Alternative Fuels Data Center, U.S. Department of Energy, Washington, DC, 2012, http://www.afdc.energy.gov/fuels/stations_counts.html.

USDOE, *Clean Cities 2012 Vehicle Buyer's Guide*, DOE/GO-102012-3314, National Renewable Energy Laboratory, U.S. Department of Energy, Washington, DC, 2012.

USDOE, *Hybrid Electric Vehicles*, Alternative Fuels Data Center, U.S. Department of Energy, Washington, DC, 2012, http://www.afdc.energy.gov/afdc/vehicles/electric_basics_hev.html.

USDOE, *Installed Wind Capacity*, U.S. Department of Energy, Washington, DC, 2012, http://www.windpoweringamerica.gov/wind_installed_capacity.asp.

USDOE, *Maps and Data*, Alternative Fuels and Data Center, U.S. Department of Energy, Washington, DC, 2012, http://www.afdc.energy.gov/data/#tab/all.

USDOE, *Plug-In Hybrid Electric Vehicles*, Alternative Fuels and Data Center, U.S. Department of Energy, Washington, DC, 2012, http://www.afdc.energy.gov/afdc/vehicles/electric_basics_phev.html.

USDOE, *Energy Savers: Tips on Saving Energy & Money at Home*, Energy.gov, U.S. Department of Energy, Washington, DC, 2013, http://www.energysavers.gov/tips/index.cfm/.

USDOE, *Sunshot*, U.S. Department of Energy, Washington, DC, 2013, http://www1.eere.energy.gov/solar/sh_basics_water.html.

USDOL, *Coal Fatalities for 1900 through 2011*, Mine Safety and Health Administration, U.S. Department of Labor, Arlington, VA, 2012, http://www.msha.gov/stats/centurystats/coalstats.asp.

USEIA, *Natural Gas Imports, Exports, and Net Imports, Selected Years, 1949–2011*, U.S. Energy Information Administration, Washington, DC, 2011, http://www.eia.gov/totalenergy/data/annual/pdf/sec6_9.pdf.

USEIA, *Natural Gas Overview, Selected Years, 1949–2010*, U.S. Energy Information Administration, Washington, DC, 2011, http://www.eia.gov/totalenergy/data/annual/pdf/sec6_5.pdf.

USEIA, *Retail Motor Gasoline and On-Highway Diesel Fuel Prices, 1949–2010*, U.S. Energy Information Administration, Washington, DC, 2011, http://www.eia.gov/totalenergy/data/annual/pdf/sec5_59.pdf.

USEIA, *Electricity Explained: Electricity in the United States—Basics*, U.S. Energy Information Administration, Washington, DC, 2012, http://www.eia.gov/energyexplained/index.cfm?page=electricity_in_the_united_states.

USEIA, *Natural Gas Explained: Use of Natural Gas*, U.S. Energy Information Administration, Washington, DC, 2012, http://www.eia.gov/energyexplained/index.cfm?page=natural_gas_use.

Velasquez-Manoff, M., Surprise: not-so-glamorous conservation works best, *The Christian Science Monitor*, November 30, 2006, http://www.csmonitor.com/2006/1130/p13s01-sten.html.

Von Hippel, F.N., Rethinking Nuclear Fuel Recycling, *Scientific American*, 298(5), 88–93, 2008.

Wald, M.L., The power of renewables, *Scientific American*, 300(3), 56–61, 2009.

Wald, M.L., What now for nuclear waste?, *Scientific American*, 301(2), 46–53, 2009.

What Is Decay Heat? Explanation of Nuclear Reactor Decay Heat, Nuclear Science and Engineering at MIT, Cambridge, MA, 2011, http://mitnse.com/2011/03/16/what-is-decay-heat/.

WNA, *World Nuclear Power Reactors & Uranium Requirements*, World Nuclear Association, London, 2012, http://www.world-nuclear.org/info/reactors.html.

WNA, *Fukushima Accident 2011*, World Nuclear Association, London, 2013, http://www.world-nuclear.org/info/fukushima_accident_inf129.html.

Nimbus-7/TOMS Version 8 Total Ozone
for Sep 30, 1979

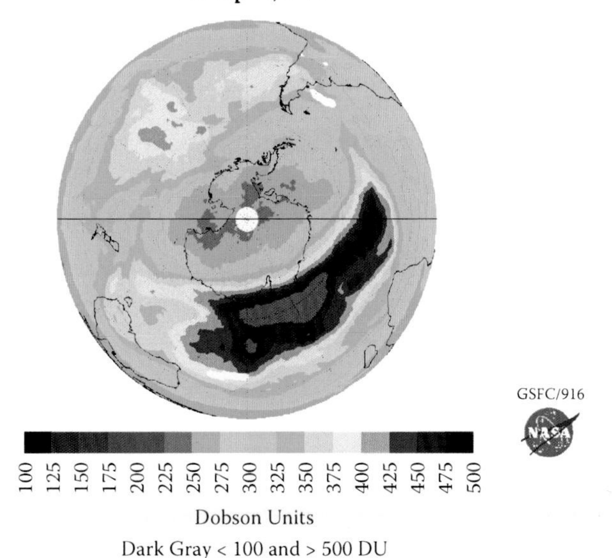

Nimbus-7/TOMS Version 8 Total Ozone
for Sep 30, 1982

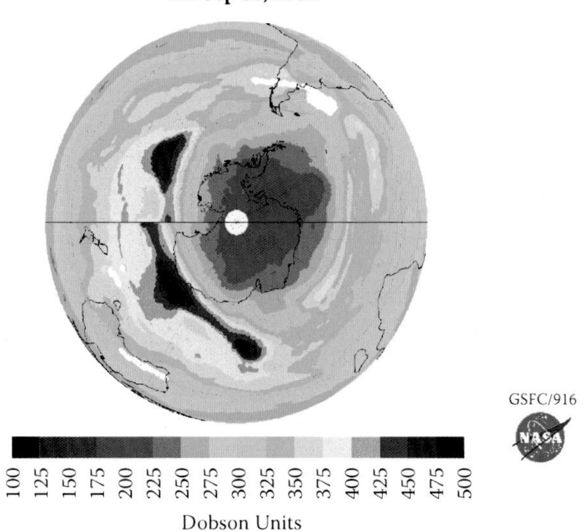

Figure 11.9 Satellite image of total ozone over Antarctica in late September of 1979 (top) and late September of 1982 (bottom). (From *Total Ozone Mapping Spectrometer*, Goddard Institute for Space Studies, National Aeronautics and Space Administration, Greenbelt, MD, http://toms.gsfc.nasa.gov/ozone/ozone_v8.html.)

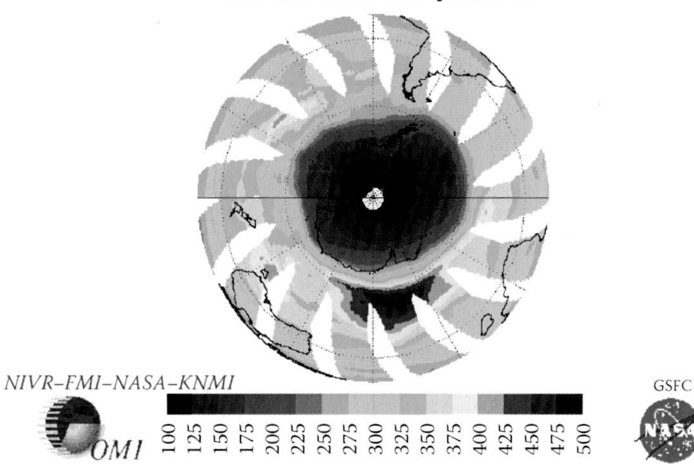

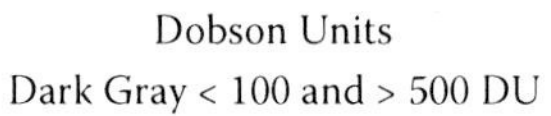

Figure 11.10 Satellite image of total ozone over Antarctica in late September of 2011. (From *Total Ozone Mapping Spectrometer*, Goddard Institute for Space Studies, National Aeronautics and Space Administration, Greenbelt, MD, http://toms.gsfc.nasa.gov/ozone/ozone_v8.html.)

CHAPTER 8

Weather and Climate

ATMOSPHERE: COMPOSITION, STRUCTURE, AND DYNAMICS

Composition

The atmosphere is a very large and nearly homogeneous mixture. No matter where in the world one samples the atmosphere, the composition of the major components, except one, would be roughly the same. The one component that varies from place to place is water. The concentration of water varies from as high as nearly 4% to air which has practically no water at all. Because the water concentration varies, the composition of the atmosphere is usually given as a percentage of dry air.

The composition of dry air was discussed in Chapter 5, where it was noted that the major components are nitrogen and oxygen. The only other gas present in a moderate concentration is argon, which, being one of the noble gases, is totally nonreactive. The remaining gases could be described as trace gases. All of these gases together make up less than 0.1% of the atmosphere. Carbon dioxide is the trace gas that is present in the highest concentration. Although a trace gas, this compound is very important in element cycles (Chapter 5) and global warming (Chapter 11). The importance of carbon dioxide in the atmosphere far outstrips its concentration. The exact composition of the atmosphere is given in Table 8.1.

Structure and Dynamics

As much as it may appear from the ground that the atmosphere is alike wherever one looks, this is not the case. We have noted earlier that the water content of air varies. Major changes take place in the nature of the atmosphere as the altitude increases. The temperature, wind patterns, and atmospheric chemistry all vary with altitude. The atmosphere is generally viewed as being made of layers that can be defined in terms of temperature, composition, or electrical properties. This chapter focuses on the layers defined in terms of temperature gradients. A **temperature gradient** is the temperature change per unit change in altitude. If the temperature

Table 8.1 Average Composition of Clean Dry Air

Gas	Percent (%)
Nitrogen	78.07
Oxygen	20.94
Argon	0.93
Carbon dioxide	0.04
Other trace gases	0.02

increases as the altitude increases, such a change is referred to as a **positive temperature gradient**. Conversely, if the temperature decreases as the altitude increases, it is referred to as a **negative temperature gradient**.

Troposphere

The **troposphere** is the part of the atmosphere with which we are most familiar. The familiarity comes from the fact that it is the part of the atmosphere in which we live. The troposphere contains 75% of the mass of the atmosphere. Within the troposphere, the air gradually cools as the altitude increases; that is, it has a negative temperature gradient. The rate at which the temperature decreases with altitude is known as the **lapse rate**. Although the lapse rate varies with location and time, it is generally about –3.6°F/1000 ft (–6.5°C/1000 m). From the lapse rate one can calculate that normally it would be about 50°F (28°C) cooler on the top of a 14,000-ft (4300-m) mountain than at sea level. The negative temperature gradient is a result of the fact that the sun does not warm the air; instead, the sun's rays go right through the atmosphere and warm the Earth. It is the Earth that warms the air next to it. For this reason, the air closest to the Earth is the warmest.

The troposphere contains all of the weather, and the negative temperature gradient in this layer is partly responsible for some of the dynamics of the weather. Warm air is less dense than cool air, and just as a hot air balloon rises so does warm air. Because warm air is constantly being generated at the lowest level, there is a constant vertical movement of air. Simultaneously, the cool air high in the troposphere, being denser, sinks toward the ground. This constant rising and sinking of air produces the vertical mixing of air that gives rise to many weather phenomena. This is discussed in a later section on weather.

At some place between 5 miles (8 km) and 10 miles (16 km) above sea level, the temperature gradient becomes roughly zero. As the altitude increases further, the temperature stays nearly constant. The beginning of this region of zero temperature change is the upper boundary of the troposphere, known as the *tropopause*. At this point the temperature is usually about –75°F (–60°C). The top of the troposphere is at a lower altitude in polar regions than in the tropics. It should be remembered that the troposphere is defined as a region of negative temperature gradient beginning at ground level and is not necessarily defined in terms of altitude.

Stratosphere

At the point where the negative temperature gradient ends, the next layer of the atmosphere begins. This layer is known as the **stratosphere**. The stratosphere can be split into two zones based on the magnitude of the temperature gradient. In the first 5 to 10 miles of the stratosphere the temperature does not change much at all, as noted earlier. Beyond this region, the temperature gradually increases with altitude. The temperature ranges from about –75°F (–60°C) at the lower altitude to about 30°F (0°C) at the upper limit. One might wonder where the heat comes from to heat the stratosphere. Certainly, this region of the atmosphere is too high above the ground to get its heat from the Earth. The region of the stratosphere from about 20 to 30 miles (30 to 50 km) contains a relatively high concentration of ozone (O_3). This region in the atmosphere is known as the *ozone layer* and is discussed further in Chapter 10. The important feature to be noted here is that the ozone absorbs certain types of solar radiation and converts them into heat, thus warming this region of the atmosphere. Because the stratosphere either has no temperature gradient or has a positive temperature gradient, there is little, if any, vertical mixing of the air. The warmer, lighter air remains above the cooler, heavier air; therefore, the air only moves horizontally in wind patterns. Most of the stratosphere is relatively calm. The strongest winds are at the tropopause, where the jet stream is found. For more details see the later section on jet streams.

Mesosphere

Above the ozone layer there is no warming mechanism for the air, and once again the temperature begins to fall as the altitude increases. This region runs from about 30 miles to 50 miles (50 km to 80 km) in altitude. The temperature falls from about 30°F (0°C) at the lowest altitude to about –120°F (–85°C) at the highest altitude. Although some parts of the **mesosphere** have relatively high temperatures, this is not a very hospitable place because the air is very thin. At the base of the mesosphere the air pressure is just 1/1000th of the air pressure at sea level.

Thermosphere

The last layer of the atmosphere is the **thermosphere**, the lower part of which is known as the *ionosphere*. This region has a positive temperature gradient, and the temperatures eventually rise to a few hundred degrees. The meaning of temperature at this altitude becomes obscure. The temperature we feel when we go outdoors is transmitted to us by the air of the atmosphere. If one took a thermometer to such a high altitude, it would probably not give a correct reading because it would not be in contact with enough air. The energy required to produce heat is provided by high-energy solar radiation that strikes the upper atmosphere, producing ionization and generating heat. The thermosphere does not really have an upper limit, but rather the air becomes thinner and thinner until there is really no atmosphere at all. Some authorities define another region called the *exosphere*, which begins at about 300 miles (500 km).

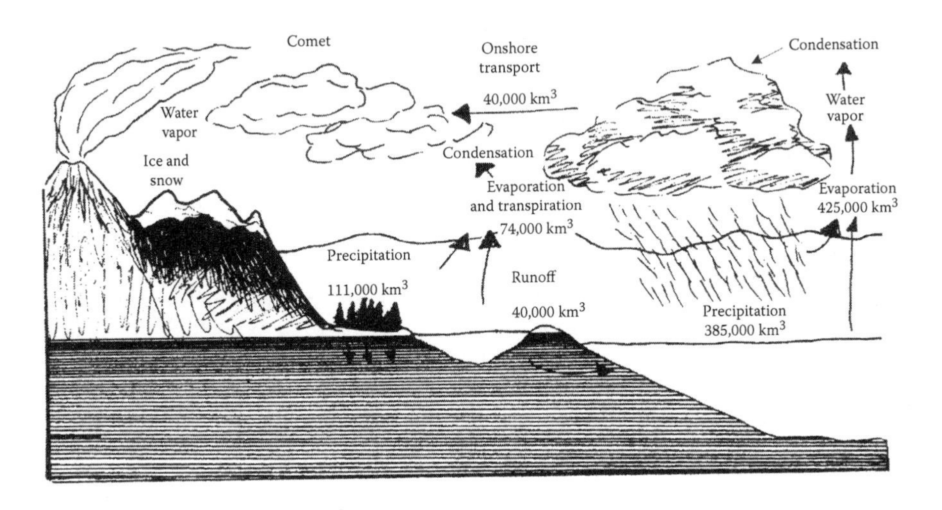

Figure 8.1 The water cycle. Total annual flows are in thousands of cubic kilometers (1 km^3 equals 264 billion gal). (Courtesy of Amy Beard.)

WATER CYCLE

The troposphere has the distinction of containing most of the water in the atmosphere. Atmospheric water and what happens to it are important features of weather dynamics. Water moves in and out of the atmosphere in what is referred to as the **water cycle**. This cycle is illustrated in Figure 8.1. Water vapor moves into the air principally by evaporation from the oceans and to a lesser extent from land sources. Land sources of atmospheric water include vegetation and surface waters. Water leaves the atmosphere most commonly as precipitation (e.g., rain or snow), most of which falls back into the oceans. Some of the water falls on the land and provides the moisture needed for vegetation. The excess water either soaks into the ground or provides the runoff that creates our rivers and streams.

WEATHER

The **weather** is the physical state of the troposphere at any given time and place. This physical state can be described by taking measurements of things such as air temperature, air pressure, humidity, clouds, precipitation, visibility, and wind. When we think of weather we are often considering the changes that occur in these physical states of the air over time and location. In other words, weather is dynamic, constantly changing from time to time and place to place.

As noted earlier, the stratosphere is a relatively calm place, whereas the troposphere is constantly in turmoil, producing wind and rain. What gives rise to the active processes of the troposphere? The fuel that drives the weather engine is the uneven heating of the Earth's surface. The four principal causes of uneven heating are as follows:

1. The variation in the angle at which the sun's rays strike the Earth from time to time and place to place
2. The difference in the rate of cooling between land and water (oceans and other large bodies of water warm up and cool off more slowly than land)
3. The uneven absorption of the sun's radiation caused by a variety of effects, including the presence or absence of clouds
4. The trapping of heat near the Earth at night by clouds

These are the reasons for uneven heating of the Earth, but not for the overall temperature. There are other factors, which are discussed in Chapter 11, that control the overall temperature of the planet.

Angle of the Sun's Radiation

As most people are aware, if one travels north in the Northern Hemisphere, the weather generally gets cooler and cooler. Traveling south usually brings one in contact with warmer weather. The reverse, of course, is true in the Southern Hemisphere. What gives rise to the different temperatures at different latitudes?

Well, on an average the sun's rays shine directly down on the Earth at the equator, whereas at any distance north or south from the equator it hits the ground at an ever-increasing angle from the perpendicular. When the sunlight hits the ground at an angle other than 90° it will be more spread out; in other words, there will be less sunlight per unit area of ground. This observation is illustrated in Figure 8.2. If one holds a flashlight directly above a surface and marks the area covered by the beam, then holds the flashlight at an angle but at the same distance, one would find that the

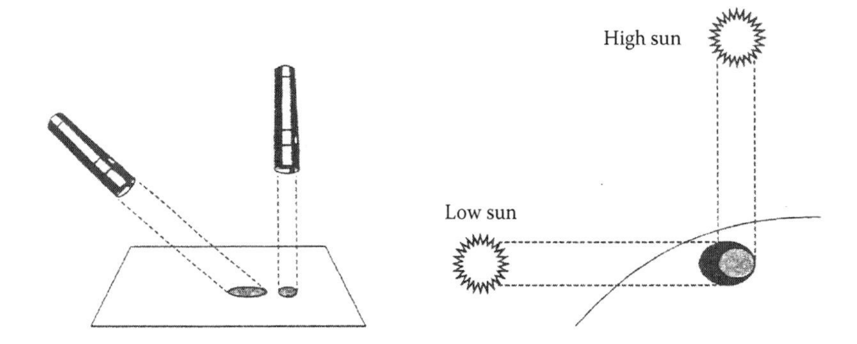

Figure 8.2 The effect of angle on the radiant intensity of a light beam. (From Ahrens, C.D., *Meteorology Today: An Introduction to Weather, Climate, and the Environment*, 5th ed., West Publishing, St. Paul, MN © 1994; Brooks/Cole, a part of Cengage Learning, Inc., www.cengage.com/permissions. With permission.)

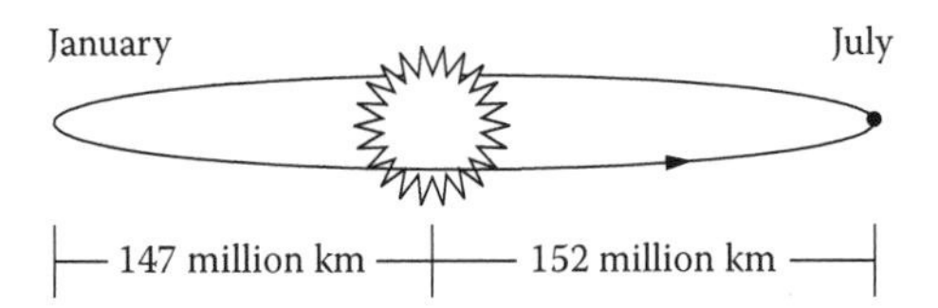

Figure 8.3 The elliptical path of the sun (highly exaggerated). (From Ahrens, C.D., *Meteorology Today: An Introduction to Weather, Climate, and the Environment*, 5th ed., West Publishing, St. Paul, MN © 1994; Brooks/Cole, a part of Cengage Learning, Inc., www.cengage.com/permissions. With permission.)

area covered by the beam is greater. If the total amount of radiation is the same in both cases and the beam at an angle covers more area, then the angled beam must have less radiation per unit area. If one thinks of the sun as an extremely large flashlight, then the same arguments can be made regarding the amount of radiant energy delivered to the Earth's surface by the sun (see Figure 8.2).

Clearly, the biggest change in our average weather conditions at a given location is caused by the seasons. To understand seasons, we need to look at the path the Earth takes as it revolves around the sun. The Earth moves around the sun in a flat elliptical path such that the distance of the Earth from the sun varies from 91 million to 94 million miles (147 million to 152 million km), as illustrated in Figure 8.3. If the Earth rotated on an axis set perpendicular to the plane of this elliptical path, there would be no seasons; however, the axis of rotation of the Earth is tilted by 23.5° from the perpendicular. Such a tilt can mean that the North Pole can be pointed toward the sun, away from the sun, or halfway in between. (The tilt can, of course, be at various intermediate positions.) From Figure 8.4, one can see that when the North Pole has its maximum tilt toward the sun, on or about June 22, the direct rays of the sun do not fall on the equator but at 23.5°N latitude. At this point in time there is a maximum amount of radiation falling on the Northern Hemisphere and a minimum amount falling on the Southern Hemisphere. This arrangement gives rise to summer in the Northern Hemisphere and winter in the Southern Hemisphere. On or about December 22, the situation is reversed and the maximum radiation falls at 23.5°S latitude, which causes winter to arrive in the Northern Hemisphere. On or about September 23 and March 21, the North Pole is tilted neither toward nor away from the sun and either spring or fall begins.

To summarize, we would expect it to be warmest near the equator and the temperature to be lower as one moves further from the equator. We would also expect the Northern Hemisphere to be at its warmest when the Southern Hemisphere is at its coolest and that each hemisphere would go through an annual cycle of warmer and cooler temperatures.

Ocean Effects

Anyone who lives near the ocean will be familiar with cool sea breezes coming from the water in the summertime. The effect is due to the fact that water warms up at a slower rate than land. The change in season from spring to summer causes the land to get heated up, but the ocean lags behind and remains cooler than the land

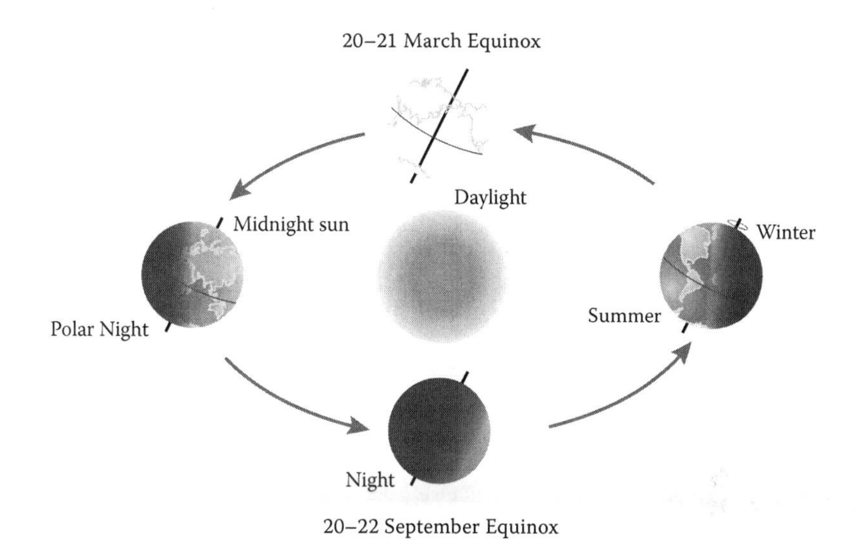

Figure 8.4 The tilt of the Earth relative to the sun for each of the seasons.

until late in the summer. Warm air rises over the land and draws cool air from the ocean. After summer, the fall and winter seasons lead to cooling of the land, but the ocean cools off much more slowly. This slow cooling leads to areas near the ocean being warmer in the winter than areas further inland. Generally, areas far inland tend to be hotter in the summer and colder in the winter than coastal areas of the same latitude. For example, the highest recorded temperature in Alaska (67°N) is the same as the highest recorded temperature in Hawaii (19°N). Alaska has areas very far from the ocean, whereas nearly every part of Hawaii is near water.

Uneven Absorption of the Sun's Radiation

When the radiation from the sun strikes the atmosphere or the Earth's surface, it is either absorbed or reflected back into space. The percentage of solar radiation that is reflected is called the **albedo**. The Earth is a very reflective planet, with an albedo of 30 compared to 17 for Mars and 7 for the Moon. The albedo has an effect on the uneven heating of the surface because the albedo of the Earth is not the same for all locations or at many locations from one time to another. For example, the albedo of a grassy field is 10 to 30, whereas that of fresh snow is 75 to 95. Fresh snow reflects most of the sunlight and keeps the temperature low, whereas a plowed field under similar conditions absorbs more sunlight and would be warmer. The clouds can keep an area cooler by reflecting the sunlight before it reaches the Earth below. Usually thick clouds have an albedo of about 60 to 90. Most of the solar energy is reflected back into space, and of the energy that is not some fails to reach the Earth's surface. All of these albedo effects will cause some locations to absorb more heat than others, leading to uneven heating of the Earth's surface.

Clouds as Blankets

Clouds behave very much as the blanket on one's bed. The clouds simply keep the heat from escaping by absorbing it and trapping it between the Earth and the cloud. Some heat will escape into space from the top of the cloud bank, but certainly much less than if the clouds were not there. Earlier, we learned that clouds reflect sunlight and that clouds hold heat near the Earth; hence, clouds generally have the effect of making it warmer at night and cooler in the daytime. Often, on very cloudy days during wintertime, the day to night temperature change may be only a few degrees.

Vertical Mixing of Atmosphere

In the discussion of the troposphere, it was noted that this lower region of the atmosphere is inherently unstable because cool, dense air is placed above warm, less dense air. As air near the Earth is constantly warmed by the ground, it expands and becomes less dense. Being less dense than the cooler air above it, this warm air begins to rise. As the air rises away from the Earth, it is no longer being warmed by the Earth and it gradually gives up some of its heat to the cooler air around it. Cooling the air causes it to contract and increase its density, until it is heavier than the warmer air below it. At this point, the air begins to sink. Such a system leads to constant mixing of the air in the troposphere.

To study this process in detail we need to consider moisture and pressure. An important observation about the moisture content of air is that cooler air cannot contain as much moisture as warmer air. This fact means that as moist air cools there will be a temperature at which the air can no longer hold all of the water vapor. At this temperature, small water droplets or ice crystals will be formed in the air, giving rise to fog or clouds.

The other issue related to vertical mixing is pressure. We all live at the bottom of a sea of air. Every square inch of surface at sea level has a column of air pushing down on it weighing about 14.7 lb (or 1.03 kg for each square centimeter). We are not aware of this pressure because we humans have always been at the bottom of this sea of air. Air pressure can change slightly due to weather conditions, so the air pressure is not exactly the same at all places on the planet. Clearly, air pressure will change as one goes to higher altitudes. This pressure drop would be expected because the column of air above one at a higher altitude would be less. You may be aware of this pressure change if you have ever driven over a high mountain pass. The human ear is fairly sensitive to pressure changes, and pressure changes due to going up and down high mountains can be felt on the eardrum.

When the Earth's surface warms the air, the air will start to rise. The rising air will have an immediate effect on the air pressure. The air pressure at the surface decreases because the rising air opposes the pressure of the air pushing down above it. At high altitudes, the air pressure increases as the rising air from below pushes up against it. This situation is illustrated in Figure 8.5 with a high pressure aloft and a low pressure at the surface.

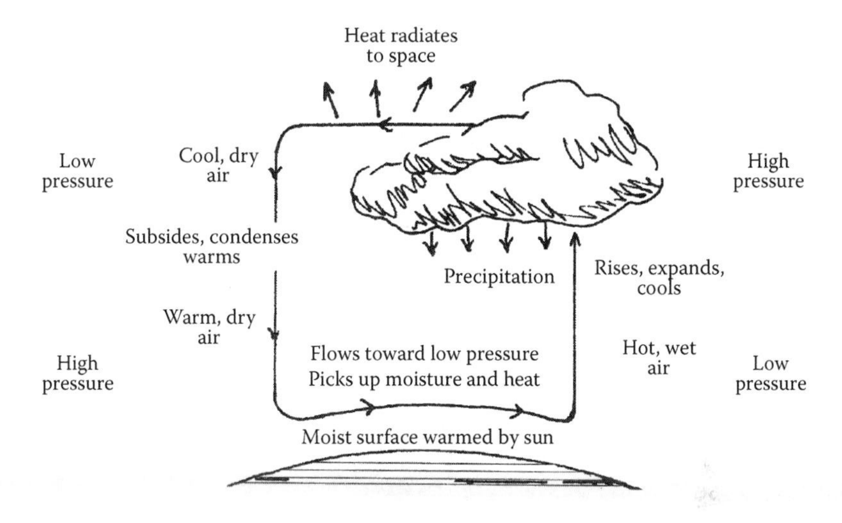

Figure 8.5 Diagram showing the movement of air as a result of heating and cooling. (Courtesy of Amy Beard.)

The rising warm moist air eventually cools and stops rising; however, if it gets cool enough the air can no longer carry all of its moisture. When the moisture is driven out of the air, clouds will be formed and perhaps it will rain. This air is now cooler and drier than when it first started out. Eventually, warmer air will begin to rise below it and this air will begin to sink. This sinking air will squeeze the air below it and raise the air pressure at the surface. The falling air will create an area of low pressure above it. As the air falls it will warm up but will remain dry as there is no water to add to it. The result is warm dry air at the surface under high pressure, as shown in Figure 8.5.

Global Air Circulation

If some simplifying assumptions are made, it is relatively easy to come up with a model for the global circulation of air in the atmosphere. Although the model is based on some fairly absurd proposals, the result is a surprisingly useful model. The assumptions are that the Earth is not rotating, the sun is directly over the equator, and the Earth is uniformly covered with water. Such a model was proposed by the 18th-century meteorologist **George Hadley**. Hadley proposed that the sun warms the air over the equator, causing the air to rise. This effect leaves a void at the surface and air flows toward the equator along the surface to replace the rising air. The air at the poles is the coldest and sinks, drawing in air at the upper atmosphere. To replace the upper atmospheric air, air flows from the upper atmosphere toward the poles. Such a circulation model leads to a circular airflow system where the air rises at the equator, flows toward the poles in the upper atmosphere, sinks at the poles, and flows toward the equator on the surface. This circulation model is shown in Figure 8.6. Each of the closed circulation loops is known as a **Hadley cell**.

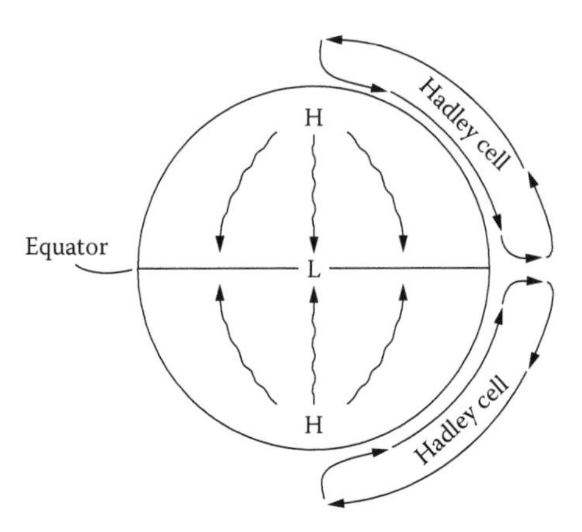

Figure 8.6 General global air circulation on a nonrotating Earth with the sun over the equator.

The fact that the Earth rotates forces us to make some modifications to the model. The Hadley cell shows the air flowing along the surface in a southerly direction toward the equator. Any air moving in a southerly direction will want to move in a straight line. The catch here is that the term *straight line* means straight in some absolute sense, but not in relation to the rotating Earth. The basic idea is illustrated in Figure 8.7. Two children are sitting on a nonrotating merry-go-round (platform A) throwing a ball to each other. The ball is thrown in a straight line and caught by the other child. If we put the merry-go-round in motion (platform B), then when the child throws the ball as he did earlier it will miss its target and the other child will be unable to catch it. The ball will go in a straight line, but the actual location of each child will have changed. To the children the ball will appear to have followed a curved path. The same is true of the air flowing away from the polar regions. From the perspective of looking down on the globe above the North Pole, the Earth rotates in a counterclockwise direction. Air moving away from the pole in an absolute straight line will appear to move in a westerly direction, as shown in Figure 8.8.

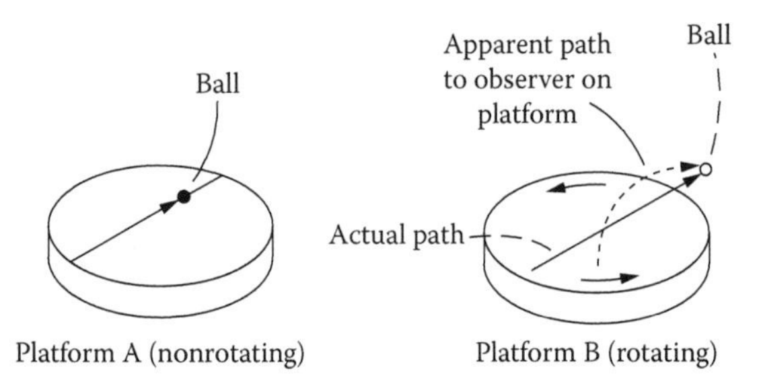

Figure 8.7 The effect of rotation on the apparent path of an object. (From Ahrens, C.D., *Meteorology Today: An Introduction to Weather, Climate, and the Environment*, 5th ed., West Publishing, St. Paul, MN © 1994; Brooks/Cole, a part of Cengage Learning, Inc., www.cengage.com/permissions. With permission.)

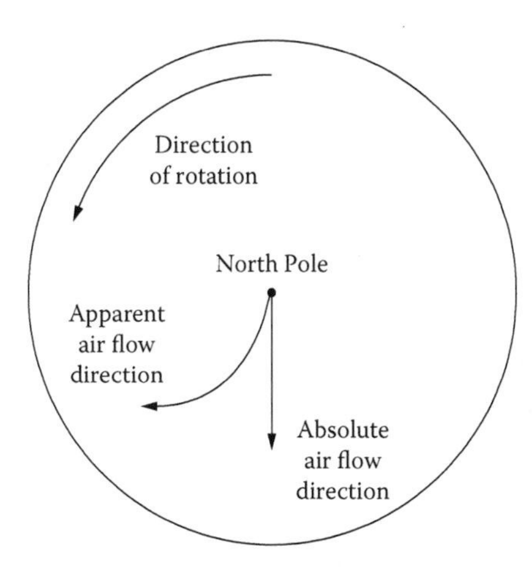

Figure 8.8 Illustration of the Coriolis effect.

Because winds are named for their direction of origin, these winds would be referred to as *easterlies*. This change in wind direction due to the rotation of the Earth is known as the **Coriolis effect**, after the French scientist **Gaspard Coriolis**, who first worked out the idea.

These winds start off headed due south and eventually end up going due west; hence, they will never reach the equator. The result is that the single-celled system of Hadley (i.e., one cell per hemisphere) needs to be replaced with the three-celled system shown in Figure 8.9. Two of the cells (one near the pole and one near the

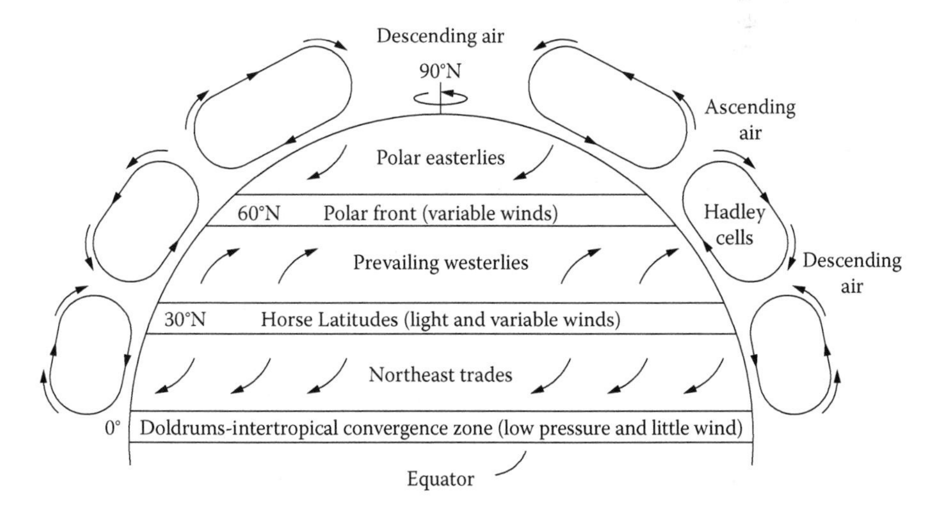

Figure 8.9 The patterns for the general circulation of air in the Northern Hemisphere. (From Cunningham, W.P. and Saigo, B.W., *Environmental Science: A Global Concern*, William C. Brown, Dubuque, IA, 1990. With permission of the McGraw-Hill Companies.)

equator) are just shortened versions of the original Hadley cell. From the Hadley cell near the North Pole we can see that it leads to air rising at about 60°N latitude (or 60°S latitude for the South Pole). The Hadley cells nearest to the equator lead to air sinking at about 30°N or 30°S latitude. As can be seen from Figure 8.9, such an arrangement gives rise to a Hadley cell in between that appears to flow in the opposite direction. The actual situation is more complicated than that. The surface winds in this cell flow north (in the Northern Hemisphere), resulting in winds that curve to the right and giving rise to prevailing westerlies.

The prevailing winds have been important in history. In the age of sailing ships the prevailing easterlies of the tropical oceans were used as a reliable means of power for commercial ships. In time, these winds came to be known as the **trade winds**. In between the bands of prevailing winds are the convergence zones, where the winds are much less predictable. At about 60°N or 60°S latitude is the polar front area where the wind may blow in nearly any direction depending on immediate weather conditions. At 30°N or 30°S latitude we find the **horse latitudes** and at the equator the **doldrums**. These areas are likely to have long periods of little or no wind. Many sailing ships met disaster by being trapped in these areas for long periods of time. The horse latitudes got their name because of the many sailing ships bringing livestock to the New World. Often the crew of a ship would throw dead horses overboard after being stuck in calm seas for too long a time.

The patterns of rising and sinking air can have dramatic effects on rainfall amounts. Near the equator in what is known as the **intertropical convergence zone** there is mainly warm moist air rising, which leads to frequent and plentiful rainfall. It is in this region that tropical rain forests are found. At the horse latitudes, there is generally dry air sinking. Although it is warmed as it sinks, the air has no moisture in it; hence, the rainfall amounts are sparse. Many of the world's deserts are found at 30°N and 30°S latitude.

Jet Streams

Near the tropopause, the winds flow toward or away from the various convergence zones (see Figure 8.9). Such airflow patterns lead to irregularities in the upper atmosphere. Because of these irregularities at the tropopause, upper atmospheric wind systems develop, called **jet streams**. The jet streams are basically rivers of air in the upper atmosphere moving around the globe at fairly high rates of speed. The air speed in these streams is usually between 100 and 250 miles/hr (mph) (160 to 400 km/hr). There are several of these jet streams, but two of them in each hemisphere are of the greatest importance to ground-level weather. One is roughly above the polar front area and is known as the *polar front jet*; the other is found above the horse latitudes and is known as the *subtropical jet*. In the temperate regions of the world (between 30° and 60° latitude), the ground-level weather is moved and directed by these jet streams. The jet streams do not tend to make a simple circular path around the globe but rather meander north and south as they circumnavigate the globe (see Figure 8.10). Under normal circumstances, the exact paths of the jet streams are constantly changing, and thus the ground-level weather directed by them is constantly

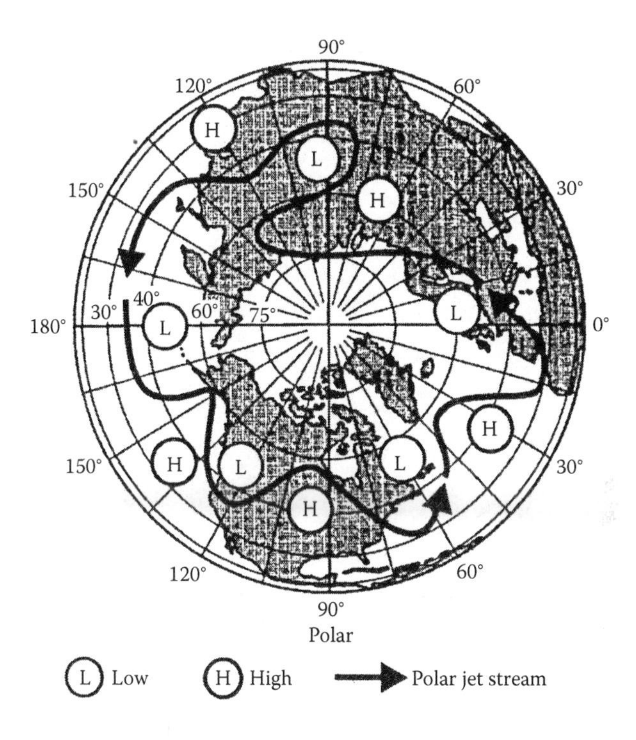

Figure 8.10 A typical path of the polar front jet stream.

changing its pattern. From time to time the jet streams can get locked into a particular pattern for a long period. When this happens the ground-level weather will be locked into a relatively fixed pattern, with some areas getting repeated storms and other areas experiencing dry weather.

Low-Pressure Cells and Cyclonic Storms

For a number of reasons there can be local areas where there is a large amount of rising air. This rising air creates low pressure. As the air rises, other air flows in to replace the air that is rising. This air does not flow directly in because it is turned by the Coriolis effect as it approaches the low-pressure center. For example, in the Northern Hemisphere air flowing south toward a low-pressure center would be turned to the west, whereas air flowing north would be turned to the east. Such a turn produces a counterclockwise spiral of air toward the center of the low-pressure area, as shown in Figure 8.11. A similar analysis of the Southern Hemisphere indicates that low-pressure systems in this region will produce clockwise winds. (Contrary to popular folklore, this analysis works only for large air masses and has no bearing on the direction water swirls down a bathtub drain.) The rising air at the center of these low-pressure centers usually leads to clouds and rainy weather if sufficient moisture is available.

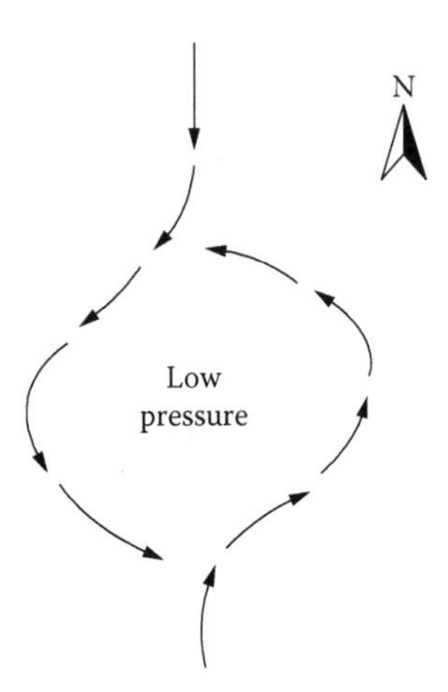

Figure 8.11 Illustration of the direction of airflow around a low-pressure center in the Northern Hemisphere.

Cyclonic storms do not form near the equator because the Coriolis effect is too weak in the region to produce air rotation. However, in the region of 5 to 20°N or S latitude, some of the most powerful cyclonic storms in the world are formed. In the Western Hemisphere, these storms are known as **hurricanes**. The storms are formed around rising warm, moist air that produces an intense low-pressure system. They occur over warm ocean waters in the late summer or early fall when the water temperature is at its maximum. These storms are usually about 300 to 400 miles (480 to 650 km) across, and as such are not as big as conventional low-pressure centers formed in the more temperate regions of the Earth. The storms are, however, much more intense. The sustained wind speeds near the center of such storms may exceed 135 mph (217 km/hr) in strong hurricanes. To be classified as a hurricane the sustained winds must exceed 74 mph (119 km/hr); otherwise, they are considered to be tropical storms or tropical depressions. Because the energy source for these storms is the warm ocean waters, hurricanes will weaken and eventually die out when they get over land or cooler ocean waters.

Hurricanes go by different names in different parts of the world. In the western North Pacific, the storms are called **typhoons**, and in Australia and India they are called **cyclones**. Certain local names have developed such as *baguio* in the Philippines or *willy-willy* in Australia. Another type of cyclonic activity that is often confused with hurricanes is the **tornado**. A tornado is a totally different type of phenomenon. These very small storms are spawned out of thunderstorms, and although the size is small they are usually vicious. The winds in a tornado run anywhere from 85 to 200 mph (137 to 322 km/hr). Not only do tornados contain high wind, but they also have exceedingly low pressure inside of them. The atmospheric pressure may drop by as much as 10%. Tornados can produce devastating destruction. Trees can be uprooted, buildings destroyed, and many heavy objects, such as home appliances, can be hurled through the air. Many items can be picked up and deposited several miles away. There are all kinds of strange stories about tornados. One time it rained frogs and toads after the creatures were sucked up out of a nearby pond by a tornado. People have reported chickens having all of their feathers removed, and pieces of straw driven into metal pipes. In one amazing story, all 85 students in a schoolhouse survived after the schoolhouse was picked up and deposited over 100 yards away.

As noted earlier, these storms are small, with the diameter varying from roughly 200 yards or meters up to about 0.25 miles (0.4 km), although an occasional tornado may have a diameter of 1 or 2 miles (1.5 to 3 km). Tornados do not generally move forward at a high rate of speed. A speed of 25 to 45 mph (40 to 70 km/hr) is common, but a few have been clocked at speeds of about 80 mph (130 km/hr).

Weather Fronts

Throughout the year, cold air is constantly trying to push toward the equator and is opposed by warm tropical air moving up from the equatorial regions. The boundary between these two air masses is known as a *front.* These frontal interactions of air masses typify the weather in the temperate parts of the world. The two most common types of fronts are the cold front and the warm front. These types of fronts tend to produce rain or snow as they pass. The amount of precipitation can vary considerably, depending on the amount of moisture available in the weather system.

In a cold front, the cold air advances against the warm air. Because the cold air is heavier than the warm air it pushes under the warm air as it moves. This action usually produces sudden updrafts of warm, moist air, often resulting in fairly narrow bands of intense precipitation, as can be seen in Figure 8.12. In warm fronts, the warm air rides up over the cool air that it has been pushing back. Because the

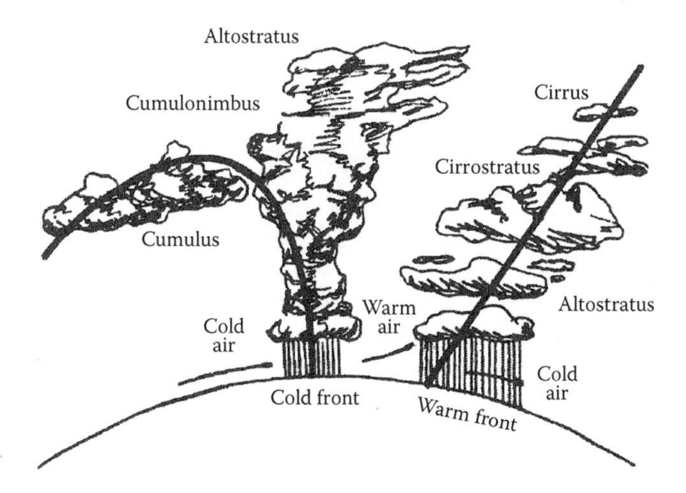

Figure 8.12 Diagram of the vertical structure of warm and cold fronts. (Courtesy of Amy Beard.)

warm air rises gradually, an extensive band of clouds is produced. The first clouds can appear hundreds of miles ahead of a warm front with a fairly large band of light to moderate precipitation mixed in. This type of front is also illustrated in Figure 8.12.

Seasonal Winds (Monsoon Winds)

The temperature differentials between land and sea often produce wind and moisture patterns. In summer when the land is warm, the warm air rises, sucking in air from over the sea, whereas in winter the air over the water is warmer and it rises, sucking in air from over the land. Therefore, in summer the breezes blow onshore, and in winter it blows offshore. In some areas of the world, this airflow pattern occurs on a giant scale and controls the seasonal weather patterns. This type of weather system exists most notably in Southeast Asia and India. In summer, the land warms and pulls in moisture-laden air from the sea. This moist air produces rain; therefore, the summer is the rainy season. In winter, the land cools and dry air flows out to the sea, producing the dry season.

Air Inversions

Probably the most important weather occurrence in environmental matters is the air inversion. In the discussion of the structure and dynamics of the atmosphere it was noted that the troposphere has a negative temperature gradient. The temperature gradient occurs because the sun warms the ground and the ground warms the air. As stated earlier, such a situation places cool air above warm air and causes continual vertical mixing of the lower atmosphere, as can be seen in Figure 8.13 (top figure).

Under certain conditions the temperature gradient near the ground can become positive, or inverted from the normal situation. The layer of air near the ground containing a positive temperature gradient (i.e., warmer air with increasing altitude) is known as an **inversion layer**. Inversion layers occur most commonly on cool, clear nights. The sun is not there to warm the ground and with no cloud cover the ground cools rapidly by emitting heat radiation into outer space. This cool ground then cools the air above it, and the air at ground level becomes cooler than the air above it. Although these inversion layers are common, they usually break up when the sun comes up the next morning. More long-lasting inversions can be caused by long periods of foggy or cloudy weather. In these situations, the sun's rays cannot reach the ground to warm it; hence, the temperature gradient may either invert or be nearly zero.

Sea breezes can often produce a temperature inversion in coastal areas. As the cool air blows in from offshore, it slides under the warmer air over the land, producing the inversion. Such an inversion can be fairly persistent if the sea breezes are strong enough and constant.

In areas of major air pollution, inversion layers are a problem because the vertical mixing of the atmosphere stops, as seen from Figure 8.13 (bottom figure). If pollutants are produced, the polluted air does not rise high enough into the atmosphere

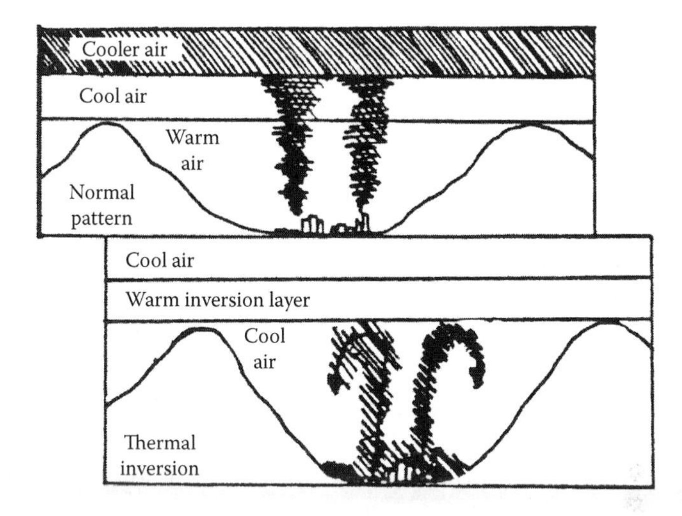

Figure 8.13 Comparison of vertical air circulation with and without an inversion layer. (Courtesy of Amy Beard.)

where the pollutants might be carried away by higher altitude winds. Instead, the polluted air stays near the ground. This stagnant air will get more and more polluted as gases and particulate matter are sent into the air by the various emission sources. Sometimes a ground-level breeze might blow the pollution away, but this can only happen if the ground is fairly flat. If the problem arises in a mountain valley, then the surface winds cannot get to the pollutants to blow them away; thus, an inversion layer in a polluted area can cause the pollutants to build up if there is no breeze or if the air is trapped by a mountain valley.

CLIMATE

When we make the comment that it has been a hot summer or a cold winter or perhaps a wet fall, we are actually talking about climate. We are stating how the weather we have been observing is deviating from the norm. When we speak of climate, we are referring to weather conditions over a long period of time, including statistical parameters such as averages, extremes, and frequency of various measurements of the state of the troposphere. **Climate** can be defined as the pattern of weather conditions over a region or regions for a long period of time.

Climate is often thought of as a relatively fixed phenomenon. When we speak of the climate in Florida being different from that in the province of Alberta in Canada, we expect that to be a fairly reliable statement that will not change from year to year. Yet, climate does change. The most obvious change in climate is from summer to winter in the temperate and polar regions of the Earth. (The summer climate in Alberta is clearly different from the winter climate.) Yet, even beyond the obvious seasonal changes, the climate of most places on the globe changes over time.

Much of our perception of climate change is anecdotal. Often, older people will comment on how much more it snowed when they were children. In the summer of 1988, many people were sure that global warming (see Chapter 11) had arrived because of the intense heat wave that covered the eastern half of the United States. It is easy to jump to conclusions about climate and climate shifts. When considering shifts in climatic data, the following precautions should be taken:

1. Do not consider data from a single year by itself to be meaningful.
2. Look at averages of data from 5 or more years at a time.
3. Consider data to be meaningful only if it fits into trends over 50 to 100 years or more.

History of the Earth's Climate

Within the life span of humans, climatic changes are usually not noticeable. Most changes in climate are geologic in their time frame (hundreds of thousands to perhaps millions of years). To better understand climatic changes, we need to look at the history of climate.

First, let us see the various sources from which the climatic information is obtained. Systematically collected data only goes back to the middle of the 19th century. There is some scattered evidence from journals and diaries for a few hundred years before the 19th century; however, for climate data from before about AD 1000, we must resort to scientific cleverness. The Earth's changing climate has left various types of evidence in the form of tree growth rings, ice bubbles in glaciers, and so on.

How precisely can we deduce the climate of, for example, 10,000 years ago? Fortunately, much information can be obtained from the fossil record. Ancient specimens of gases and water can be obtained by drilling into the permanent ice in Antarctica and Greenland. This information reveals the temperature and the concentrations of gases in the atmosphere. The exact method of deducing the temperature is beyond the scope of this chapter, but it is related to the analysis of various isotopes of oxygen in the water sample. Similar information can be obtained by analyzing the layers of ocean sediment. The biological fossils, including pollen, reveal what type of life flourished that would imply something about the climate. Also, isotope analysis of the ocean sediment can be done.

For most of Earth's history, the planet was 15 to 18°F (8 to 10°C) warmer than it is today. During most of this time, the Earth had little or no permanent ice. This relatively warm state has been interrupted at least three times by periods of glaciation. Two of the periods were 600 and 300 million years ago, and the third period of general glaciation, which started about 2 to 3 million years ago, is still going on.

Near the end of the period when dinosaurs still roamed the Earth, which was about 65 million years ago, the planet was still quite warm. By about 55 million years ago, a long cooling trend had begun. Gradually, polar ice began to appear, thicken, and spread out. Finally, about 2 to 3 million years ago large ice sheets were firmly established at both poles.

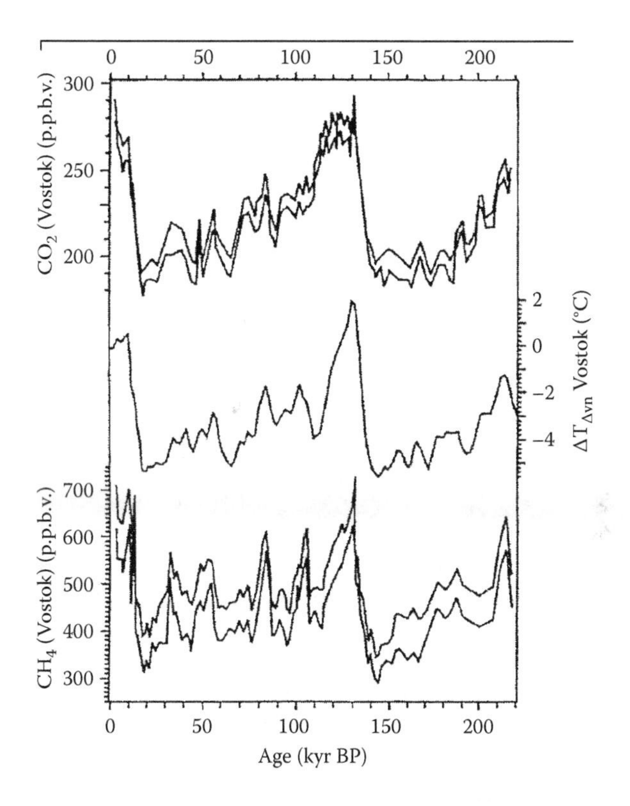

Figure 8.14 Carbon dioxide concentration (top), temperature (middle), and methane concentration (bottom) from analyses of ice cores from Vostok, Antarctica. (From Lorius, C. et al., *Nature*, 347(6289), 139–145, 1990. With permission of Macmillan Publishers, Ltd.)

About this same time, something interesting began to happen. The polar ice sheets began to spread out and recede in periodic cycles. At first, the cycles were about 40,000 years in length. About 800,000 years ago the cycles became longer with a cycling time of about 100,000 years. Each of these cycles started with a long period of slow cooling, called the *glacial period*. After this period the climate began to become warm. When compared to the rate of cooling, the warming was fairly rapid. The glacial period lasted for 80,000 to 90,000 years, whereas the period of warming, called the *interglacial period*, lasted only about 10,000 years. The graph in Figure 8.14 shows that at the present time we are very near the end of an interglacial period. The Earth should start to cool soon, but because the cooling is gradual and the time span is so large, it is difficult to pinpoint exactly when the cooling should begin. Other factors (which are discussed in Chapter 11) may actually cause warming of the globe.

Although the past 10,000 years have been a period of gradual warming, it has not been a time when there was a smooth general increase in temperature. The gradual increase in temperature continued until about 5000 years ago. The Earth at that

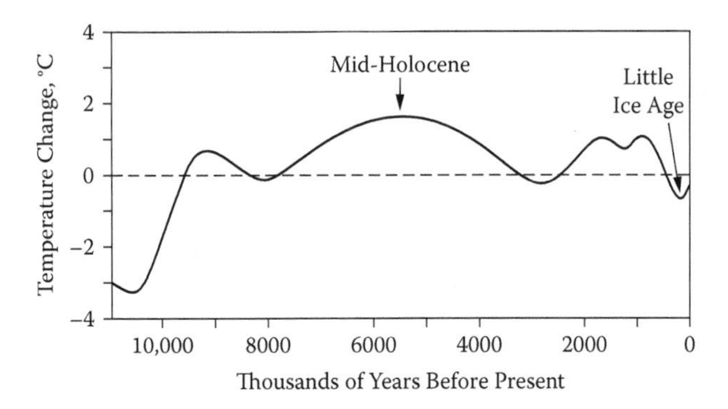

Figure 8.15 Variation in global temperatures over the past 10,000 years. (From Hileman, B., *Chemical and Engineering News*, 70(17), 11, 1992.)

time was as warm as it has been in the past 10,000 years. Figure 8.15 shows that the Earth cooled slightly and then began to warm again, shortly after the beginning of the Christian era. After this, two warm periods occurred with a short cool period in between. The second warm period is known as the Medieval Warm Period (see Figure 8.16). During this time, vineyards flourished in England and the Vikings were able to colonize Iceland and Greenland. About AD 1400 the climate began to cool and stayed cool for some time. The temperatures began to drop seriously about AD 1550 and stayed low for about 150 years. In this time period the vineyards in England vanished, farming became all but impossible in northern Europe, and the Viking colonies in Greenland were lost to colder temperatures. This time came to be known as the Little Ice Age. The temperatures later became moderate, but in the late 1800s the Earth began to warm more rapidly and has continued to warm up to the current time. Currently, the average temperature of the Earth is nearly as high as it was about 5000 years ago.

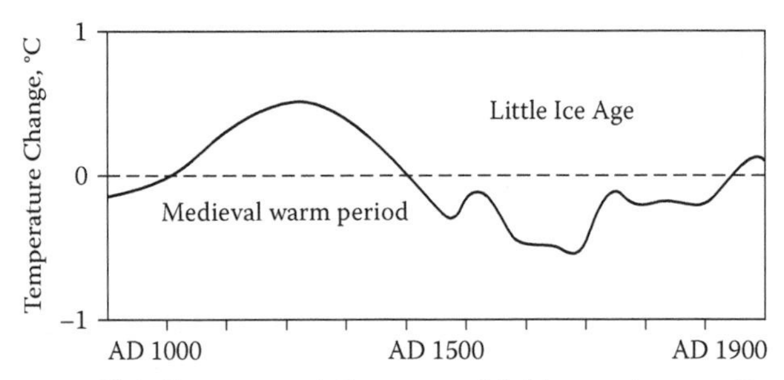

Note: Zero represents the average global temperature near the beginning of the 20th century.
Source: UN Intergovernmental Panel on Climate Change

Figure 8.16 Variation in global temperature over the past 1000 years. Zero represents the average global temperature near the beginning of the 20th century. (From Hileman, B., *Chemical and Engineering News*, 70(17), 11, 1992.)

External Factors Affecting the Earth's Climate

Many factors within the atmosphere, such as water vapor levels, cloud cover, the amount of dust, and carbon dioxide levels, have an effect on climate. Furthermore, ocean currents can also change climate in a major way. These internal factors are discussed in Chapter 11, but there are also important external factors. The warmth of the Earth comes from the sun, which does not have a constant output; however, currently we have no good method of predicting solar output. Other factors affecting the amount of solar radiation reaching the Earth's surface are more predictable. Some of these factors are the tilt of rotation of the Earth on its axis and the path of the Earth around the sun. There are cyclic changes in both of these factors over a long time span of 20,000 to 100,000 years. The Serbian mathematician **Milutin Milankovitch** was able to relate these cyclic changes to the amount of solar radiation striking various parts of the planet, and in his honor these cycles are referred to as the **Milankovitch cycles**. They have been checked against the climatic changes of the past and have been found to correlate well with the actual temperature shifts.

Although our understanding of the forces that drive climatic change is certainly imperfect, it is clear that we do understand some of the factors. The external factors just noted and the atmospheric factors that are discussed in Chapter 11 give some insight into the physical forces that control the climate.

DISCUSSION QUESTIONS

1. Why is the composition of air always given in terms of dry air?
2. What is a temperature gradient and how are the layers of the atmosphere defined using this concept? Why would adjacent layers of the atmosphere be required to have opposite temperature gradients in terms of being either positive or negative?
3. Explain why weather exists in the troposphere and not in other layers of the atmosphere.
4. How does the ozone layer explain the positive temperature gradient in the stratosphere?
5. Why is temperature such a difficult concept in the upper reaches of the atmosphere?
6. From the information in Figure 8.1 explain why a widespread increase in the amount of permanent ice would lead to a decrease in sea level.
7. How is the temperature of air related to its density? How does the relationship between air density and air temperature explain the rising and falling of air?
8. What are the causes of uneven heating of the Earth's surface and how is this related to the weather?
9. George Hadley originally conceived of only one cell in the Northern Hemisphere, with air rising at the equator and air sinking at the poles. Explain how the Coriolis effect breaks a Hadley cell into three cells.
10. Explain why the Coriolis effect is so much weaker near the equator than it is nearer to the poles.
11. Explain why there is so little wind at the horse latitudes and in the intertropical convergence zone (doldrums). Why are so many of the world's deserts at the horse latitudes, while rain forests are often at or near the intertropical convergence zone?

12. Explain why the winds around a low center in the Northern Hemisphere rotate in a counterclockwise fashion.
13. Compare hurricanes and typhoons to conventional low-pressure system in terms of size and wind speed. Why do hurricanes and typhoons tend to die out when they get over land or cool water?
14. Why are details about hurricanes easier to predict than tornados?
15. Differentiate the type of weather associated with a cold front vs. the type associated with a warm front. Explain why.
16. Explain why inversion layers often cause pollutants to build up in the air in a local area.
17. Differentiate between climate and weather.
18. What kind of data would you need to determine that climate change is occurring? Would one hot summer or one cold winter allow the conclusion? What about two or five or fifty?
19. Describe the general temperature changes that occur as a function of time in one complete ice age cycle.
20. What are the factors that go into Milankovitch cycles?

BIBLIOGRAPHY

Ahrens, C.D., *Meteorology Today: An Introduction to Weather, Climate, and the Environment*, West Publishing Company, St. Paul, MN, 1994, pp. 58, 59, 232.

Christianson, G.E., *Greenhouse: The 200-Year Story of Global Warming*, Walker & Company, New York, 1999.

Cunningham, W.P. and Saigo, B.W., *Environmental Science: A Global Concern*, Wm. C. Brown, Dubuque, IA, 1990, pp. 289, 318–320.

Diamond, J., *Collapse*, Penguin Books, London, 2005.

Hileman, B., Web of interactions makes it difficult to untangle global warming data, *Chemical and Engineering News*, 70(17), 7–19, 1992.

Lorius, C., Jouzel, J., Raynaud, D., Hansen, J., and LeTreut H., The ice-core record: climate sensitivity and future greenhouse warming, *Nature*, 347(6289), 139–145, 1990.

Lutgens, F.K. and Tarbuck, E.J., *The Atmosphere*, Prentice-Hall, Upper Saddle River, NJ, 2007.

Schneider, S.H., The changing climate, *Scientific American*, 261(3), 70–79, 1989.

Schoen, D., Learning from polar ice core research, *Environmental Science and Technology*, 33(7), 160A–163A, 1999.

CHAPTER 9

Air Pollution

Of all the substances taken into the human body, air is probably the most critical. People can survive for days, often weeks, without food. Although water must be consumed more frequently, one can still go on for a few days without it and survive. But, this is not so with air. The amount of time one can survive without air to breathe is measured in minutes. This factor has an immediate consequence when we are discussing air pollution. Air contamination is the type of pollution over which we as individuals have the least control. If the food is bad, we can choose not to eat it. Similarly, if the water is bad, we can choose not to drink it. But, if the air is bad, we have to breathe it whether we like it or not.

Air pollution is not new. Bad air is a problem that has plagued humankind since the discovery of fire. One can easily envision the caveman starting a fire in his cave and being driven out by the smoke. The problem became more serious with the development of cities. In cities, many fires were concentrated in a small area because of the high population density. If a city happened to be in a valley and the conditions were right, the smoke would linger, resulting in air pollution. The use of coal seemed to make the situation worse, so much so that King Edward I banned the burning of coal in medieval England. The king was responding to the increased use of coal that had resulted from a shortage of wood. In the 14th century, when the Black Death swept through Europe, the population fell and the fuel crisis disappeared; however, by the 16th century wood shortages reappeared, and coal became more appealing as a less expensive, alternative fuel. London, in particular, became more and more smoke ridden. Even at its worst, however, the problem was basically local and transient. Generally, air pollution was not seen as a very serious problem.

This attitude was to change with the advent of the English Industrial Revolution. By the 17th century, a severe shortage of firewood had developed, and coal was being burned in homes for heating and cooking. The development of the steam engine led to the use of coal to power it, as well. All types of factories began to use coal to power their equipment and often as a source of heat for certain processes. As the coal consumption increased, the amount of air pollution also increased. At first, the coal smoke was basically seen as a nuisance—that is, something that one had to endure as a part of "progress." Then, in the late 19th century, the situation became much more ominous.

In 1873, during a period of foggy, cool weather (which is typical of London during certain times of the year) an event occurred that was but a harbinger of things to come. The air became acrid and smoky because there was little wind, if any. By the time the fog lifted and the winds returned 700 people had died either directly or indirectly from the effects of the air. The fog and smoke were so thick that 15 people drowned in the Thames River at Northside docks because they lost their way and fell into it. Many of the cattle at an exhibition called "The Great Show at Islington" died.

This phenomenon was to be repeated many times in London. Other episodes occurred in 1880, 1892, 1911, 1948, 1952, 1956, 1957, and 1962. In 1905, Dr. H.A. Des Voeux of the Coal Smoke Abatement Society coined a term that is still used today. He referred to the mixture of smoke and fog that the people of London were forced to breathe as **smog** (**sm**oke + **fog**). In the early years of the 20th century, England and the industrial valleys of Germany suffered the most from this type of unhealthy air, but gradually, as the United States became industrially powerful, it too began to suffer a similar fate.

The town of Donora, Pennsylvania, is an industrial city set in a valley south of Pittsburgh. In 1948, it sported steel mills, zinc smelters, and a sulfuric acid plant. An inversion layer (discussed in Chapter 8) developed over the city which lasted for 5 days. During this period 6000 people fell ill, but miraculously the death toll was not nearly what one might have expected. Twenty people died directly from the pollution.

The greatest smog incident occurred in London in December of 1952. The weather before this event had been favorable, and then as if by some sinister design the fog began to settle in and the winds ceased. The weather remained this way for 3 days as Londoners and their industries continued to push more and more smoke into the air. When it was over, it was estimated that 4000 people had died due to the smog.

CLASSICAL AIR POLLUTION

The first part of this chapter uses the common understanding of air pollution, and foul, hazy, cough-producing, unhealthy air is considered. In reality, this type of air pollution deals with trace gases in the atmosphere. None of these polluting gases is present in the atmosphere in concentrations remotely approaching the concentrations of major or even minor atmospheric constituents. Why then are the gases so important? It is because of their high toxicity to humans and other life forms. Because this is the type of air pollution first observed by humans, it is referred to as **classical air pollution**, which is defined as the constituents of air that are clearly unhealthy and that have been traditionally considered as a part of polluted air. It turns out, however, that the pollution development process is a bit more complicated than simply releasing gases into the atmosphere. Many of these gases react with the gases already present in the atmosphere. As a result, chemical reactions occur that produce new pollutants. The materials released into the air initially are referred to as **primary air pollutants**. Those that appear due to chemical reactions are known as **secondary air pollutants**.

INDUSTRIAL SMOG

Industrial smog was the type of air pollution to which the term *smog* was originally applied. It is so named because it appeared in heavily industrialized cities starting with the beginning of the Industrial Revolution. The most obvious component of industrial smog is **smoke**. What is smoke? Smoke is not a gas, but is instead very finely divided solid particles. Smoke will eventually settle out of air, and it is this property that makes industrial smog so dirty. Smoke causes houses, cars, streets, etc. to be covered with a black film.

Smoke generally becomes a serious problem only when one burns coal. It can be a problem with other types of fuel but usually is not. To understand the formation of smoke we need to investigate how coal burns. The burning of coal is a type of oxidation reaction known as **combustion**. Combustion is the rapid combination of a material with oxygen accompanied by the evolution of considerable amounts of heat and gases. When coal undergoes combustion the following reaction occurs:

$$C + O_2 \rightarrow CO_2$$

Under ideal conditions carbon dioxide is the only product. First, the principal component of coal is carbon, and, if all is well, the carbon will combine with the plentiful supply of oxygen in the air. As is true of many things we try to do, the ideal is difficult to achieve.

There are two major problems with the way coal burns. Both of them are related to the supply of oxygen. Because coal is a solid, it is very difficult to get air into all of the parts uniformly; hence, some parts of it do not get sufficient oxygen. When this happens, some of the coal either burns incompletely or does not burn at all. Second, when coal burns, it is broken down into fine particles, which if unburned are lifted into the air by the hot flame. These small particles of unburned coal are known as **soot**. They vary in size, and, as noted in Chapter 6, the smaller they are the more dangerous they are. The smoke also contains **fly ash**. Fly ash results from the fact that coal is not absolutely all carbon and contains some materials that do not burn. These materials are known as ash, and some of the ash particles are very small. These smaller particles are carried up the chimney as a part of the smoke. The smoke then contains soot and fly ash particles of various sizes, some of which remain suspended in the air for relatively long periods of time.

As mentioned earlier, if there is insufficient air, then some of the coal would burn incompletely. The lack of oxygen leads to the formation of poisonous carbon monoxide:

$$2C + O_2 \rightarrow 2CO$$

Note the effect of the lack of oxygen. Carbon normally combines with two oxygen atoms, whereas in an oxygen-deprived situation it combines with only one oxygen atom. Because it is very difficult to get sufficient air into all parts of the coal, it is also very difficult to avoid the production of carbon monoxide. One can force air in, under pressure, and thereby get more complete combustion, but the fire will be hotter

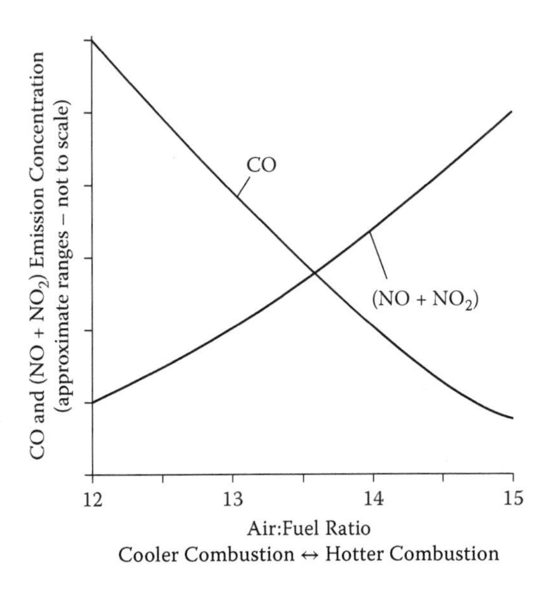

Figure 9.1 Idealized variation of CO and NO_x output from combustion as a function of temperature.

and will consume the fuel faster. Unfortunately, this solution to the carbon monoxide problem only creates another problem, the formation of oxides of nitrogen. At very high temperatures, nitrogen and oxygen will react to give nitric oxide:

$$N_2 + O_2 \rightarrow 2NO$$

Hence, a "Catch 22" type of situation is created. At higher temperatures, more NO is produced and less CO. At lower temperatures, NO production is minimized, but CO production increases (see Figure 9.1). Which is the preferable situation?

The chemistry of nitrogen oxide pollutants includes the production of secondary pollutants. Once nitric oxide (NO) is placed in the air, it is slowly converted into nitrogen dioxide (NO_2) by reaction with the oxygen in the air:

$$2NO + O_2 \rightarrow 2NO_2$$

Nitrogen dioxide is a much more dangerous and toxic chemical than nitric oxide. This very important secondary pollutant goes on to produce nitric acid (HNO_3) by reaction with the water in the air. Eventually, the acid is removed from the air in rainwater, and this is one of the processes by which acid rain is produced:

$$NO_2 + H_2O \rightarrow HNO_3 \text{ (not balanced)}$$

It has already been noted that these nitrogen chemicals are produced in the atmosphere naturally when lightning strikes. The difference between the situation during smog conditions and that occurring naturally is that smog may produce localized high concentrations that may be dangerous. The nitrogen pollutants are difficult to avoid because only high temperature is necessary to produce these hazardous gases. The raw materials are always there, as nitrogen and oxygen are present in the atmosphere.

The most important primary pollutant in industrial smog has not been discussed yet. Interestingly, it is produced not from the main constituent of coal, but from an impurity. Because coal is a fossilized plant material from millions of years ago, at one time it contained all of the elements associated with living plants. Although some of these elements have been removed over time, most coal contains various amounts of sulfur. When the coal is burned, the sulfur contained in it burns as well:

$$S + O_2 \rightarrow SO_2$$

The material produced is the unpleasant and poisonous gas sulfur dioxide. Early in the 17th century, the English chemist Robert Boyle noticed a corrosive component in the air in and around London, but it was not until 1744 that the French chemist Rouelle found this to be sulfur dioxide. It is this gas along with particulate matter that has been implicated as the villain in major episodes of industrial smog throughout the world. Levels of sulfur dioxide have been recorded as high as 38 ppm (Meuse River Valley of Belgium in 1930, with 60 deaths); however, levels of 0.5 to 3 ppm in polluted air are more common.

The story, however, does not end there. Sulfur dioxide produces a series of secondary air pollutants. As was the case with nitric oxide, sulfur dioxide reacts slowly with the oxygen in the air to produce sulfur trioxide:

$$2SO_2 + O_2 \rightarrow 2SO_3$$

Although sulfur trioxide is a quite toxic and corrosive gas, it is so reactive with water that it will immediately react with the water in the atmosphere. As a result, sulfur trioxide seldom presents much of a direct threat; however, the reaction of this material with water produces a very strong acid, sulfuric acid:

$$SO_3 + H_2O \rightarrow H_2SO_4$$

This is certainly not the type of material one would like to inhale on a regular basis. One further reaction needs to be added to complete the picture, as sulfur dioxide also reacts with water to produce a weaker acid, sulfurous acid:

$$SO_2 + H_2O \rightarrow H_2SO_3$$

The production of these acids and sulfur trioxide will become more important in the discussion of acid rain; however, in direct industrial pollution, sulfur dioxide is the main sulfur-containing pollutant.

Industrial smog has most commonly been associated with coal combustion and that continues to be the case. There is some sulfur in oil; therefore, oil-fueled industrial plants may produce some sulfur dioxide. If burned properly, oil is less likely to

Table 9.1 Industrial Smog

Primary pollutants	Smoke, carbon monoxide (CO), nitric oxide (NO), and sulfur dioxide (SO_2)
Secondary pollutants	Nitrogen dioxide (NO_2), nitric acid (HNO_3), sulfur trioxide (SO_3), sulfurous acid (H_2SO_3), and sulfuric acid (H_2SO_4)
Weather conditions	Any weather conditions producing an inversion layer in a valley or with no winds

produce smoke, but the production of the gaseous pollutants may still be a problem. Another major source of sulfur dioxide is ore smelting. Many metal-containing ores occur as sulfur compounds. If they are heated strongly in the air, the pure metal can be obtained. But, as the law of conservation of matter would indicate, the sulfur has to go somewhere. It ends up as sulfur dioxide, which is often released into the atmosphere.

The weather condition that usually concentrates industrial pollutants is the *inversion layer*. Chapter 8 described the formation of an inversion layer and the conditions under which such a layer leads to the buildup of air pollutants. This type of weather condition is important in smog formation because wherever air pollutants are being produced it is the inversion layer that makes the problem worse by concentrating the contaminants near the ground. See Table 9.1 for a summary of industrial smog.

PHOTOCHEMICAL SMOG

Nearly everyone has heard that historically Los Angeles has had a serious smog problem. What most people likely are unaware of is that this type of smog has very little in common with industrial smog, for which the term was developed. The use of the term *smog* to describe Los Angeles air pollution is probably not correct; however, the term is so often used to describe air pollution of this type that the word smog has essentially been redefined. The sources are different, the chemical nature is different, and the conditions under which this type of air pollution will form are very different.

The west coast of the United States is not significantly different from anywhere else in the world in that air pollution has always been around to some extent in this area. In 1542, when the Spanish explorer Juan Rodriguez Cabrillo was investigating the California coast, he observed a heavy haze over San Pedro Bay, which led him to call the bay the "Bay of Smokes." Some of the earliest records of eye irritation caused by air pollution in the Los Angeles basin date back to 1868; however, at that time the problem was transient and did not have serious effects.

California has always been a place of drastic change and to many people a far-off land of golden dreams. In 1840, California was not even a part of the United States, yet by 1850, due to the Mexican War and the California Gold Rush, it was an American state with a population of nearly 100,000. People were drawn by gold, new land, and the climate. The gold and the land are no longer factors, but the climate still is. California is warmed by prevailing ocean winds that keep the state unusually warm for its latitude in the winter. The population steadily increased until the 1940s. Following World War II, many things about American culture began to change. Americans became more mobile. Many defense plants were established in southern California, and with the lure of the climate Americans began to move to California in droves. The state added 3.5 million people in the 1940s and another 5 million in the 1950s. From 1940 to 1960 the population of the state more than doubled.

It was during this period that the people of southern California began to notice that there was something wrong with the air. The air often became hazy and sometimes caused coughing and eye irritation. Some plants began to develop yellow spots on their leaves. Rubber products began to crack and wear out prematurely. This air pollution seemed unlike any that had been encountered elsewhere. Where was it coming from? Why was it so different?

One of the leading suspects was the automobile, but they had cars in New York and Chicago, and they did not have this type of smog. In experiments with mixtures of gases from car exhausts, nothing happened until someone had the idea to shine an intense light on the mixture. At that point, the mixture of gases began to behave just like smog. This light-induced smog came to be called **photochemical smog** (the prefix photo means "light").

It is now known that three conditions are necessary for the production of photochemical smog:

1. Air containing a mixture of gases such as those contained in auto exhaust gases, particularly nitrogen oxides and hydrocarbons
2. Sunlight
3. Temperatures above 65°F (18°C)

One additional condition is necessary to concentrate it:

4. An inversion layer along with calm or an inversion layer in a mountain valley

These criteria make it quite clear why many urban centers of the United States are not chronically plagued by this type of smog and Los Angeles is. Nearly all urban areas will fit into the first criterion. It is with points 2 and 3 that differences will appear. Large areas of the United States are too cold during significant months of the year to produce this type of smog. The second criterion for photochemical smog is sunlight, and of course Los Angeles has this in abundance. It is ironic that part of southern California's great appeal, its sunshine, causes such a daily problem for Los Angeles residents. Yes, photochemical smog can develop in New York. Such smog is very likely on a warm sunny day, but there are many days when it cannot occur, either because of clouds or cool temperatures. Even if photochemical smog develops in New York, it does not build up without calm and an inversion layer. This does not mean, however, that photochemical smog cannot be a serious problem in New York City. It can be and sometimes is. In fact, serious problems related to photochemical smog are spreading across the country in cities such as Houston, Charlotte, and Atlanta, among others.

It should be noted that photochemical smog is becoming a more and more serious problem in many places throughout the world. Less-developed countries are seeing more and more automobiles on the road in more and more densely populated cities. Places such as Mexico City, Mexico, and São Paulo, Brazil, have serious air pollution problems, of which photochemical smog is a considerable part.

Photochemical smog has its beginnings, as is true for industrial smog, in the process of burning fuel. In the case of photochemical smog we are considering the combustion of gasoline in various motor vehicles. Gasoline is a mixture of low-boiling compounds obtained from petroleum. Nearly all of these compounds are hydrocarbons. Under ideal circumstances, burning gasoline produces carbon dioxide and water. Among the compounds that make up gasoline are a group of compounds of the formula C_6H_{14}, called *hexanes.* Using this group of compounds as an example, the burning of gasoline occurs according to the following equation:

$$2C_6H_{14} + 19O_2 \rightarrow 12CO_2 + 14H_2O$$

The same problems that we encountered with coal occur in this case also. There are problems with incomplete combustion due to lack of oxygen. The fuel either is burned incompletely or perhaps is not burned at all. Additionally, the heat of combustion will cause the production of nitric oxide.

If there is insufficient oxygen, several things can happen. The most likely occurrence is the production of carbon monoxide. This is a well-known feature of auto exhaust as noted earlier. Less oxygen is consumed in producing carbon monoxide rather than carbon dioxide:

$$2C_6H_{14} + 13O_2 \rightarrow 12CO + 14H_2O$$

In some cases, the product of the reaction may be soot (carbon). This type of pollution is usually easily avoidable in the standard gasoline engine by keeping one's car well tuned; however, due to the nature of operation, the emission of carbon particles is much more difficult to avoid in a diesel engine. These diesel emission particles contain large quantities of hydrocarbons, many of which are likely to be carcinogenic. The California Air Resources Board has listed diesel exhaust particulate matter as a toxic air contaminant.

Ultimately, as is the case with coal, some of the fuel in any engine remains unburned. Before the advent of pollution control equipment, a considerable amount of fuel vapor would escape from the fuel tank and from the carburetor. The rest of the unburned fuel released into the air came from that which made it through the engine and out of the exhaust without being consumed. The amount of carbon monoxide and unburned fuel can be decreased easily by increasing the oxygen-to-fuel ratio, but this makes the engine hotter and, as discussed in the earlier section on industrial smog, causes increased production of nitric oxide (NO).

Therefore, at this point we have three fairly ordinary primary air pollutants (CO, NO, and hydrocarbons), two of which have been discussed earlier. What is so different here? It is this. These three materials along with sunlight produce a chemical soup of incredible complexity. The process begins with the conversion of nitric oxide (NO) to nitrogen dioxide (NO_2) in the same way that it occurred in the case of industrial smog, except that sunlight seems to speed up the process. The nitrogen dioxide produced is a brown-colored gas, which will absorb solar radiation. When nitrogen dioxide absorbs radiation, it reacts to produce nitric oxide and an oxygen atom:

$$NO_2 + \text{Sunlight} \rightarrow NO + O$$

A free oxygen atom is very different from an oxygen molecule. Oxygen atoms pair up because this makes a more stable arrangement. When an oxygen atom is unpaired, it is very reactive. It is so reactive that it will make a bond with a large number of other substances. The oxygen atom is so reactive that it never lasts very long in the air, reacting almost as soon as it is formed.

One of the things that the oxygen atom reacts with quite often is an oxygen molecule to make ozone:

$$O + O_2 \rightarrow O_3$$

Ozone (O_3) is not very stable and tends to revert to molecular oxygen (O_2) over time:

$$2O_3 \rightarrow 3O_2$$

Hence, if ozone is not continually put into the air its concentration will tend to decrease because ozone reverts to oxygen over time. The problem in photochemical smog is that ozone is continuously produced.

To our "soup" of CO, NO, and hydrocarbons, the chemical species of NO_2, O, and O_3 have been added, but that is not the end of the recipe. The hydrocarbons in the air are subject to attack by very reactive chemical species produced in the air such as atomic oxygen or ozone to give aldehydes:

$$C_6H_{14} + (O \text{ or } O_3) \rightarrow \text{Aldehydes}$$

Aldehydes are a collection of similar compounds. They do not vaporize as easily as hydrocarbons and as a result tend to form tiny droplets in the air which produce the haze generally associated with photochemical smog. The aldehydes are moderately toxic but by themselves would not cause any great problem. Unfortunately, the aldehydes are very reactive and go on to produce a collection of compounds which, along with ozone (O_3), are known as oxidants.* Most of these compounds are toxic, nasty materials that produce much of the damaging effects of photochemical smog.

One particularly well-known member of this group is a material known as *peracetyl nitrate* (PAN), which is produced when a particular aldehyde known as acetaldehyde reacts with oxygen and nitrogen dioxide in the presence of sunlight:

$$\text{Acetaldehyde} + \text{Sunlight} + O_2 + NO_2 \rightarrow \text{PAN}$$

* The oxidants are analyzed as a class of compounds by their ability to oxidize iodide ion (I^-) to iodine (I_2).

PAN is one of those materials that make the smog so immediately unpleasant. It causes tearing and burning in the eyes. Chemicals that cause the eyes to produce tears are known as *lacrimators*. Not only is PAN a lacrimator that affects humans but it also has deleterious effects on plants and animals. PAN is the most toxic of any of these photochemical pollutants to plants. Luckily, it is also present in the lowest concentration of any of the major components of smog.

One unfortunate byproduct of the conversion of hydrocarbons to aldehydes is the production of NO_2 from NO:

$$C_6H_{14} + NO + O_3 \rightarrow \text{Aldehydes} + NO_2$$

The problem here is that NO_2 is produced, which in the presence of sunlight will make more ozone to replace the ozone that was consumed. In fact, in some cases, it is possible to produce more NO_2 molecules per molecule of hydrocarbon because the preceding reaction can occur more than once in chewing up a large hydrocarbon. Because

$$NO_2 + \text{sunlight} \rightarrow NO + O$$

and

$$O + O_2 \rightarrow O_3$$

these reactions will either maintain the ozone level or increase it.

The last of the secondary pollutants from photochemical smog is nitric acid (HNO_3), which is formed in the same way as it is in industrial smog. This acid is very water soluble and as such represents the major way in which nitrogen dioxide is eventually removed from the atmosphere.

In Los Angeles and many other urban areas of the world, there is a definite daily cycle of buildup and dissipation of air pollution. As seen from Figure 9.2, various pollutants start out at fairly low levels in the morning, but each rises and

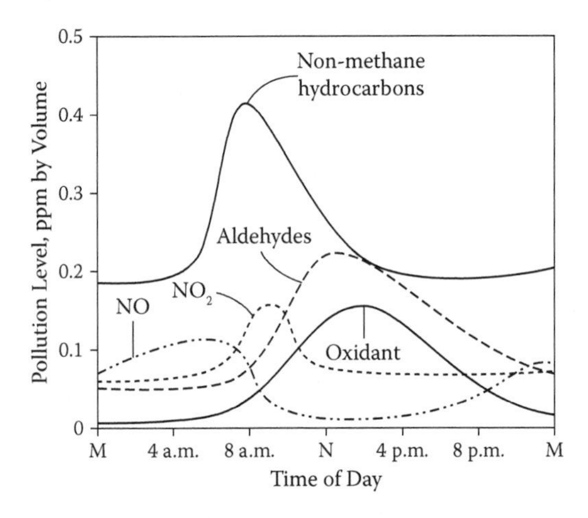

Figure 9.2 General daily variation of photochemical smog pollutants. (From Manahan, S.E., *Environmental Chemistry*, Lewis Publishers, Chelsea, MI, 1991. With permission.)

Table 9.2 Photochemical Smog

Primary pollutants	Carbon monoxide (CO), hydrocarbons (unburned fuel), and nitric oxide (NO)
Secondary pollutants	Nitrogen dioxide (NO_2), nitric acid (HNO_3), ozone (O_3), aldehydes, PAN, and other oxidants
Weather conditions	Usually occurs in sunny, warm weather, with concentration by an inversion layer in a valley or due to lack of wind

falls at different times in a pattern that is very predictable. As the rush hour begins in the morning, automobiles release unburned hydrocarbons, carbon monoxide, and nitric oxide (NO). The concentrations of these chemical compounds are the first to increase. (Carbon monoxide is not on the graph because it is not involved in any reaction to form secondary pollutants.) Because sunlight is an important factor in smog production and during the morning rush the sun is still low in the sky, initially fewer reactions occur except for the conversion of nitric oxide (NO) to nitrogen dioxide (NO_2).

When the sun gets high enough in the sky, around 8 or 9 a.m., the concentrations of hydrocarbons and nitrogen dioxide begin to fall as they are photochemically converted into aldehydes and oxidants such as ozone (O_3), PAN, and a host of other compounds. There is no comparable increase in the levels of hydrocarbons or nitrogen oxides during the evening rush hour because the sun is high in the sky and they are converted into aldehydes and oxidants almost as quickly as they appear. As the traffic decreases and the sun gets lower in the sky, the pollution levels gradually decrease to their early morning levels. See Table 9.2 for a summary of photochemical smog.

REGIONALIZATION OF AIR POLLUTION

Ozone

As much as we usually associate air pollution with urban areas, it is unreasonable to expect that these pollutants will remain trapped in one area; for example, particulate matter consistent with European and Russian sources has been found on the islands of the Canadian Arctic. An issue of prime importance is establishing a pollutant's effects as it spreads out over a region. Many of the pollutants cease to be damaging as they are diluted by larger and larger amounts of air and are slowly removed from the atmosphere by various means. Oddly, one of the pollutants that remains a problem regionally is ozone (O_3).

Ozone might not be expected to cause regional problems, as it is an unstable gas that reverts to regular oxygen (O_2) over time; however, ozone can exist at significant enough concentrations in the atmosphere to survive more than a day of transport. As a result, it can be carried far from its point of origin. Additionally, air pollutants involved in the production of ozone (ozone precursors), such as NO_2 and hydrocarbons, can be transported by air currents to remote locations as well. At any time, these ozone precursors can produce more ozone. Historically, it has been estimated that the world background levels for ozone were about 10 to 20 ppb. Following the introduction and widespread use of the internal combustion engine, the world background levels have gradually increased to the current 30 to 35 ppb. By the mid-1980s, nearly all rural areas of the United States had ozone levels above the world background levels. When the rural ozone levels for each state were averaged, these levels varied from 40 ppb in North Dakota to 63 ppb in California. Such averages, however, can be misleading.

Prevailing winds can produce high ozone levels downwind from any major urban area. The higher the ozone levels in the city, the higher they will be downwind. It is often the case that 50 to 60% of the ozone in the greater Toronto area originates in the Detroit, Michigan, area. Furthermore, it has been demonstrated that much of the Toronto ozone load finds its way to Hastings, Ontario, a rural site about 85 miles (140 km) to the east. Transport of ozone over 60 to 300 miles (100 to 500 km) occurs frequently over the eastern part of the United States.

Ozone can have a deleterious effect on plant life even at the lower levels often found in rural areas. Ozone interferes with photosynthesis at relatively low concentrations; hence, crops and forests are affected. It has been estimated that crop losses in case of sensitive crops such as peanuts, soybeans, and wheat range from 5 to 20%.

One question that is difficult to answer is the effect of ozone on forests. Forests are complex ecosystems that respond in subtle ways to many different external influences. The list includes disease, insects, animals, drought, wind, fire, and temperature extremes. Ozone is simply another external influence. In extreme cases, it can be shown to cause direct damage. This has been the case in the San Bernardino and Angeles national forests near Los Angeles where high ozone concentrations cause pine needles to turn yellow and fall off the trees. The effect of lower ozone concentrations on forests is unclear, but because ozone does inhibit photosynthesis the effect is probably not good. One fairly common view is that ozone weakens trees, after which they become more susceptible to death from other agents such as insects.

Because of the mobility of ozone, controlling its production becomes a political issue. The general approach to ozone control requires an ozone-emitting urban area to limit emissions of nitrogen oxides (NO and NO_2). This approach is necessary because ozone is not a directly produced primary pollutant, but rather is created from other materials in the air. The general assumption is that the nitrogen oxides (NO_x) are the main compounds involved in the production of ozone; therefore, if NO_x production is limited, then ozone production will be limited as well. Many states are now required by the U.S. Environmental Protection Agency (EPA) to tighten their controls on NO_x emissions, if it appears that ozone and other constituents of smog from a state are adversely affecting other states. Not only can this be an issue within a country, but ozone can cross international boundaries, as well. The United States and Canada have engaged in talks to control NO_x emissions in each country to decrease the levels of ozone in both countries. Much of the smog in New England comes from Canada, and the United States is responsible for much of the smog in Ontario, New Brunswick, and Nova Scotia.

Acid Rain

The other pollutants of regional concern come from coal smoke. One of the oldest ways to cut down on air pollution from coal smoke is to build a high chimney. This solution was recognized as early as 1377, when in a London lawsuit an armor maker was sued for not having a chimney as high as good practice required. The main concern with coal smoke until the late 19th century was the particulates in the smoke. For the purpose of removing these particulates the chimneys worked well. If they were tall enough, the smoke went high in the air where it could be picked up by winds higher up in the atmosphere. An extreme example of this is the "superstack" built at a smelter in Sudbury, Ontario, Canada, which is 1300 feet (400 m) high, or the height of over four American football fields. These stacks remove the effluent from the breathing public on the ground and the pollutant is diluted and carried somewhere else.

The use of the tall smokestack to get rid of pollutants has some interesting and unintended consequences that are related to the law of conservation of mass, which states that everything goes somewhere and that there is no "away." All of the major gases produced from industrial coal burning, with the exception of carbon dioxide (CO_2), carbon monoxide (CO), and nitric oxide (NO), are very water soluble, which means that most gases are usually removed from the atmosphere in rainwater. Therefore, these pollutants have an effect not only on the air we breathe but also on the water that nourishes our fields, gardens, and streams.

The oxides of nonmetals such as sulfur and nitrogen tend to produce acid when dissolved in water, and as a result any rainwater in which they are dissolved would be acidic as well. The term **acid rain** is applied to such rainwater. The term is not new, having been used by **R. Angus Smith** in his book *Air and Rain* in 1872, but general public interest in the subject probably extends back to only 60 years at the most.

To discuss acid rain, first it needs to be defined. One important observation is that ordinary, unpolluted rainwater is slightly acidic, having a pH of about 5.6. This phenomenon is largely due to the carbon dioxide in the atmosphere, which is slightly soluble in water. When carbon dioxide dissolves in water it produces carbonic acid:

$$CO_2 + H_2O \rightarrow H_2CO_3$$

Carbonic acid is a very weak unstable acid, which causes the fizzing (bubbles of carbon dioxide) when one opens a bottle of carbonated beverage. The amount of carbonic acid dissolved in rainwater is very small, but there is enough of this weak acid to give rainwater the slightly acidic pH. Additionally, there are varying amounts of naturally occurring organic acids and sulfur-based acids that may affect the pH slightly; therefore, to allow for some variability, acid rain may be defined as any rainwater with a pH of less than 5. (Remember that pH is an inverse scale, so a value less than 5 would represent greater acidity.) There can also be acid snow and acid fog. Acidic gases can be directly absorbed by the soil, groundwater, and vegetation. All of these modes for depositing acid from the air taken together are known as acid deposition.

Earlier in the chapter we pointed out that industrial smog produces three acids as the end products of the chemical pollution process: nitric acid (HNO_3), sulfuric acid (H_2SO_4), and sulfurous acid (H_2SO_3). Coal burning tends to have much more severe acid deposition effects than auto emissions because coal smoke pollution will produce all three acids.

The production of sulfur trioxide (SO_3) by oxidation of sulfur dioxide (SO_2) in the air tends to intensify the acidification process. First, sulfur trioxide is an active molecule with a great affinity for water. It not only dissolves in water but also reacts vigorously with any water it can find. Second, sulfuric acid (H_2SO_4) produced when sulfur trioxide reacts with water is one of the strongest acids known. As a result, the sulfur oxide type of pollution from coal burning tends to produce more severe acid deposition problems than photochemical smog.

Photochemical smog will give rise to acid, but only nitric acid. Nitric acid is one of the three strong acids and can produce very acidic solutions. In the case of nitric-acid-contaminated rain, fortunately the concentrations are very low and the pH values produced are generally not outside normal limits.

As seen from the first map in Figure 9.3, in 1994 the eastern half of the United States had pH values much lower than the western half of the country. (Remember that a decrease of 1 pH unit increases acidity by 10-fold.) This situation is related to the fact that most of the high-emitting coal-fired electric power plants are in the east. Many of these plants are located in the Ohio River Valley.

On reviewing the 1994 map, low pH values of about 4.5 can be seen. It may be noted that these values occurred in the northeastern part of the country, and it is reasonable to assume that these values extended up into Canada. Because it was generally believed that these emissions originated in the United States and were carried to Canada by the prevailing winds, the United States needed to control these emissions not only to protect itself but also to avoid doing harm to its neighbor as well. Other indications of the international nature of this problem can be found in Europe, where evidence suggests that most of the acid rain that fell on Norway originated in England.

In the later section on air pollution and the law we will review some of the controls that have been put into place in the last couple of decades. Clearly, from the pH maps shown in Figure 9.3, considerable improvement in the situation was made between 1994 and 2009. In the second map, we find that pH values in the eastern part of the United States were generally up around half a pH unit.

Effects of Acid Deposition

What are the effects of acid deposition? Generally, they can be grouped into the following four categories:

1. Structures
2. Lakes
3. Vegetation
4. Human health

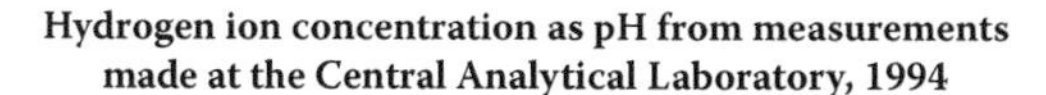

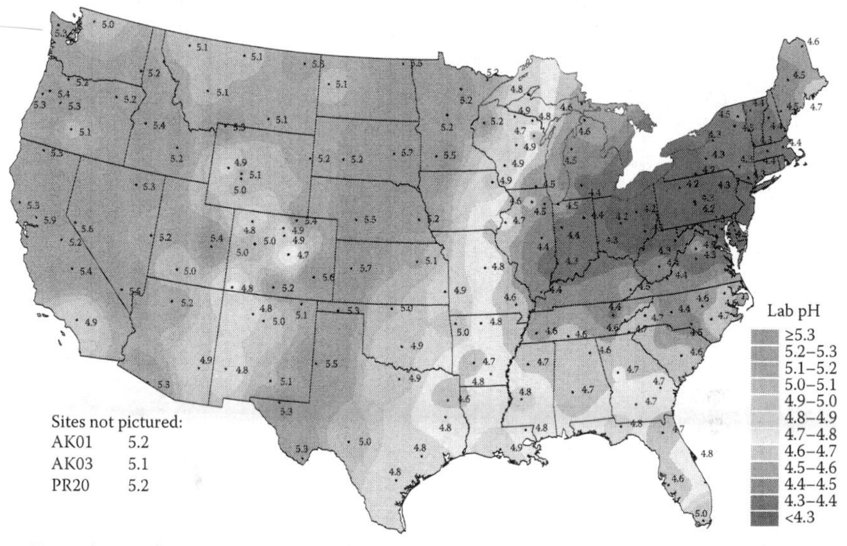

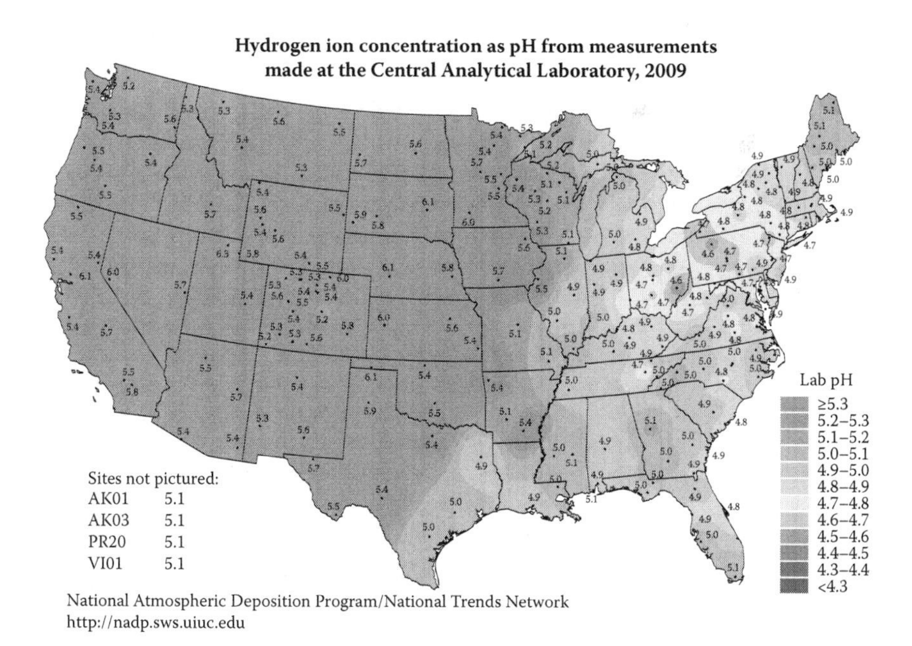

Figure 9.3 The average pH value for precipitation over the United States for 1994 (top) and 2009 (bottom). (Courtesy of the National Atmospheric Deposition Program, Old-Style Isopleth Maps, NADP Map Archive, http://nadp.sws.uiuc.edu/maplib/archive/NTN/.)

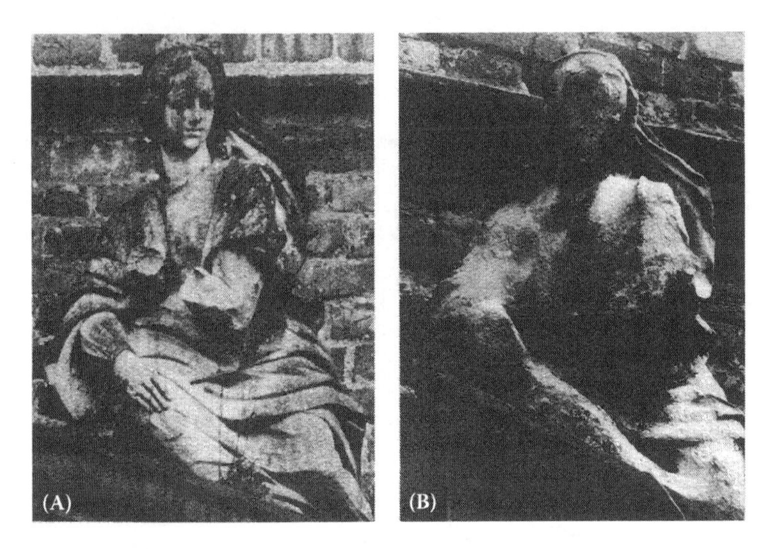

Figure 9.4 Photographs of a statue in New York City taken in (A) 1908 and (B) 1969. (From Chang, R., *Chemistry*, McGraw-Hill, New York, 1991. With permission of The McGraw-Hill Companies.)

Effects on Structures

The effect of acid deposition on structures is strikingly visible in the photographs in Figure 9.4. The first photograph, taken in 1908, shows a statue with considerable fine detail; yet, over a span of 60 years it deteriorated to the point where it is scarcely recognizable. Some of the structures that are affected most severely by acid deposition are those that are constructed of limestone or marble. Limestone and marble are two forms of the same chemical, calcium carbonate ($CaCO_3$). All carbonates share the same chemical property—they react with acid to produce carbon dioxide and water. This can be illustrated by using sulfuric acid (H_2SO_4) as an example:

$$CaCO_3 + H_2SO_4 \rightarrow CaSO_4 + CO_2 + H_2O$$

Whether it is sulfuric acid or one of the other acids associated with acid deposition, the limestone or marble is changed into another substance, which either flakes off or dissolves in the rainwater. Metal structures are, of course, also damaged by acid deposition, as many of the more reactive metals, including iron, react with acid. In the early part of the 20th century, for example, a girder collapsed in a railway station in London. On analysis, it was found that the girder was made up of 9% ferrous sulfate ($FeSO_4$) from reaction with sulfuric acid in the air:

$$Fe + H_2SO_4 \rightarrow FeSO_4 + H_2$$

Although other reactions are associated with this process, this reaction gives a general indication of how metal is attacked. Other metals used in construction react in similar ways.

Effects on Lakes

Much of the concern about acid rain has involved many of the lakes in the scenic areas of upstate New York, New England, and eastern Canada. Interestingly, lakes behave very differently from one another in the presence of acid deposition. Some lakes become noticeably acidic, whereas others seem to remain unchanged. The difference is related to our old friend limestone, which was discussed earlier. Underlying most lakes is some type of bedrock, most commonly either limestone or granite. As noted earlier, acid reacts with limestone to produce products that are not acids. This means that the acid is removed by reaction with the limestone as soon as it enters the lake; hence, these lakes seem unaffected by acid precipitation. Granite, however, does not react with acids, and this leaves the acid to build up in granite-based lakes. Of course, it takes some time before a lake can become appreciably acidic. When acidic rainwater first enters a lake it is diluted by the lake water and the effect is minimal. It is only after many years that the cumulative effect begins to appreciably affect the pH of the entire lake. Some of the areas of the world where granite-based lakes have developed high acidity (low pH) include eastern Canada, northeastern United States, and Scandinavia. For most lakes, a "normal" pH would be in the range 6.5 to 7. Many of the lakes in these areas had pH values in the range of 5 or less. Over the last 20 years the acidity situation in many lakes and streams has improved markedly. Over half of the lakes and streams monitored in the Adirondack and Catskill Mountains and in the northern Appalachian Plateau have shown movement toward a general decrease in excess acidity.

Lakes affected by high levels of acidity often have a deceptive appearance. The lake may look beautiful, clear, and tranquil. This is particularly true of those few lakes that have pH values of about 4. At this pH, nearly all of the normal aquatic life such as trout, perch, and pike die. These lakes are very clear because no action takes place in them to stir them up. They are beautiful dead lakes.

Interestingly, one of the major factors in fish kills in acidic lakes is related to increasing concentrations of aluminum ions in the water. Water in the usual pH range would tend to dissolve very little aluminum from rocks and soil. Hence, the levels of dissolved aluminum in most lakes are quite low. Acidic waters will increase considerably the quantities of aluminum found dissolved in a lake. Aluminum in high concentrations tends to kill fish by clogging their gills with gelatinous aluminum hydroxide ($Al(OH)_3$), thereby suffocating the fish.

Effects on Vegetation

The effect of acid deposition on vegetation has been more difficult to access. In 1980, the federal government set up the National Acid Precipitation Assessment Program (NAPAP) to undertake a 10-year study of acid precipitation and its effects. This study was reauthorized and continued under the 1990 amendments to the Clean Air Act. The study considered all aspects of acid precipitation, including the effect on vegetation. The main foci of the consideration of vegetation were the forest ecosystems.

Any visit to the high-altitude forests of the northeast of United States or the southern Appalachian Mountains reveals a scene of devastation. A sense of the hand of death over the trees is difficult to escape. What killed these trees? The study done by NAPAP clearly indicates that acid precipitation was involved in the demise of these forests. The major effect of acid precipitation seems to be on the various nutrients that the trees need to obtain from the soil for good growth. Many necessary nutrients are washed out of the soil by the acidic waters, while other materials that may be toxic to the trees are released from the soil and are taken up by the trees. These imbalances weaken the trees so much that they can easily be killed by other factors (insects, cold, viruses, bacteria, and fungi).

A significant impact of acid precipitation on lower altitude forests has not been observed. Some disturbance in nutrients has been noted in some cases, but no clear deterioration in the health of the trees has been tied to these changes. So why are the more extreme effects observed at high elevation? The difference is related to what is known as **acid fog**. Because the highest peaks of a mountain often shoot up above the cloud level, they are often bathed in fog. For reasons that are too complex to go into here, fog tends to be much more acidic than rain under similar circumstances. The pH values will often go down to the vicinity of 2. Because many of the high-altitude trees are bathed in this acidic fog for longer periods of time than trees of lower altitude, one would expect the effects to be more severe.

Effects on Human Health

The effects of acid deposition on human health are almost totally related to the gases and particulate matter that are related to the production of the acids. Some of these gases are sulfur dioxide (SO_2), sulfur trioxide (SO_3), and nitrogen dioxide (NO_2). The health effects of these materials are discussed in Chapter 6. The acid precipitation itself has very little direct effect on human health.

AIR POLLUTION AND THE LAW

Attempts to control air pollution by the use of law go back at least as far as the 13th century, but the advent of the Industrial Revolution in the 18th century made air pollution a serious problem. Many attempts were made to control smoke by force of law in the 19th and early 20th centuries, but all to little avail. It was difficult to even agree on a clear definition of smoke. In the London Public Health Act of 1891, it was stated that "any chimney sending forth black smoke might be deemed a nuisance." One company avoided penalty under the law by successfully arguing that their smoke was actually dark brown. Only after the Great Smog of 1952 were really effective air pollution control laws passed in England.

In the United States even less attention was paid to the regulation of air pollution. The United States was a large country without the high population density of England, and except for a few high-population locations there was no pressure to regulate air pollution. Only after the beginning of the environmental movement of the late 1960s and early 1970s did the federal government become seriously interested in the control of air pollution. The first American attempt was the **Clean Air Act** of 1970, amended in 1977 and heavily rewritten and expanded in 1990. This law has always been controversial, with industry arguing that the law is too restrictive and too expensive to comply with and environmentalists charging that it clearly does not go far enough.

The Clean Air Act generally separates air pollutants into two categories: **criteria air pollutants** and **hazardous air pollutants** (**HAPs**, also known as **air toxics**). Criteria pollutants would be those pollutants that are common and are found in many areas of the United States. For these pollutants, the **Environmental Protection Agency** (**EPA**) is required to produce a **criteria document** on each pollutant before setting out to control its emissions. Criteria documents contain scientific data on the health effects, environmental effects, and property effects of the pollutants in question. So far, criteria documents have been issued for sulfur dioxide (SO_2), particulate matter, carbon monoxide (CO), ozone (O_3), nitrogen oxides (NO and NO_2), and lead (Pb).* From these documents the EPA drafts **National Ambient Air Quality Standards** (**NAAQS**), which set the permissible limits for each of the pollutants in air. However, when the law was passed some areas of the country were already well below the NAAQS. Some people were concerned that the air quality might be allowed to "deteriorate" up to the limits of the NAAQS. The Sierra Club filed suit in federal court to require the EPA to protect the current air quality in regions that were then below the NAAQS. The result was that the EPA was forced to adopt the **Prevention of Significant Deterioration (PSD) policy**. This policy in its simplest form says that the EPA cannot allow areas with air cleaner than national standards to deteriorate. As a result, the country was divided into attainment areas and nonattainment areas. For each of the nonattainment areas, schedules were set up to bring them into compliance. The fact that after more than 20 years some nonattainment areas still exist suggests that the original Clean Air Act has not been a complete success, but the air is certainly better than it would have been without the Act.

The NAAQS include both a *primary* and a *secondary standard* for each criteria pollutant. The two types of standards differ in terms of what they are designed to protect. **Primary standards** are pollution limits based on human health effects, whereas **secondary standards** are pollution limits based on environmental and property effects such as damage to structures, plants, and animals or reductions in visibility. The NAAQS are given in Table 9.3.

Hazardous air pollutants are not generally found in the atmosphere over most urban areas, but there may be some significant risk of large-scale releases in some specific locations. In some cases, such a release could be a one-time large-scale release, such as the 1984 methyl isocyanate release in Bhopal, India, that killed approximately 4000 people. More commonly, these releases are of smaller quantities of air pollutants over a longer period of time. In the 1990 amendments to the Clean Air Act, Congress listed 189 specific chemicals to be controlled under the Act. The

* The USEPA originally set standards for hydrocarbons but revoked them in 1983. Also known as *volatile organic compounds* (VOC), they are still tightly controlled in many states.

Table 9.3 National Ambient Air Quality Standards

Pollutant	Time Frame	Primary Standard Value	Secondary Standard Value
Carbon monoxide	8-hour average	9 ppm (10 mg/m³)	—
Carbon monoxide	1-hour average	35 ppm (40 mg/m³)	—
Nitrogen dioxide	Annual arithmetic mean	0.053 ppm (100 μg/m³)	0.053 ppm (100 μg/m³)
Nitrogen dioxide	1-hour average	0.100 ppm (189 μg/m³)	—
Ozone	8-hour average	0.075 ppm (147 μg/m³)	0.075 ppm (147 μg/m³)
Lead	Rolling 3-month average	0.15 μg/m³	0.15 μg/m³
Particulate <10 μm	24-hour average	150 μg/m³	150 μg/m³
Particulate <2.5 μm	Annual arithmetic mean	15 μg/m³	15 μg/m³
Particulate <2.5 μm	24-hour average	35 μg/m³	35 μg/m³
Sulfur dioxide	1-hour average	0.075 ppm (196 μg/m³)	—
Sulfur dioxide	3-hour average	—	0.50 ppm (1300 μg/m³)

Act also allows the EPA to add other chemicals to this list as necessary. The control of HAPs, given the large number of chemicals and variability in the paths of release, is much more complex than the control of criteria pollutants.

An important feature of the 1990 version of the Act was the incorporation of **permits** and **offsets**. Under the Act, all major emitters of HAPs into the atmosphere are required to obtain a permit to release the pollutant or pollutants. The EPA sets in the permit each year the amount of the pollutant or pollutants that may be emitted legally. Generally, the permitted quantities are decreased each year until the emitter is in compliance.

The offset procedure was set up to provide relief to an individual plant that might temporarily need to increase the emission of a pollutant while allowing the EPA to move toward the goal of achieving an overall decrease in emission of that pollutant. If a permit for emission of a pollutant at one facility is to be increased, then the company will need to find either within their own company or elsewhere another facility that will reduce the emission of that pollutant by more than is required in their permit. The added reduction must equal or "offset" the amount that the other facility will go over the permitted emissions. This process works because the net amount of the pollutant in the area is decreased. Under this program, offsets cannot be bought and sold between companies. Any offset transactions must be without monetary reward.

In the 1990 amendments, the Act was expanded to include acid rain and chemicals that destroy stratospheric ozone. Chemicals that destroy stratospheric ozone are discussed in Chapter 11. The section on acid rain focuses on limiting the emission of the gases SO_2 and NO_x, which lead to acidic precipitation. The legislation was designed mainly to decrease these emissions from stationary sources such as electric power plants, smelters, and other coal-fired boilers. Although the legislation provides for specific reductions in SO_2 and NO_x relative to 1980 emissions of these gases, there are significant differences in how the program operates in each case.

The SO_2 program was the most innovative approach at the time and the most clear cut of the two programs. The program basically called for a 10 million-ton reduction in SO_2 emissions compared to 1980 levels by 2010. Each year the allowable total emissions would be decreased until 2010, when the total annual SO_2 output was to be capped at 8.95 million tons. This program used a new variation on the offset concept. On an annual basis, each facility was issued a certain number of SO_2 **emission allowances** to be used that year or in future years. Each allowance was worth 1 ton of SO_2 emissions, and these allowances were transferable to other facilities. As a result, unused allowances could be sold on the open market. Hence, if a company did very well in emissions control, the company could either bank the unused allowances for future use or sell them to another company that was not doing well at controlling emissions. Because the total number of allowances issued each year continued to decrease through 2010, total emissions of SO_2 had to decrease when averaged over time. This approach was known as a **cap and trade program**.

In phase I of the SO_2 emissions control program, only the large coal-fired electric power plants were included. In 2000, all large-scale stationary-source sulfur dioxide emitters were brought under the regulations. As seen from Figure 9.5, phase I of the SO_2 emissions program went very well, with a nearly 25% decrease in total SO_2 emission from 1990 to 1995. After 1995, the total SO_2 output increased for a while before

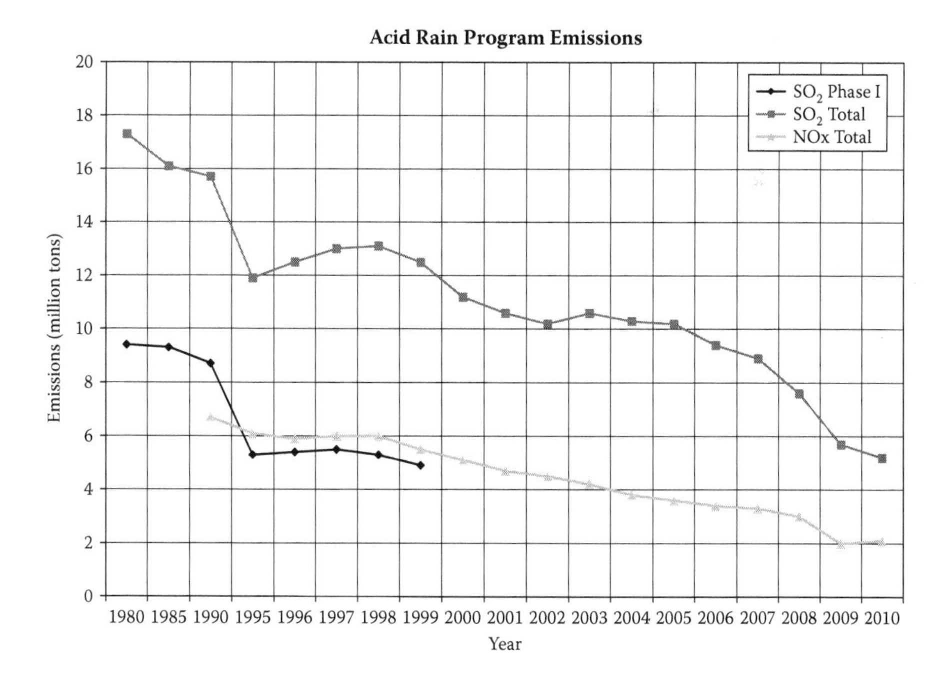

Figure 9.5 Total U.S. phase I emissions of SO_2 and total U.S. emissions of SO_2 and NO_x (compiled from USEPA data).

decreasing again, which was probably related to the fact that many sources were not controlled until 2000. Recently, the output has fallen steadily to the point where in 2010 the total was about 5.2 million tons, well under the goal of 8.95 million tons.

The approach to control NO_x emissions was much more traditional and considerably more modest. The goal was a 2 million-ton reduction in emissions by 2000, and the process got off to a good start with emissions declining in 1996 well ahead of the target. The emissions have continued to decline in a smooth manner as seen from Figure 9.5. The NO_x emissions under the Acid Rain Program, however, have no national cap, which leads to concern that the total emission levels may increase despite this legislation. Without a cap there is really no way to factor in new facilities that may emit NO_x or to consider the large amount of NO_x emitted from mobile sources such as cars, trucks, buses, and planes. The large amount of NO_x emissions from car, trucks, buses, etc. is a key reason why there is no national cap. These sources add to the total emissions but are very difficult to control because there are such a large number of small emitters. Any cap would be arbitrary under these circumstances and difficult to guarantee.

Because, as it turns out, nitrogen oxides not only cause acid rain but are also important precursors for the production of ground-level ozone, there has been considerable interest in controlling NO_x emissions. Much of the standard approach to controlling NO_x emissions was done on a city-by-city basis to meet NAAQS ozone standards; however, nitrogen oxides produced in one area can easily find their way into another area or another state. The EPA became concerned by this situation and in 2003 set up a cap and trade program for a limited number of eastern states designed specifically to decrease the amount of NO_x produced that was likely to find its way into another state. With fewer nitrogen oxides less ozone should be produced. This program has been operated under three different EPA programs, the most recent of which is the Cross-State Air Pollution Rule. These programs are not nationwide and operate with a NO_x cap for the states involved. These EPA programs not only try to control total nitrogen oxide emissions but also look specifically at NO_x production during the ozone season (May to September). All of these programs together should help control acid rain and ground-level ozone pollution.

POLLUTION REDUCTION

The business of pollution reduction and control has many different faces ranging all the way from elegant, highly technical solutions to simple solutions such as "stop doing that." It is far outside the intent and scope of this book to go into many of the technical solutions in detail. Instead, I will try to point out where technical devices exist to correct a problem and where they do not.

Probably the oldest form of pollution control would be smoke abatement. Smoke was and is the most obvious of air pollutants. As mentioned earlier, a traditional solution was simply to build a higher smokestack, which directed the polluted air somewhere else. This is but one of several general strategies for the control of air pollution emissions. Generally, pollution control can be grouped into three categories as follows:

1. Change the energy source (i.e., the fuel).
2. Carry out the process differently so as not to produce the pollutant.
3. Remove the pollutant before it enters the air.

In terms of smoke production, coal is one of the dirtiest fuels available. There are many cleaner energy sources, including oil and natural gas. Coal, however, is often the least expensive and at times the most readily available. It is frequently true that price and availability become the driving factor. This phenomenon was discussed in detail in Chapter 7. Switching to anthracite coal (see Chapter 7) tends to reduce smoke, as anthracite burns with less smoke than the more common, softer bituminous coal. Unfortunately, anthracite coal is much less available and therefore more expensive.

A number of devices exist for removing smoke from stack gases. Figure 9.6 shows the remarkable ability of these devices to remove soot and ash from stack gases. The issue of sulfur dioxide is somewhat different from the problem with smoke. The sulfur that gives rise to this pollutant is an impurity in the coal. One can use coal with

Figure 9.6 Stack emissions (A) before and (B) after installation of pollution control devices. (From Enger, E.D. et al., *Environmental Science: The Study of Interrelationships*, 3rd ed., William C. Brown, Dubuque, IA, 1989. With permission of The McGraw-Hill Companies.)

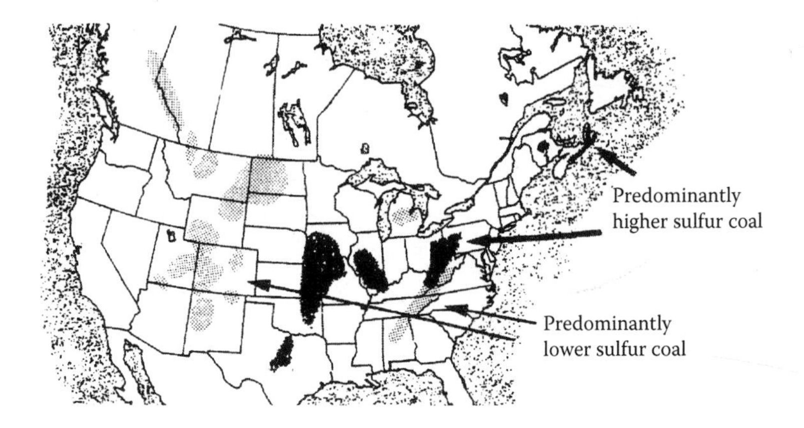

Figure 9.7 High- and low-sulfur coal-producing regions of North America. (From Bunce, N., *Environmental Chemistry*, 2nd ed., self-published, Guelph, Ontario, Canada, 1994.)

lower sulfur content, but, depending on location, this may or may not be practical. If the plant is at a long distance from low-sulfur coal, transportation costs will run up the price. The map in Figure 9.7 shows the origin of many high- and low-sulfur coals. Generally, coal from the western United States is for the most part low in sulfur content and is often desired despite the larger transportation costs in some places. Alternatively, a plant can try to remove the sulfur from the coal. Various physical and chemical techniques exist for removing sulfur from coal; however, these processes cost money and will run up the cost of the coal.

Because sulfur dioxide is a very reactive chemical, removing it from stack gases is relatively easy. Such technology has been adopted at many sites around the world. Major progress has been made in limiting the nitrogen oxides from automobile exhausts by the use of a catalytic converter. Catalytic converters were originally designed to remove unburned gasoline from the exhaust gases. They have since been redesigned to remove nitrogen oxides, as well. Catalytic converters are now required in all cars sold in the United States. To understand how they work, consider a hypothetical mixture of carbon monoxide (CO), nitric oxide (NO), and the hexanes (C_6H_{14}) from gasoline. Catalytic converters contain catalysts that allow reactions to occur at lower temperatures and with greater speed than they would in the absence of the catalysts. These catalysts are very expensive metals, such as palladium, platinum, and ruthenium; hence, they have become a fairly expensive addition to our cars.

The converters are built in two sections. In the first section of the converter, the oxides of nitrogen react with the pollutant carbon monoxide:

$$NO_2 + CO \rightarrow NO + CO_2$$

$$2NO + 2CO \rightarrow N_2 + 2CO_2$$

The products are the relatively harmless gases nitrogen and carbon dioxide. In the second stage of the converter, a reaction similar to regular combustion occurs, except that it occurs at a much lower temperature.

$$2C_6H_{14} + 19O_2 \rightarrow 12CO_2 + 14H_2O$$

As a result, the emission of pollutants from the exhaust pipe is greatly reduced.

These converters have created one additional benefit. The catalysts used in the devices are rendered useless by contamination with lead; hence, when catalytic converters were introduced lead had to be eliminated from gasoline. Lead, in the form of tetraethyl lead, had been added to gasoline for years to improve its performance. As discussed in Chapter 6, lead is quite toxic to humans. When the use of leaded gasoline was at its maximum in the early 1970s, over 200,000 tons of lead were released into the atmosphere each year in the United States. Since that time, the atmospheric deposition of lead onto lands and waters has virtually ceased in the United States.

DISCUSSION QUESTIONS

1. Compare and contrast air pollution and water pollution in terms of their general characteristic. How might an ordinary individual cope with each?
2. Why did the first serious air pollution episodes occur in England? Why did air pollution problems in Germany and the United States only occur later? Air pollution is a serious problem in China. Why has this air pollution occurred mainly in the last 50 years?
3. Chemical reactions occur in the air continuously. How do these reactions differ from the reactions that produce secondary air pollutants?
4. Describe what a typical day with heavy industrial smog might be like. What are some of the adverse effects of industrial smog that might not be so obvious?
5. Which of the following cities is likely to be afflicted by photochemical smog in the summertime under normal circumstances? Which would normally be afflicted year around? Indicate why in each case.
   a. Chicago, Illinois
   b. Denver, Colorado
   c. Edinburgh, Scotland
   d. Los Angeles, California
   e. Mexico City, Mexico
   f. Seattle, Washington
6. Why is sunlight so important in the production of photochemical smog?
7. Considering its source, why is photochemical smog so much more difficult to control than industrial smog?
8. Describe what a typical day with heavy photochemical smog might be like. What are some of the adverse effects of photochemical smog that might not be so obvious?
9. Because the term *smog* is a contraction for the words smoke and fog, explain why it is applied to the photochemical air pollution of California which involves neither smoke nor fog.

10. Explain why Salisbury, North Carolina, which is roughly 40 miles northeast of the major metropolitan city of Charlotte, often has ozone levels as high as Charlotte itself.
11. Explain why unpolluted rainwater has a slightly acidic pH of 5.6.
12. How do we get from sulfur in coal to acid rain? Indicate all of the various chemicals involved in the production of acid rain.
13. Give some examples of the law of unintended consequences. Indicate why this law is so important in environmental studies.
14. Assume that you are a sculptor and plan to carve a statue out of rock. What type of rock would you use if you wish to avoid any impact over time from acid rain?
15. What is the difference between criteria pollutants and HAPs under the U.S. Clean Air Act?
16. Let's assume that the sulfur dioxide level in Sioux Falls, South Dakota, is 7.5 ppb and has been historically for a number of years. If I wanted to build a plant in Sioux Falls that would likely increase the level to 30 ppb, would that be legal given that the NAAQS for sulfur dioxide is 75 ppb?
17. Under cap and trade for sulfur dioxide emissions, the total allowances for sulfur dioxide decrease each year. Given that situation, how could sulfur dioxide emissions increase, as they did from 2002 to 2003?
18. In the acid rain program, why is emission for sulfur dioxide controlled in a different manner than the nitrogen oxides (NO_x)?
19. The first catalytic converters were designed to convert hydrocarbons and carbon monoxide into carbon dioxide and water. Why was this type of converter found to be inadequate and replaced by the two-stage type?

BIBLIOGRAPHY

Altshuller, A.P. and Bufalini, J.J., Photochemical aspects of air pollution: a review, *Environmental Science & Technology*, 5(1), 39–64, 1971.

Brimblecombe, P., *The Big Smoke: A History of Air Pollution in London Since Medieval Times*, Methuen, London, 1987.

Bufalini, M., Oxidation of sulfur dioxide in polluted atmospheres: a review, *Environmental Science & Technology*, 5(8), 685–700, 1971.

Bunce, N., *Environmental Chemistry*, 2nd ed., self-published, Guelph, Ontario, Canada, 1994, p. 161.

Chang, R., *Chemistry*, 4th ed., McGraw-Hill, New York, 1991, p. 661.

Enger, E.D., Lormelink, J.R., Smith, B.F., and Smith, R.J., *Environmental Science: The Study of Interrelationships*, 3rd ed., Wm. C. Brown, Dubuque, IA, 1989, p. 400.

Manahan, S.E., *Environmental Chemistry*, Lewis Publishers, Chelsea, MI, 1991, p. 324.

National Atmospheric Deposition Program, *NTN Maps*, 2012, http://nadp.sws.uiuc.edu/ntn/maps.aspx.

USEPA, *Acid Rain and Related Programs: 2007 Progress Report*, EPA-430-R-08-010, U.S. Environmental Protection Agency, Washington, DC, 2009.

USEPA, *NO_x Budget Trading Program: Basic Information*, U.S. Environmental Protection Agency, Washington, DC, 2009 (http://www.epa.gov/airmarkets/progsregs/nox/sipbasic.html).

USEPA, *Clean Air Interstate Rule, Acid Rain Program and Former NO_x Budget Trading Program: 2010 Progress Report—Emission, Compliance, and Market Analyses*, U.S. Environmental Protection Agency, Washington, DC, 2011 (http://www.epa.gov/airmarkt/progress/ARPCAIR_downloads/ARPCAIR10_analyses.pdf).

USEPA, *National Ambient Air Quality Standards (NAAQS)*, U.S. Environmental Protection Agency, Washington DC, 2011 (http://epa.gov/air/criteria.html).

USEPA, *Clean Air Interstate Rule, Acid Rain Program and Former NO_x Budget Trading Program: 2010 Progress Report—Environmental and Health Results*, U.S. Environmental Protection Agency, Washington, DC, 2012 (http://www.epa.gov/airmarkt/progress/ARPCAIR_downloads/ARPCAIR10_environmental_health.pdf).

USEPA, *Clean Air Interstate Rule: Basic Information*, U.S. Environmental Protection Agency, Washington, DC, 2012 (http://www.epa.gov/cair/basic.html).

USEPA, *Cross-State Air Pollution Rule: Basic Information*, U.S. Environmental Protection Agency, Washington, DC, 2012 (http://www.epa.gov/crossstaterule/basic.html).

USEPA, *The Plain English Guide to the Clean Air Act*, U.S. Environmental Protection Agency, Washington, DC, 2012 (http://www.epa.gov/air/peg/).

CHAPTER 10

Air Inside

There has been much written and said about the quality of air; however, such discussion is usually about the local atmosphere. In other words, it is a discussion on outdoor air. The fact is that modern humans spend only a minority of their time outdoors. Most of us spend much of our lives inside some type of a building. This situation, of course, varies from individual to individual, but even those individuals who work outdoors come home in the evening and spend 10 to 12 hours a day in their homes. Because we spend so much of our time indoors, we need to examine the indoor air we breathe.

Breathing indoor air can have advantages and disadvantages. One of the great advantages is that we have more control over it. In our own homes, we even have considerable personal control over it. We can open or close the windows, turn down or turn up the heat, turn up or turn down the air conditioner, or install any of a variety of air purifying systems. Another advantage is that when the outdoor air is particularly bad, the air inside our homes or offices can be a refuge. Often during a pollution alert in large cities, the elderly and infirm are advised to stay indoors.

The biggest problem with indoor air is that there is too little of it. Compared to the atmosphere, even office buildings have very little air in them. As a result, any pollutants that are produced indoors will have little air in which to be dispersed. Even a low-level emission of a contaminant inside a structure can lead to significant air quality problems, and indoor air can end up dirtier than outdoor air.

SOME BACKGROUND ON INDOOR AIR QUALITY

Indoor air pollution is probably not a new problem. The caveman mentioned in Chapter 9 may have created indoor air pollution by building a fire in his cave; however, through most of history, indoor air quality has been only a minor issue. There are some early references to the air quality of mines, but for the most part indoor air quality is a very recent concern.

There are several reasons why indoor air quality has become a hot topic in the past 40 years. First, we started looking for it. Outdoor air began to improve, and we wondered what the indoor air might have in it. Second, we developed better methods for looking for contaminants; therefore, we looked, and when we looked we found them. It is quite possible that many of the pollutants had been there all along, but we simply did not know about them.

Another factor was the series of energy crises in the 1970s. With the shortage of oil everyone became energy conscious, and one way to cut down on energy loss from heated homes and businesses was to make them as airtight as possible, which had the effect of sealing up the air inside the structure. Any pollutants produced intentionally or unintentionally were trapped and would build up in the air. This is an example of a situation where the solution to one problem aggravates another. Both indoor air quality and energy conservation are important environmental concerns, yet they run at cross purposes here.

There are other factors that conspired to make this indoor air quality problem worse. One has to do with post-World War II changes in Western society. As soldiers returned home after the war, there was a great effort to reabsorb them back into civilian life. The postwar economy began to boom, and there was great optimism. Huge housing developments were started with the hope that most American families could have their own home. However, as this great building effort was speeding up, a shortage of natural building materials occurred. The chemical industry came to the rescue with synthetic materials. Most of these materials worked well, but they often contained volatile substances that slowly leaked into the air.

Other changes have been taking place in the period between World War II and today. We have been gradually becoming a more urban and technological society, which means that we are spending more of our time indoors. As a result, it is indoor air that we breathe most of the time. In the past 30 years there have been drastic changes in what we do indoors, in both the modern office and the home. In the early 1960s, most offices had a typewriter, a telephone, a dictating machine, an intercom, some paper, carbon paper, pencils, and erasers. None of these materials would contaminate the air. There might have been a copying machine or a duplicating machine somewhere, but only in a very large office complex. Today, we have photocopy machines, computers, and laser printers in almost every office and in many homes. Many of these modern technological devices that make our lives easier also send fumes into the air that have an adverse effect on the air quality.

CLASSIFICATION OF INDOOR AIR CONTAMINANTS

Indoor air contaminants will generally fit into one of the following categories:

1. Respiratory gases
2. Combustion products
3. Volatiles
4. Non-combustion-related particulates
5. Odors
6. Radioactive substances
7. Biological agents

This classification system sorts various air contaminants into groups related to their nature and sources. Although the last group, biological agents, is a very important class of contaminants, it will not be discussed in this chapter, as they are outside the scope of this book. We will, however, take a look at each of the other groups.

Respiratory Gases (CO_2, H_2O)

One thing that everyone does when they are in a building is breathe. We all inhale air and transfer oxygen into our bloodstream; however, when we exhale we release carbon dioxide and water vapor into the air. Such activity is normal and not usually viewed with concern, particularly with only a few people in the house or a small office staff. But, with large groups of people in a poorly ventilated enclosed space, moisture and carbon dioxide levels can rise to fairly high levels. The moisture would do no more than raise the humidity to an unpleasant level, but the carbon dioxide can have potentially serious consequences. Carbon dioxide is not technically a toxic substance, but at very high levels it can act as an asphyxiant by excluding oxygen. At carbon dioxide levels above 800 ppm some symptoms may be present such as fatigue and the inability to concentrate. These symptoms occur because high levels of carbon dioxide interfere with the ability of the lungs to rid themselves of the carbon dioxide. In some very tightly sealed office buildings, levels as high as 2000 ppm have been measured.

Combustion Products

Kerosene Space Heaters (CO, NO, NO_2, CO_2, H_2O)

One of the most common ways to warm a home that has inadequate central heating is with a kerosene space heater. They are not very expensive (usually costing about $150 for a new unit) and at the same time are inexpensive to operate. A few dollars worth of kerosene can heat an average size room for quite some time; however, some serious issues are associated with their use. The potential to start house fires can be a problem, but if used properly newer models do not pose a serious fire hazard.

One key problem is that, unlike a central heating unit, a kerosene heater does not send the combustion products up a chimney to the outside. The combustion products remain within one's house. In a normal family living space, the main products, carbon dioxide and water vapor, are not an issue. Unfortunately, as seen from earlier discussions, these are not the only products. It is very difficult to burn kerosene and not produce some carbon monoxide from incomplete combustion. Because carbon monoxide is a colorless, odorless, and extremely toxic gas, it could easily be a silent killer. Traditionally, this gas has been a major concern for these heaters. When using a kerosene space heater, one should install a carbon monoxide detector to alert the occupants of the house to any buildup of the gas. In the early 1970s, a catalytic device was added to all new heaters to allow the conversion of any carbon monoxide to carbon dioxide, although installing a detector would still be a good idea. One problem remains, though: the production of oxides of nitrogen (NO and NO_2) in the hot air of the kerosene flame. The hotter the flame, the more they are produced. If the heater is operated for a long time in a room with inadequate ventilation, the levels of the two nitrogen oxides can build up to unacceptable levels.

Gas Heaters and Ranges (CO_2, H_2O, NO, NO_2)

Natural gas has always been clean burning compared to the other common fuels, as it produces primarily carbon dioxide and water when burned. Hence, one would expect this to be a perfect fuel for use in the house, and it is as perfect as it gets. But, it still suffers from the fact that it is a hot flame—in fact, a very hot flame. As a result, it can produce nitrogen oxides, perhaps in quantities greater than those produced in kerosene heaters. Acceptable standards for nitrogen oxides are often exceeded near gas ranges or gas heaters.

Furnaces

Generally, furnaces provide central heating by burning coal, oil, or natural gas. Technically, they do not figure into indoor air pollution if they are functioning properly. The combustion products that usually would include some carbon monoxide are sent up the chimney or flue to the outside. These devices become contributors to indoor air pollution only when they malfunction due to a leak somewhere in the combustion gas venting system. If the gases are leaked into the home, then carbon monoxide can rise to unacceptable levels. In homes with central heating systems, no matter the type of fuel burned, it would be wise to install carbon monoxide detectors to be sure that carbon monoxide levels never reach dangerous levels. It is also wise to have the furnace inspected annually by a professional.

Environmental Tobacco Smoke (CO, CO_2, NO, NO_2, Other Vapors, Many Particulates)

Cigarette smoke contains about 4500 compounds which makes it one of the most complex of indoor air pollutants. The smoke contains many things such as carbon monoxide, nicotine, and benzo(α)pyrene. These compounds and many others are known to have negative health effects. Carbon monoxide and nicotine are deadly poisons, and benzo(α)pyrene is a known carcinogen. In fact, 60 of the compounds found in tobacco smoke are known or suspected carcinogens.

The health risks of direct tobacco smoking are well known and hence have not been discussed in this chapter. The significant environmental issue seems to be that the smoker turns out not to be the only one to inhale the smoke. The smoke mixes with the air and is inhaled by others in the vicinity of the smoker. Such inhalation of smoke by others is referred to as *passive smoking*. This type of smoke is known as **environmental tobacco smoke (ETS)**, which is really a combination of two types of smoke: **sidestream smoke** and **mainstream smoke**. Mainstream smoke is the one that the smoker has pulled through the cigarette and exhaled, whereas sidestream smoke comes directly from the end of the cigarette between drags. Of the two types, the sidestream smoke is of the most concern. First, sidestream smoke is not filtered, but comes directly from the burning cigarette into the air, whereas mainstream smoke goes through the cigarette, the filter, and probably the smoker's lungs. Second, most of the cigarette smoke produced is sidestream smoke. On average, cigarettes are actively smoked for eight to ten 3-second periods during the 12 minutes that they are lit.

In 1992, the U.S. Environmental Protection Agency (EPA) released a report on the health effects of passive smoking and concluded that at that time approximately 3000 cases of lung cancer per year in the United States were the result of ETS. Other estimates considering all diseases related to ETS suggest that smoke has caused 46,000 deaths per year from various cancers and heart disease. In children, the effects are even more severe. It is estimated that 150,000 to 300,000 cases annually of bronchitis and pneumonia in children up to 18 months of age can be attributed to ETS. ETS also plays a clear role in aggravating childhood asthma, with 200,000 to 1,000,000 children showing an increase in asthma symptoms.

In 2006, the Surgeon General of the United States issued a report entitled *The Health Consequences of Involuntary Exposure to Tobacco Smoke*. The report came to six major conclusions:

1. Secondhand smoke causes premature death and disease in children and in adults who do not smoke.
2. Children exposed to secondhand smoke are at an increased risk of sudden infant death syndrome (SIDS), acute respiratory infections, ear problems, and more severe asthma. Smoking by parents causes respiratory symptoms and slows lung growth in their children.
3. Exposure of adults to secondhand smoke has immediate adverse effects on the cardiovascular system and causes coronary heart disease and lung cancer.
4. The scientific evidence indicates that there is no risk-free level of exposure to secondhand smoke.
5. Despite substantial progress in tobacco control, millions of Americans, both children and adults, are still exposed to secondhand smoke in their homes and workplaces.
6. Eliminating smoking in indoor spaces completely protects nonsmokers from exposure to secondhand smoke. Separating smokers from nonsmokers, cleaning the air, and ventilating buildings cannot eliminate exposures of nonsmokers to secondhand smoke.

The simple fact that smoking remains a common human activity implies that avoiding ETS means avoiding smokers. Unfortunately, this may be easier said than done. In your home, you can make the rules, but in public places it is more difficult. It becomes an issue of the public's right to pollution-free air vs. the smoker's right to enjoy a cigarette. Many ideas have been tried, such as separate smoking and nonsmoking areas in restaurants. These separate areas do not work well, as such an approach is similar to having a non-chlorinated section in a swimming pool. In the United States, indoor smoking in public building is becoming less and less common.

Volatiles

Building Construction Materials and Interior Furnishings

Most people would be surprised at the number of chemicals that can be emitted from common building materials. A simple item such as plywood can emit as many as 11 different chemicals. Where do these chemicals come from, and why are they there? These chemicals as a group are known as **volatile organic compounds** (**VOCs**). These compounds start out as solvents and raw materials in the manufacturing process. In the case of plywood, which is made up of very thin layers of wood glued together, the VOCs come from the glue that holds the plywood together. As the glue bonds, the solvent evaporates, but not all of it. Some of the solvent is slowly emitted from the wood for a long period of time. This slow solvent loss is often referred to as **outgassing**. Some similar products that also outgas are chipboard and particleboard.

There are more obvious examples of products containing solvents. These are items that are intended to lose solvent into the environment, such as adhesives, caulking compounds, paints, stains, and varnishes. The VOCs evaporating from these materials contain a very long list of chemicals. At the time of application one would expect the concentrations of these chemicals to be high; however, emissions from the material can continue for months or even years.

The other sources of VOCs are synthetic materials used in building and furniture making. One common agent used in making many of these synthetic materials is formaldehyde. Formaldehyde-based materials are used for everything from foam insulation to bonding agents (e.g., glue). Items that may emit formaldehyde include caulking compounds, carpeting, ceiling tiles, particleboard, draperies, wall coverings, upholstery, and some types of insulation. The formaldehyde actually arises in a couple of ways. If excess formaldehyde was used in the manufacture, it may be trapped in the polymer and leak out slowly over the years. The formaldehyde may also come from slow deterioration of the polymer if moisture and a little acid are present.

The most severe problems with formaldehyde emissions have been encountered in mobile homes, for a couple of reasons. Mobile homes were traditionally constructed with a high proportion of formaldehyde-emitting materials in them. One particularly bad offender was urea formaldehyde foam insulation (UFFI), which is no longer used in the construction of mobile homes. In 1985, the Department of Housing and Urban Development (HUD) set formaldehyde emission limits for any material used in the construction of prefabricated and mobile homes.

Another issue with mobile homes is their small size compared to a regular house. Any emissions will be trapped in the much smaller space of a mobile home, which leads to higher concentrations inside the mobile home. Fortunately, UFFI has been banned for so long that any remaining insulation of this type has most likely outgassed nearly all of its formaldehyde.

Formaldehyde is probably best known for its use in preserving biological tissue in laboratories and as an embalming fluid. The fluids used for these purposes are solutions of formaldehyde in water called *formalin*. Formaldehyde itself is a gas that at high levels produces headaches, drowsiness, and nausea and may also produce some toxic effects at much lower levels. It is suspected of being a mutagen and carcinogen. In addition to the normal toxic effects of formaldehyde, many people can become severely allergic to even very low concentrations of the gas.

Household Pesticides

We humans use chemicals to kill pests, but anything that kills one species may be somewhat toxic to another, including ourselves. Pesticides can spread through the air, but outdoors the pesticide quantities are small enough and the volume of air large enough that air contamination is seldom a problem. When insecticides are sprayed in the home, however, the chemicals are dispersed in a smaller quantity of air. If the amount of pesticide is large enough, then the small quantity of air in the house will lead to high concentrations of the chemical in the air, which the occupants breathe.

The occasional spraying of small areas with an aerosol can is not likely to cause any serious problem as long as none of the occupants is hypersensitive to the insecticide. A more serious problem is the spraying of basement or crawlspace areas with large amounts of pesticide to combat termites. The insecticide is soaked up by the soil or the concrete, which then slowly releases the insecticide over a long period of time. Some homes that have been commercially treated for termites have had pesticide levels 20 to 40 times those found outdoors. The long-term health effect of these pesticide levels is still unknown.

Appliances, Office Equipment, and Supplies

Appliances and office equipment are hidden sources of indoor air pollution. An office with a lot of electronic equipment is not generally associated with air pollution; yet, a computer with an older CRT display may emit as many as 24 separate volatile compounds. Various duplicating machines contain a cocktail of assorted volatile solvents. Then there are the supplies used in an office. Some of those with volatile solvents in them are obvious—for example, rubber cement or various types of markers; however, something as simple as preprinted paper forms may also emit a large number of chemicals. The preprinted form is printed with various inks that contain an assortment of chemicals. When this form is fed through a photoduplicating machine at 355°F (180°C), many of these compounds will be roasted right out of the paper. An interesting source of volatile chemicals is carbonless copy paper. This type of paper has, trapped within it, various chemicals that can be released by the pressure of a pen or pencil. The chemicals from the two sheets of paper mix and react to develop a color. Many of these chemicals are volatile and are released into the air during the writing process.

Non-Combustion-Related Particulates

Everyone has probably observed dust particles in the air as sunlight streams through a window in the late afternoon. Air is full of a large collection of particles, most of which occur naturally. These particles include pollen, carpet dust, paper dust, and many other related materials. Most of these airborne materials are not a problem unless one is allergic to them. An allergic response to one or several of these airborne materials can make a person's life miserable and may make medical attention necessary. Because many of these particles are natural, their complete removal from the air is impractical in most situations. Even materials such as paper dust, metal dust, and airborne particles from frayed materials are nearly impossible to exclude from indoor air while still allowing a normal lifestyle.

Luckily most of the non-combustion-type airborne materials are not hazardous to our health, except for allergenic responses. One well-known exception to this is **asbestos**. Asbestos fibers are well known for their insulation properties and ability to withstand high temperatures. Because the material was less expensive than other comparable materials, it was used for everything from firefighters' insulated suits to spray-on insulation to oven mitts. In the 1950s and 1960s, it was demonstrated that miners who worked in asbestos mines had higher incidence of certain health problems. The conditions included **asbestosis**, lung cancer, and **mesothelioma**. Asbestosis is a serious but benign chronic degenerative lung condition, whereas mesothelioma is an extremely rare cancer. By the 1970s, the U.S. government had become seriously concerned about asbestos in public buildings. Spray-on asbestos insulation had become very popular in the 1950s and 1960s and had been used in many public buildings, including schools.

Asbestos is a complex set of silicon- and oxygen-containing minerals existing in at least six different forms. The feature that made asbestos appealing for many purposes was that one form of it could be drawn out into fibers. This type is referred to as **chrysotile**. The other forms, collectively known as *amphibole* forms, do not form long fibers.

Although there are health concerns clearly associated with exposure to asbestos, the risks are on the order of four cases in a million. There is some uncertainty concerning the health risks of the various types of asbestos. All types of asbestos have been shown to cause lung cancer and asbestosis, but the amphibole forms are more likely to cause mesothelioma than the more common commercially available chrysotile form. There is clearly an enhanced risk of lung cancer from asbestos exposure for persons who smoke. This is a synergistic effect and is greater than the sum of the risks from smoking and asbestos alone.

Another issue related to asbestos is removal of the material. In the 1970s and 1980s, as the asbestos fear began to peak, many buildings, including schools, had the asbestos insulation ripped out. At that time, this approach seemed the sensible thing to do; however, it was not long before people began to wonder about the dangers faced by asbestos removal workers. Studies were conducted that suggested that undisturbed, intact asbestos insulation did not release asbestos particles into the air.

Officials in both the public and private sectors were then faced with a decision as to whether it was best to leave asbestos alone or to remove it. It was found that asbestos that is intact and showing no signs of deterioration probably should be left alone. This is particularly true if monitoring shows no signs of asbestos in the air. Removal of asbestos is an expensive proposition. The main reason for the expense is protection of the removal workers. Once asbestos insulation is disturbed, large amounts of asbestos dust are sent into the air. Elaborate precautions have to be taken to see that the workers do not inhale any of the asbestos dust.

Odors

Chemical odors can produce the most intense psychological reaction of any of the indoor air contaminants. Unfortunately, the intensity and disagreeable nature of an odor are not very closely related to the toxicity associated with the chemical causing the odor. Some very foul-smelling compounds are relatively harmless and some other pleasant-smelling or odorless compounds can be quite lethal. For example, hydrogen cyanide, used in the often used in gas chambers, has the slight smell of bitter almonds and is very lethal, whereas butyric acid has the smell of rancid butter but is relatively harmless. Because most of the chemical pollutants that cause odors fit into one of the other chemical classes, they are best discussed in these categories.

Radioactive Substances

Radon Gas, the Main Source of Radioactive Material in Air

Although there are many radioactive materials in the environment, this chapter discusses only one, radon gas, which makes up 54% of natural background radiation. This high level alone would make radon important, but the gas has also been found to be especially important with regard to indoor air. To talk about radiation levels, first we should discuss how they can be measured. Radiation intensity is commonly measured in one of two units. We will use a unit derived from the curie (Ci). One curie is 3.7×10^{10} disintegrations per second (counts/sec) or 2.2×10^{12} disintegrations per minute (counts/min). The curie is too large a unit for most purposes; therefore, when taking measurements in household air use of the picocurie (pCi) is most appropriate. One picocurie is 2.2 counts/min or one trillionth of a Curie ($1 \text{ pCi} = 10^{-12} \text{ Ci}$). One disintegration per second is also known as a becquerel (Bq); hence, one curie is equal to 3.7×10^{10} becquerel ($1 \text{ Ci} = 3.7 \times 10^{10} \text{ Bq}$).

Uranium is one of the radioactive elements that were present when the Earth was formed; the disintegration of uranium produces radon gas. Uranium is relatively plentiful for a radioactive material and is found in many soils and rock formations. Uranium produces radon through a series of reactions, which are as follows:

	Half-life
$^{238}_{92}U \rightarrow ^{4}_{2}He + ^{234}_{90}Th$	4.51×10^9 yr
$^{234}_{90}Th \rightarrow ^{4}_{2}He + ^{230}_{88}Ra$	1 sec
$^{230}_{88}Ra \rightarrow ^{0}_{-1}e + ^{230}_{89}Ac$	1 hr
$^{230}_{89}Ac \rightarrow ^{0}_{-1}e + ^{230}_{90}Th$	<1 min
$^{230}_{90}Th \rightarrow ^{4}_{2}He + ^{226}_{88}Ra$	80 yr
$^{226}_{88}Ra \rightarrow ^{4}_{2}He + ^{222}_{86}Rn$	1622 yr

With all of the products that appear in this series one might wonder why radon is more problematic than the others. The answer to this question is related to the chemical and physical properties of these elements. All of the product elements in the series are reasonably reactive and will react with their surroundings to form compounds, if they exist there long enough. The compounds formed will be solids, and if the product elements do not react the elements themselves are solids. Radon, however, is an unreactive gas, one of the noble gases that rarely form compounds. This fact is important because the other decay series products remain where they are produced, whereas radon, being a gas, will leave the site of its production. In addition, radon gas has a half-life (3.82 days), which allows it to exist long enough to move around.

Radon gas can arise from any place where uranium can be found. Some of the sources, and their total radiation, can be found in Table 10.1. It is important to understand the table correctly. These numbers represent all of the emissions from each of these sources, but the intensity of the emission at a given location may be greater. There are not many places where one may find uranium tailings, but they may put out a lot of radiation for the amount of material present. Therefore, the question comes down to how much uranium is present in the soil, water, etc. at a given location.

Table 10.1 Sources of Radon Gas and Their Total Annual Contribution to Radon Radiation

	Radiative Emissions per Year	
Source	**Millions of Curies ($Ci \times 10^6$/yr)**	**Trillions of Becquerels ($Bq \times 10^{12}$/yr)**
Soil	2400	89,000,000
Groundwater	500	18,000,000
Oceans	34	1,300,000
Phosphate residues	3	100,000
Uranium mill tailings	2	70,000

Concern about radon gas dates back to the 1930s when it was known to be a problem for miners. In the 1950s, the Atomic Energy Commission reaffirmed this concern, and by the 1970s there was some concern that radon gas might be leaking into homes built on reclaimed land over uranium mill tailings or old phosphate mines. This understanding remained the generally accepted theory until December 1985, when Stanley Watras, an engineer at the Limerick Nuclear Power Plant in eastern Pennsylvania, set off plant monitors for radiation on his way into work one morning. A check of the Watras' home found radiation levels of 2700 pCi/L in the air in the basement, or about 100 counts per second per liter. Other homes in the area were also found to have high radiation levels, although none was nearly so high as the Watras' home. A puzzling aspect of this problem was the fact that this area was not close to any type of uranium or phosphate mining. If radon gas could be a problem in an area with no apparent high-risk factors, then the problem could, indeed, be widespread.

The area in which these homes are situated is over a geologic formation known as the Reading Prong. It is located in eastern Pennsylvania, northern New Jersey, and a small part of southern New York State. The bedrock in this geologic formation is known to be rich in uranium, which explains why radon gas production is a problem in this area. The rock formations also suggest that there are many other areas where similar conditions probably exist.

Radon has obviously been seeping out of the ground for hundreds of millions of years, so why is it such a problem now? One obvious reason is that for most of human history we did not know that it was present because we had no way to detect it. Even if we had been able to detect radon long ago, it is doubtful that the gas would have been a serious problem. Most radon comes to the surface in open areas and is blown away by winds, and the atmosphere dilutes the radon to a harmless level. Even radon gas emitted into homes was less likely to create a problem because homes did not used to be as airtight as they are today. The gas was more likely to leak out and be blown away.

A phenomenon known as the *stack effect* aggravates radon release into homes. This term refers to the fact that air tends to rise inside a home much as air rises up a smokestack. There are two principal reasons for this. First, during part of the year we warm the air in our homes. This warm air is usually generated in the basement or on the ground floor and rises, as it is lighter than the cool air present in the house. Second, most homes have many devices that remove air from the home, thus resulting in air moving up from the basement or crawlspace to replace the air being removed. Such air removal devices include bathroom vent fans, stove hood vent fans, and clothes dryer vents.

The stack effect causes radon gas to rise up into reasonably airtight homes, because the rising air will pull gases out of the soil under the house. If the house does not have exposed soil (crawlspace), then the gases will be pulled through cracks in the concrete foundation or the concrete floor of a basement. If the gases in the soil contain radon, then radon will be among the gases drawn into the house.

Health Hazards of Radon Gas

Out of about 160,000 deaths due to lung cancer each year in the United States, approximately 20,000 can be attributable to radon exposure, whereas 80 to 90% can be attributed to cigarette smoking. (Some deaths can be attributed to both smoking and radon.) Considering the average radon concentration in the American home, the risk of developing lung cancer due to radon is similar to the risk of dying in a fire or from a serious fall at home. Of course, actual homes have either more or less radon than the average. To get some perspective, let us note that many of the homes in eastern Pennsylvania near Stanley Watras had radon emission levels of about 200 pCi/L (7.4 Bq/L). A radon level of 20 pCi/L (0.74 Bq/L) would pose a risk equivalent to smoking two packs of cigarettes per day. Generally, the EPA suggests that action should be taken whenever the level exceeds 4 pCi/L (0.15 Bq/L). One additional fact is worth noting. There is a synergistic effect between smoking and radon exposure. Smokers seem to be more sensitive to the effects of radon gas than nonsmokers.

Interestingly, the health effects of radon gas are not caused by radon alone. The effect is caused by what are known as radon daughters or radon progeny (in other words, the radioactive decay products from radon). Radon itself is an inert gas, which if inhaled will be exhaled in the next breath. Should the radon decay in the lungs, it produces α particles, which have very little penetrating power; therefore, those produced in the lung airways will probably cause very little damage to lung tissue. The decay products, however, are not inert gases and as a result will settle down in the lung tissue or react with it. Those daughter isotopes that are α emitters will cause great damage at this close range. The series of decay steps gives rise to two deadly α emitters in a relatively short period of time. The steps can be seen in the series of reactions as follows:

	Half-life
${}^{222}_{86}\text{Ra} \rightarrow {}^{4}_{2}\text{He} + {}^{218}_{84}\text{Po}$	3.82 days
${}^{218}_{84}\text{Po} \rightarrow {}^{4}_{2}\text{He} + {}^{214}_{82}\text{Pb}$	3.05 min
${}^{214}_{82}\text{Pb} \rightarrow {}^{0}_{-1}e + {}^{214}_{83}\text{Bi}$	27 min
${}^{214}_{83}\text{Bi} \rightarrow {}^{0}_{-1}e + {}^{214}_{84}\text{Po}$	20 min
${}^{214}_{84}\text{Po} \rightarrow {}^{4}_{2}\text{He} + {}^{210}_{82}\text{Pb}$	0.0002 sec
${}^{210}_{82}\text{Pb} \rightarrow\rightarrow\rightarrow {}^{206}_{82}\text{Pb}$	21 yr

Although the decay of lead-210 to lead-206 leads to the production of one more α particle, this particle generally does not cause a problem, as lead-210 has a long enough half-life to allow it to be excreted from the body before emission of the α particle. Lead-206 is not radioactive and represents the end of the series.

REMEDIES FOR INDOOR AIR CONTAMINATION

Generally, there are two approaches to the reduction of indoor air contamination: (1) stop the production of the offending contaminant, or (2) try to remove it. The first approach is probably preferable if it is practical. For example, in most public buildings smoking is prohibited, thus eliminating the production of cigarette smoke. The spraying of insecticides could be reduced by opting for the use of a fly swatter. We might switch from kerosene heaters to electric heaters and avoid combustion products. (See Chapter 7 for an indication of why it might not be such a good idea from an energy perspective.) For many reasons it may not be practical or even possible to eliminate the production of all indoor air contaminants. Some of the issues that might make elimination impractical include the fact that radon comes out from the ground, offending furniture and building materials may be expensive to replace, and office equipment and supplies are needed in a modern office.

The other option is to remove the pollutants once they are formed. The removal of indoor air pollutants largely depends on air exchange rates. As seen from Figure 10.1, increasing the air exchange rate from relatively low values to a moderately high value has a drastic effect on the levels of indoor gases. Carbon dioxide is used in this example, but the dynamics are roughly correct for any gases except that the outdoor reference levels would be different. The graph seems to indicate that increased ventilation rates up to about one air change per hour have a striking effect on gas levels, but above this rate little more is accomplished.

Historically, humans have held varied opinions about the amount of ventilation that is desirable. Benjamin Franklin recognized that some ventilation produced an improved quality for indoor air. About 1860, Florence Nightingale was a proponent

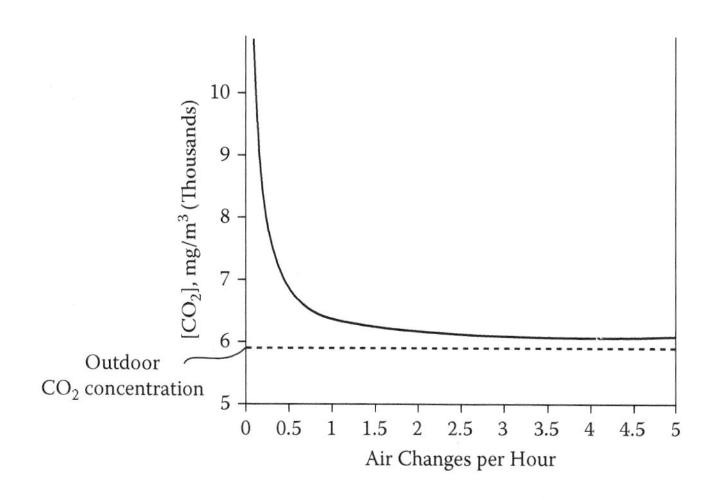

Figure 10.1 Indoor pollutant concentration as a function of air exchange rate at production rates of 10 kg/hr. (From Bunce, N., *Environmental Chemistry*, 2nd ed., self-published, Guelph, Ontario, Canada, 1994.)

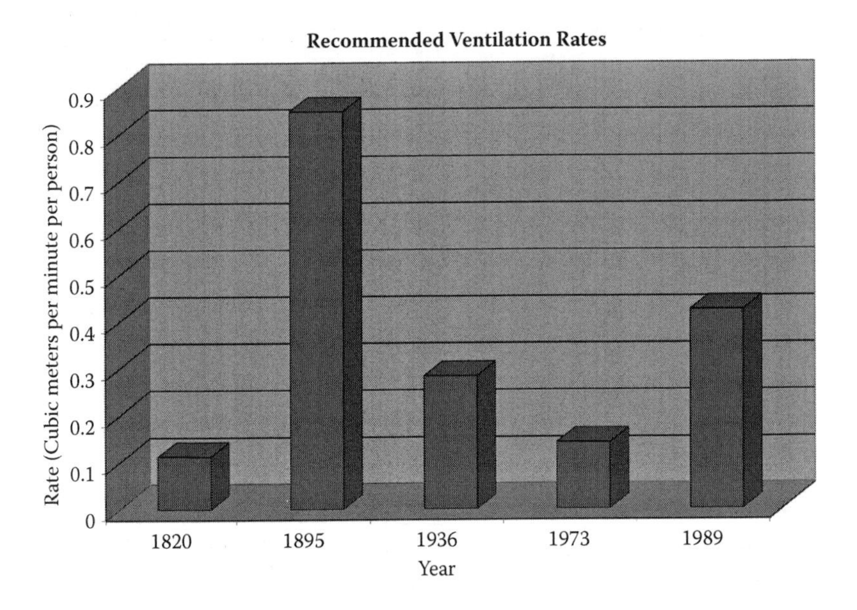

Figure 10.2 Ventilation rate recommendations: 1820 to 1991. (Compiled from data in Brooks, B.O. and Davis, W.F., *Understanding Indoor Air Quality,* CRC Press, Boca Raton, FL, 1992.)

of good ventilation, particularly in hospitals and sanatoria. Recommendations of various kinds can be found dating back to 1820; Some of these can be seen in Figure 10.2. The variations can largely be traced to the relative concerns for energy vs. indoor air quality. In 1895, people became more aware of infectious diseases and believed that fresh air was really good for health. Fuel was relatively inexpensive and not of a great concern. During the energy crisis of the early 1970s, the primary concern was energy conservation to the exclusion of all else; hence, the recommendations were for very low exchange rates. By the late 1980s, it became apparent that both considerations were important and that some compromise was needed. This led to the intermediate recommendation.

Radon is somewhat unique in that it is normally the only indoor contaminant that seeps out of the ground. Furthermore, because radon is colorless, odorless, and tasteless, it must be tested for to be aware of its presence. There are two common ways to test the presence of radon. One is an activated charcoal detector, which on exposure will absorb many contaminants from the air. One of the materials that it will absorb is lead-210, which is a radioactive decay product of radon. The amount of lead-210 is directly related to the amount of radon present in the air. Because lead-210 is radioactive, we can determine how much radon is present by measuring the radiation emitted from the charcoal. Generally, the charcoal is exposed for 3 to 7 days, after which it is sent to a laboratory for analysis. The cost is about $25 to $35. This test is relatively quick, but because radon levels can vary with the time of year the test may have to be done more than once.

The Track Etch Monitor is a piece of plastic on which emitted α particles will leave tracks when they strike the plastic. The plastic strip is left exposed to the air for up to a year, after which it is sent to a lab. The strip is treated with acid to reveal the α particle tracks. This method is slower than the other one but gives a good average reading, which is probably a better reflection of the overall radon exposure risk.

Solving the radon problem can vary from the simple to the elaborate. The general cost is usually about $1000 to $2000, although in the extreme case of Stanley Watras the cost was $32,000. If a house is built over a crawlspace, then a crawlspace fan will usually solve the problem; however, this approach may run up heating costs in the winter by pulling cold air under the house. There are other crawlspace approaches that waste less heat energy. The solution is usually more difficult for houses built on concrete slabs or over basements, where the radon is probably coming in through cracks in the foundation. Sealing the cracks may help but may not completely solve the problem. Often, pipes have to be installed under the concrete to allow air from the soil under the house to be pumped directly to the exterior.

DISCUSSION QUESTIONS

1. List the reasons why indoor air pollution has become more of an issue in the past 40 years.
2. Respiratory gases include carbon dioxide and water vapor. Explain why the buildup of these indoors could be a problem. What would be the signs that one of these gases might be at too high a level?
3. Compare kerosene space heaters, indoor gas heaters, and furnaces. Discuss the gases produced by each and the likely dangers of each. Why is it important to have a furnace checked regularly?
4. Discuss the smoking of tobacco products. Where and under what circumstances should smoking be allowed? Should tobacco smoking ever be allowed in indoor spaces? Does the presence or absence of children affect your opinion?
5. Why is smoking perhaps more dangerous for persons near the smoker than for the smoker? Which is more dangerous—sidestream smoke or mainstream smoke? Why?
6. Why are building construction materials and interior furnishings a concern for indoor air pollution?
7. Explain why radon is not a pollutant in the classical sense and why radon only became a concern during and after the 20th century.
8. Of the various radioactive isotopes produced in the decay of uranium-238, why is radon the only one to cause serious health problems?
9. Explain how the stack effect works.
10. In what way does radon affect health? (*Hint:* What are radon daughters or progeny?)
11. Using the graph in Figure 10.1 and assuming an initial concentration of CO_2 of 11 mg/m^3, what is the percentage decrease in the CO_2 concentration after 1 hour at 0.5 air changes per hour? What would be the percentage decrease at 1 air change per hour? What would be the percentage decrease at 2 air changes per hour? Above what rate does increasing the air change rate cease to matter?
12. Describe the two types of radon monitors and explain how they work.

BIBLIOGRAPHY

ALA, *Lung Cancer Fact Sheet*, American Lung Association, 2013, http://www.lung.org/lung-disease/lung-cancer/resources/facts-figures/lung-cancer-fact-sheet.html.

Anderson, D.J. and Hites, R.A., Chlorinated pesticides in indoor air, *Environmental Science & Technology*, 22(6), 717–720, 1988.

Austin, B.S., Greenfield, S.M., Weir, B.R., Anderson, G.E., and Behar, J.V., Modeling the indoor environment, *Environmental Science & Technology*, 26(5), 851–858, 1992.

Baker, R.R. and Protor, C.J., The origins and properties of environmental tobacco smoke, *Environment International*, 16, 231–245, 1990.

Brody, A.R., Samuels, S.W., Nicholson, W.J., Johnson, E.M., Harington, J.S., Melius, J., Landrigan, P.J., and Crump, K.S., Series of four letters on "asbestos, carcinogenicity, and public policy," *Science*, 248, 795–799, 1990.

Brooks, B.O. and Davis, W.F., *Understanding Indoor Air Quality*, CRC Press, Boca Raton, FL, 1992, p. 39.

Bunce, N., *Environmental Chemistry*, 2nd ed., self-published, Guelph, Ontario, Canada, 1994, p. 116.

Jaakkola, J.J.K., Reinikainen, L.M., Heinonen, O.P., Majanen, A., and Seppanen, O., Indoor air quality requirements for healthy office buildings: recommendations based on an epidemiologic study, *Environment International*, 17, 371–378, 1991.

Janerich, D.T., Thompson, W.D., Varela, L.R., Greenwald, P., Chorost, S., Tucci, C., Zaman, M.B., Melamed, M.R., Kiely, M., and McKneally, M.F., Lung cancer and exposure to tobacco smoke in the household, *New England Journal of Medicine*, 323(10), 632–636, 1990.

Kuller, L.H., Garfinkel, L., Correa, P., Haley, N., Hoffmann, D., Preston-Martin, S., and Sandler, D., Contribution of passive smoking to respiratory cancer, *Environmental Health Perspectives*, 70, 57–69, 1986.

Löfroth, G., Burton, R.M., Forehand, L., Hammond, S.K., Seila, R.L., Zweidinger, R.B., and Lewtas, J., Characterization of environmental tobacco smoke, *Environmental Science & Technology*, 23(5), 610–614, 1989.

Schaefer, V.J., Mohnen, V.A., and Veirs, V.R., Air quality of American homes, *Science*, 175, 173–175, 1972.

Sexton, K., Petreas, M.X., and Liu, K.-S., Formaldehyde exposure inside mobile homes, *Environmental Science & Technology*, 23(8), 985–988, 1989.

Somersall, A.C. and Natural Wellness Group, *Fresh Air for Life: How to Win Your Unseen War against Indoor Air Pollution*, Natural Wellness Group, Toronto, 2006.

USCPSC, *What You Should Know about Space Heaters*, U.S. Consumer Product Safety Commission, Washington, DC, 2001.

USDHHS, *Toxicological Profile for Asbestos*, U.S. Department of Health and Human Services, Public Health Service, Agency for Toxic Substances and Disease Registry, Atlanta, GA, 2001.

USDHHS, *The Health Consequences of Involuntary Exposure to Tobacco Smoke: A Report of the Surgeon General—Executive Summary*, Public Health Service, U.S. Department of Health and Human Services, Rockville, MD, 2006.

USEPA, *A Citizen's Guide to Radon*, EPA 402-K02-006, U.S. Environmental Protection Agency, Washington, DC, 2005.

USEPA, *Consumer's Guide to Radon Reduction*, EPA 402-K-06-094, U.S. Environmental Protection Agency, Washington, DC, 2006.

USEPA, *Care for Your Air: A Guide to Indoor Air Quality*, EPA 402/F-08/008, U.S. Environmental Protection Agency, Washington, DC, 2008.

USEPA, *Health Effects of Exposure to Secondhand Smoke*, U.S. Environmental Protection Agency, Washington, DC, 2011, http://www.epa.gov/smokefree/healtheffects.html.

USEPA, *The Inside Story: A Guide to Indoor Air Quality*, U.S. Environmental Protection Agency, Washington, DC, 2012, http://www.epa.gov/iaq/pubs/insidestory.html.

USEPA, *Why Is Radon the Public Health Risk That It Is?*, U.S. Environmental Protection Agency, Washington, DC, 2012, http://www.epa.gov/radon/aboutus.html.

CHAPTER 11

Global Atmospheric Change

As discussed in Chapter 8, air pollution is generally considered to be a local problem. We can discuss the air quality in Los Angeles, New York, London, or Mexico City, but some gases produce problems on a global rather than on a local scale. Why is it that some gases produce local problems and others have global consequences? Recall that the molecules in our atmosphere do not remain there indefinitely. Residence times vary greatly. Most of the local pollutant gases have residence times of only a few days and as a result are removed fairly quickly before they can get very far. Those gases that are of global importance, however, have residence times that are measured in years. These gases remain in the atmosphere long enough to get mixed in all parts of the atmosphere.

Another interesting feature of most of these globally important gases is that on a local scale they are generally considered to be harmless. Before the latter part of the 20th century, the generally accepted wisdom was that these gases would never become an issue, because many of them had been present since the early evolution of the Earth without causing any apparent problems. Other gases had been carefully developed to minimize negative effects; yet, somehow when they were released into the atmosphere over a long period of time such effects became apparent.

What went wrong? Nothing and everything is one way of putting it. What we will find is that the natural systems of the universe are a great deal more complicated than we generally assume. It is very difficult for humans to predict all of the consequences of their actions. Some people refer to this as the **law of unintended consequences**. It is important to understand this law. The human race did not get together one day and decide to pollute the planet. It appears to have happened as an unexpected side effect of many very well-intended activities.

GASES AS INSULATORS: GREENHOUSE EFFECT

Perhaps you have heard of the **greenhouse effect** and wondered, "What is it, and why should I care?" The greenhouse effect is really no more mysterious than what happens in a car on a sunny day. One may have noticed that after being in the sun the interior of a car becomes very hot even on a somewhat cool day. The sun's energy enters the car as visible light through the car's glass windows. Much of this light is absorbed by the objects and materials inside the car. As these things warm up, they reemit this energy as heat (infrared) radiation. Heat radiation has one key difference from visible light, and that is that this radiation will not pass through glass. Hence, as more and more visible light comes in and is converted into heat energy, the car's interior grows hotter because the heat cannot escape easily. The glass of a car stops the heat radiation by absorbing it, and as a result the glass warms up. Once it warms up, it will reemit the heat energy as radiation, some to the outside and some back into the car.

Certain gases in the atmosphere are capable of behaving in the same way as the glass of the car. The sun's radiation reaches the surface of the Earth mainly as visible light that is absorbed or reflected by various surfaces. Most of that which is reflected will find its way back into outer space. Of the radiation that is absorbed by the surface, most will be converted into heat, as was the case in the car. As the surface warms up, it will reemit the energy as heat radiation. If there were no atmosphere, all of this energy would escape back into space; however, certain atmospheric gases absorb a large portion of this heat radiation in the same way as the glass of the car does. These gases will reemit the heat energy, with some significant portion of the energy being sent back toward the surface. Light can get in, but heat has trouble getting out.

GLOBAL WARMING: CONCEPT

The jump from gases as insulators to the concept of **global warming** is not a long one. It is no more complicated than the idea that if one wants the bed to be warmer then simply put on more blankets (i.e., insulators). If we blanket the Earth with more greenhouse gases, it will get warmer. No one should get the idea that this is all bad. As we know, some greenhouse warming is essential for life. It is estimated that if there were no greenhouse gases, the mean temperature of the Earth would be –3°F (–19°C); yet, the average temperature of the Earth is close to 60°F (33°C), which one must admit is much warmer. The greenhouse effect is definitely real. The concern is about how large the greenhouse effect may become. To examine this question we must look at the gases that produce the effect.

Which gases are involved? There are several of them. Some of them have been with us since the formation of the planet and others are man-made. The natural greenhouse gases include water vapor, carbon dioxide, **methane** (CH_4), **nitrous oxide** (N_2O), and ozone (O_3). The man-made greenhouse gases of importance include **chlorofluorocarbons**, which are known as **CFCs** or by their trade name, Freon.

Carbon Dioxide

Probably the most important of these gases is carbon dioxide. As pointed out in Chapter 5, carbon dioxide is one of the key ingredients in the carbon cycle. As we know, the existence of carbon dioxide in the atmosphere is absolutely essential for life. Plants remove carbon dioxide from the air during photosynthesis to create the compounds of life on which both plants and animals depend. Not only is there no direct toxic effect of carbon dioxide, but it is also essential for the existence of life. Carbon dioxide is returned to the air by both plants and animals during respiration. Decomposition of dead plant and animal matter also releases carbon dioxide into the atmosphere. Finally, any burning of organic material also releases carbon dioxide.

The level of carbon dioxide in the air has increased dramatically since preindustrial times when the level was about 280 ppm. As of 2012, the atmospheric concentration was approaching 395 ppm. Where did all of this carbon dioxide come from? It came largely from extensive burning. The burning of what? The answer is fossil fuels. Coal, oil, and natural gas all contain considerable amounts of carbon. When these fuels are burned, carbon dioxide is produced. Fossil fuels are the principal energy source used in our society; that is, they are the energy source on which a large majority of things run. The U.S. emissions of carbon dioxide account for 18% of the world's carbon dioxide emissions and represent per capita emissions of nearly 18 metric tons for every man, woman, and child in the United States.

Another major source of additional carbon dioxide is deforestation. Today, systematic destruction of the rain forests is occurring across the world. One major effect of widespread "slash and burn" of the rain forests is the dumping of large amounts of carbon dioxide into the atmosphere. This represents an increase in carbon dioxide levels in the air because the plant life that replaces the forest contains much less carbon.

The trends in atmospheric carbon dioxide levels are very evident in Figures 11.1 and 11.2. In Figure 11.1, the atmospheric levels of carbon dioxide are measured continuously at one location, and the trend seems unmistakable. Even more striking is

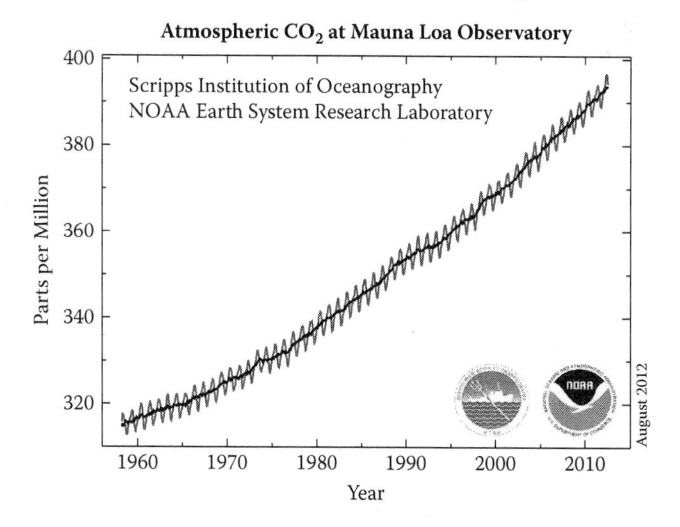

Figure 11.1 Atmospheric concentrations of carbon dioxide as measured at Mauna Loa Observatory, Hawaii. (From NOAA, *Trends in Atmospheric Carbon Dioxide*, Global Monitoring Division, Earth System Research Laboratory, National Oceanic and Atmospheric Administration, Boulder, CO, 2013, http://www.esrl.noaa.gov/gmd/ccgg/trends/co2_data_mlo.html.)

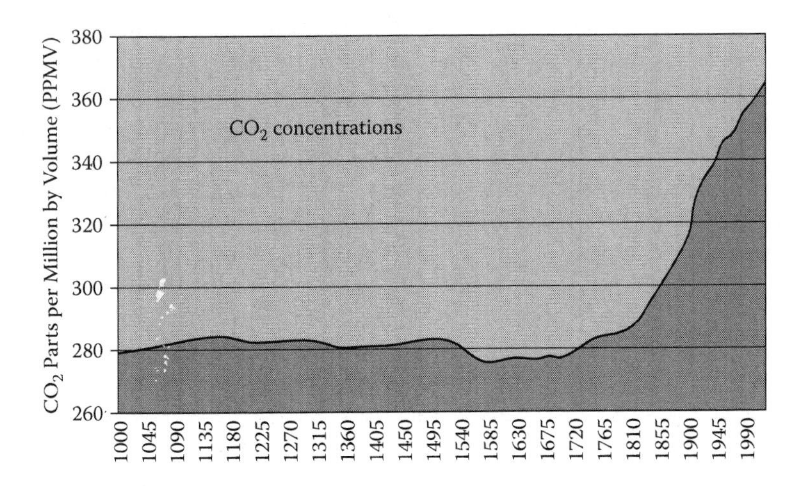

Figure 11.2 Carbon dioxide concentrations for the past 1000 years, derived from measurements of air bubbles in the ice cores in Antarctica and from atmospheric measurements since 1957.

the data found in Figure 11.2. This graph looks at carbon dioxide levels over the past 1000 years. For most of the millennium, the concentrations were reasonably stable. Then, in about 1810, the levels began to increase drastically and have continued to increase at a fairly rapid rate ever since. Of course, the trend observed here does not necessarily predict the future levels of this gas, but what it does suggest is that if things continue as they are there is very little reason to believe that this increase in carbon dioxide will not also continue.

Methane

Another important greenhouse gas is methane. Methane arises from several different sources, including solid waste decomposition, the digestive tracts of livestock, coal mining, rice paddies, and natural gas and petroleum production. For methane there is good as well as bad news. The bad news is that methane is 21 times more effective as a greenhouse gas than carbon dioxide, but the good news is that its concentration is less than 1% that of carbon dioxide. On a comparative basis, methane represents a greater greenhouse warming effect than would be predicted from its concentration. For example, based on emissions of greenhouse gases in the United States for 2009, the methane emitted represented about 12.5% of the insulating effect for all of the carbon dioxide emitted, yet this gas represented only 0.2% of the weight compared to the weight of the emitted carbon dioxide.

Until recently, concentrations of methane have been increasing and have apparently more than doubled over the past 200 years. Most of this increase is believed to be due to human activity; however, the rate of increase has slowed over the past

30 years. The annual rate of increase has fallen from 20 ppb in the late 1970s to about 10 ppb in 1990. In the 1990s, the rate of increase became erratic, varying from 0 to 15 ppb/year. From 2000 to 2006, the concentration basically leveled out at about 1775 ppb but in 2007 the concentration began to rise again. At this point, it is unclear as to whether the recent increase in methane levels is part of a larger pattern or just a temporary jump. The driving force behind methane levels is not well understood; therefore, long-term predictions of methane concentrations are difficult. More research in this area will certainly be required.

Chlorofluorocarbons

The CFCs as a group are powerful greenhouse gases, but they are generally studied because of their effect on the ozone layer. (We will deal with this effect later in the chapter in the section on ozone depletion and chlorofluorocarbons.) The potential of these materials as greenhouse gases may be seen in the fact that in 2010 carbon dioxide was present at an atmospheric concentration of over 700,000 times that of CFC-12, but the global warming effect of the carbon dioxide in the atmosphere was less than 50 times that of CFC-12. Many of these gases have leveled off and are beginning to decline, because their emission has been made illegal by international agreement due to their effect on the ozone layer. One troubling factor is that many of the CFC replacement gases are also powerful greenhouse gases.

Nitrous Oxide

A somewhat interesting gas that has a fairly powerful global warming potential is nitrous oxide (N_2O). Nitrous oxide is a colorless, sweet-tasting gas with very little odor. It has long been known for its anesthetic properties and is often used in dentist offices; it is often referred to as "laughing gas" because when inhaled in small quantities it produces a type of hysteria. Nitrous oxide is present in the atmosphere in concentrations much too low to be toxic for humans. The level in 2010 was 323 ppb and has been increasing at about 0.8 ppb/year since 1988. Nitrous oxide is a very stable gas with an average residence time in the atmosphere of about 120 years. It appears that this gas is principally produced by denitrifying bacteria that break down nitrogen-containing wastes; however, there are a number of anthropogenic sources, including the breakdown of fertilizers, combustion of fossil fuels, production of some chemicals (e.g., adipic acid and nitric acid), breakdown of manure from agriculture, and breakdown of human sewage.

We can conclude that global warming is a result of all of these greenhouse gases. These gases as a group are increasing in concentration, and because they all act as insulators their increased presence in the atmosphere will result in higher average temperatures. This idea is no more mysterious than putting more blankets on a bed. The more blankets one adds, the warmer the bed will be. The more greenhouse gases one puts into the atmosphere, the warmer the Earth will be.

IS GLOBAL WARMING IMPORTANT?

Demonstrating that global warming exists does not necessarily make it important. Bear in mind that many phenomena exist which when examined are irrelevant. The sun is burning away a little bit of itself every day. This can be shown to be true, but it is also unimportant to our lives. It is unimportant because the effect is so small. Therefore, the great debate about global warming rages on, not about whether it exists but rather about whether it is important. It seems we need to begin by getting a clear picture of just what a significant global temperature change would be. When considering climatic temperature change, however, it is important to consider the time span over which it occurs. If temperature change is slow enough, then plant and animal species will have time to adjust, and the biological effects would at least be minimized.

The temperature record indicates that the greatest temperature swings in this period of general glaciation are about 14°F (8°C) from the end of an ice age to the beginning of an interglacial period. These changes, however, have occurred over a period of 8000 to 12,000 years, and although drastic overall they do not amount to much on a per-year basis. Such changes average out to 0.0012°F (0.0007°C) per year or 0.12°F (0.07°C) per century.

The Medieval Warm Period at its maximum was about 1°F (0.5°C) warmer than it was at the beginning of the 20th century, whereas the Little Ice Age was about 1°F (0.5°C) cooler at its worst. These changes occurred over a century or two. Because the climate of the Earth is always changing, temperature changes of 1°F (0.5°C) or less that occur over a century may fall within the realm of normal.

Since 1880, the carbon dioxide level in the atmosphere has increased from about 290 ppm to about 390 ppm (currently). Over this time global temperatures have been gradually increasing. As seen from Figure 11.3, the increase has been steady for the most part except for a strange, and unexplained, cooling period between 1940 and 1975. The temperature change since 1880 is an increase of about 1.6°F (0.9°C). This seems to be a little larger than one would usually expect for normal climatic variability, as this rate of change figures out to be approximately 1.3°F (0.7°C) per century. Could this temperature increase be due to the concurrent increase in carbon dioxide concentrations? It seems reasonable but is very difficult to prove. However, the preponderance of evidence regarding what we know about the isolating properties of greenhouse gases would strongly suggest that carbon dioxide and other greenhouse gases are causing the warming of the atmosphere. Using the data we have, let us look at some computer projections and consider the consequences they suggest.

The one thing we know for sure is that if we put enough greenhouse gases into the atmosphere then global warming will occur. What concentration of these gases will it take to produce a significant warming effect? With a computer we can attempt to simulate our future world. The programs are very complex and it takes a supercomputer to run them, but they are the only way we have to peek at the world as it would be if .... Without putting anybody at risk, we can build a virtual world with any level of carbon dioxide or methane or CFCs and see what it could be like.

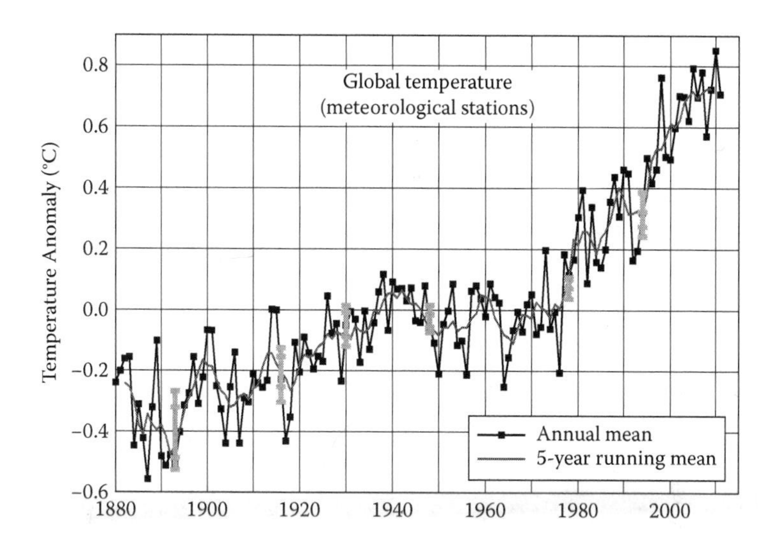

Figure 11.3 Variation of mean global temperatures since 1880. The dashed line connects annual mean global temperatures, and the solid line connects 5-year-mean global temperatures. (From *GISS Surface Temperature Analysis (GISTEMP)*, Goddard Institute for Space Studies, National Aeronautics and Space Administration, Greenbelt, MD, http://data.giss.nasa.gov/gistemp/graphs/.)

It's not a simple exercise, though; for example, how can we know or even estimate the concentrations of greenhouse gases in the year 2020, 2040, 2050, 2070, or 2100? Figure 11.4 shows some projections for carbon dioxide levels into the future. What is clear is that there may be drastically different carbon dioxide levels in the year 2030 depending on which projection proves to be most accurate.

Looking into the future requires considering all of the ways in which heat is introduced into our atmosphere and all means by which it is moved around in our atmosphere. The computer must track the movement of air, the heating of air and the Earth, the evaporation and precipitation of moisture, the effect of clouds, and the effects of vegetation, among many other parameters.

Focusing on the atmosphere and the Earth's surface, however, overlooks a major player in the climate control system—the oceans, which can hold tremendous amounts of heat and dissolve incredible amounts of carbon dioxide. Furthermore, the oceans contain great ocean currents that move more water than all of the world's rivers combined. These great ocean currents move incredible quantities of heat between the surface and the depths of the ocean and from place to place within the ocean. The major global ocean currents are seen in Figure 11.5.

The effect of clouds can also be difficult to figure out in climate modeling. Clouds reflect sunlight and cool the Earth, in addition to trapping heat near the surface like a blanket. It appears that the higher, thinner-type clouds tend to warm the Earth as these clouds transmit sunlight more than they reflect it while trapping heat near the Earth. In contrast, low thick clouds tend to have a net cooling effect in the daytime as such clouds reflect most of the sunlight that strikes the cloud tops. At night, all

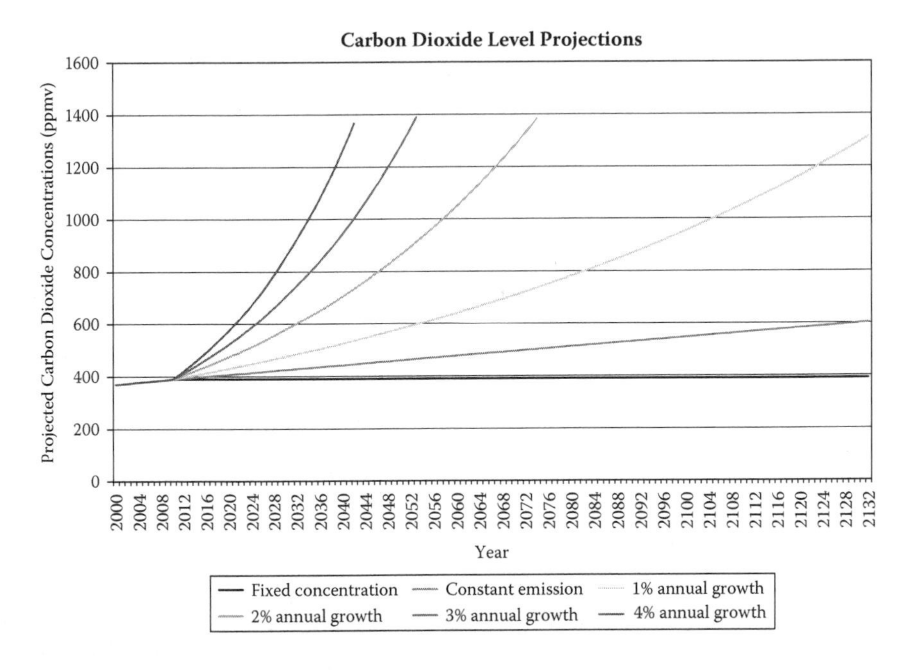

Figure 11.4 Projections of carbon dioxide levels into the future based on various growth assumptions. (Adapted from Schneider, S.H., *Global Warming*, Sierra Club Books, San Francisco, CA, 1989.)

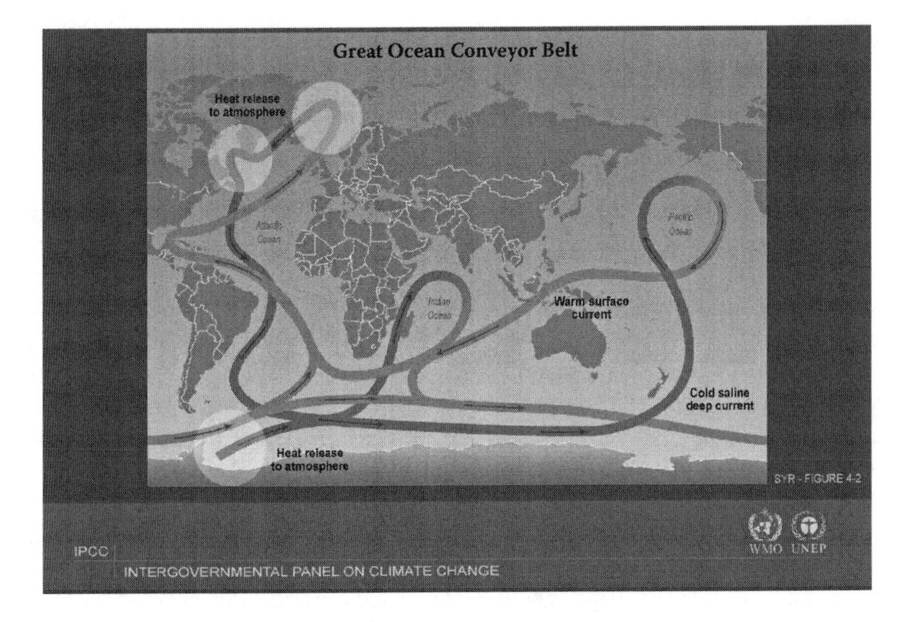

Figure 11.5 The ocean circulation conveyer belt. (From Watson, R.T. et al., *Climate Change 2001: Synthesis Report—Summary for Policymakers*, Intergovernmental Panel on Climate Change, Geneva, Switzerland, 2001.)

clouds are warming because they tend to trap heat. The overall effect of clouds, however, appears to be a net cooling one. For computer modeling of clouds, therefore, one must consider whether warming the Earth causes more clouds or fewer clouds to form and what type they would be. Would changes in cloud patterns cause the Earth to warm even faster or cause the Earth to cool back down a little?

Vegetation can have a significant effect on climate. For one thing, various types of ground cover can affect the albedo or reflectivity of the Earth (see Chapter 8), and therefore change the amount of solar radiation absorbed at a particular location. The amount of solar radiation absorbed at a particular location would be related to the amount of heat available at that location; however, increases and decreases in vegetation on a large scale can affect the amount of the carbon dioxide in the atmosphere. Growing vegetation will remove carbon dioxide and cool the climate, whereas dying vegetation will release carbon dioxide and warm the Earth.

Despite all of these uncertainties, computer projections used by the **Intergovernmental Panel on Climate Change (IPCC)** (for more information on this group, see later discussion on international agreements on global warming) suggest temperature increases from 2.0 to 11.5°F (1.1 to 6.4°C) by the year 2100. Is this temperature increase unprecedented? It most certainly is. As discussed earlier, a temperature increase of maybe 1°F (0.5°C) in a century might be within the realm of normal, but there is no historical indication of a temperature rise of 2.0°F (1.1°C) over a century. In other words, the world is now the warmest it has been in the past 1300 years and another 2.0°F (1.1°C) would probably make it warmer than at any time in the past 125,000 years. Now, if we look at the other extreme of 11.5°F (6.4°C), then there is no question that the Earth would be warmer than it has been in the past 4 or 5 million years.

GLOBAL WARMING: EFFECTS

One may be tempted on a cold day in the wintertime to think, "Global warming would not be so bad. I am cold and a Florida-type climate here would certainly warm me up." Well, think again. There are a number of problems with such a development. We often forget that as *Homo sapiens sapiens* we are probably the most adaptable of any species on the planet. We live in the tropics, we live in the Arctic, we live in wet places, and we live in dry places. We use our brains to figure out how to survive in very inhospitable places. Most animals, and certainly plants, do not have that option. They all live in a particular habitat that is partly defined by climate. If the climate changes, the habitat might also change and the species be put in jeopardy. Animals can sometimes react by migrating, but this is not always feasible. Plants cannot migrate in the usual sense, but instead move gradually when old plants die out in the old habitat and new plants take root in a new habitat. This change works well if the change in habitat is slow in comparison to the lifespan of the plant.

Forests are particularly at risk because trees are some of the longest living organisms on the planet. The formation of new forests takes time, and a tree that finds itself in an inhospitable habitat can only struggle to survive despite the conditions.

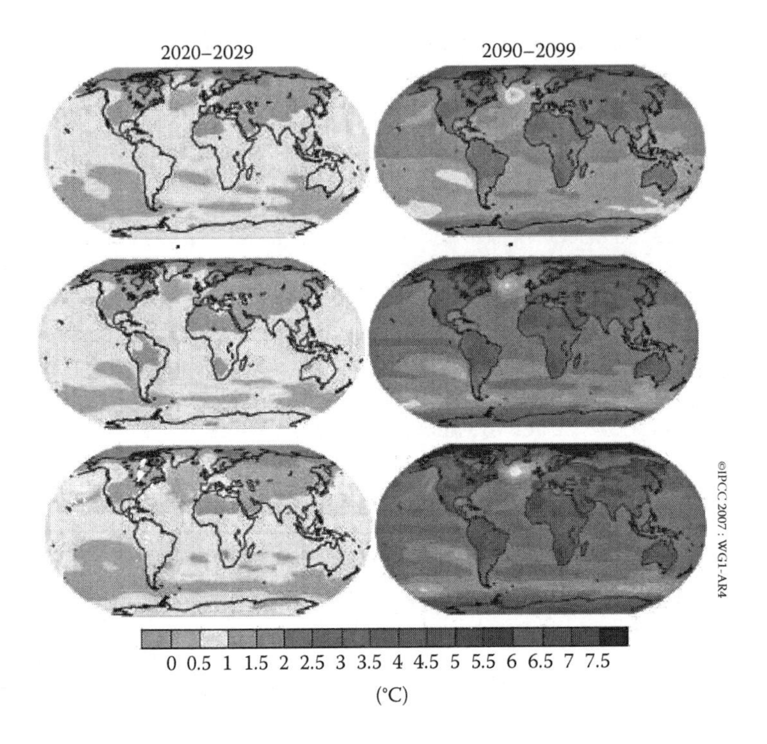

Figure 11.6 Likely temperature increases in three different scenarios projected to either 2020–2029 or 2090–2099. (From Solomon, S. et al., Eds., *Climate Change 2007: The Physical Science Basis*, Intergovernmental Panel on Climate Change, Geneva, Switzerland, 2007.)

Some of the weaker trees will probably not survive. These changes can have economic consequences; for example, maple syrup is produced in fairly large quantities in New England but global warming projections suggest that the best habitat for maple trees will move northward into Canada by 2070.

One problem with all of these conjectures is trying to get a realistic idea of what global warming would mean to you and me right here in Anytown, Anywhere, World. Let's assume that in the year 2050 the global mean temperature has increased by 2°F (1°C). How will the planet's inhabitants cope with this? I am sure that some people would think, "Well, let's see. Two more degrees would make it about 95°F (35°C) in the summer and 35°F (2°C) in the winter. That doesn't sound all that bad." Unfortunately, the situation is not that simple. In any global warming scenario, the amount of heating at various places on Earth is uneven. Some places will warm very little or moderately, whereas others will experience extreme temperature increases. Figure 11.6 gives some idea of how this might work out. In any of these scenarios, the short-term projections (2020 to 2029) show most of the warming occurring in the Northern Hemisphere and being most intense in the Arctic regions. Generally, global warming is most intense over land because oceans are difficult to heat and can move the heat away from the surface, as noted earlier. The Northern Hemisphere has proportionately more land than the Southern Hemisphere, which results in more rapid warming. The Arctic regions, paradoxically, are ocean regions but are experiencing the most extreme heating. The mechanism here is entirely different. Currently, the Arctic Ocean is covered to a large extent by permanent ice. With global warming the ice melts around the edges, exposing more and more ocean water. The albedo or reflectivity of polar ice and ocean water differs greatly. Polar ice will reflect most of the sunlight that strikes it; hence, very little of the solar radiation remains to warm these icy regions. When the water is exposed, the amount of solar radiation adsorbed goes up dramatically and the area will begin to warm up. This phenomenon gives rise to positive feedback because as the area warms due to increased carbon dioxide the ice melts and exposes ocean water, which cause even more warming. The average Arctic ice cover has been shrinking at about 2.7% per decade and may disappear in the late summer altogether by the latter part of the 21st century. Such an occurrence would be devastating to the polar bear and many marine mammals that depend on the Arctic ice for their habitat.

Across Alaska and northern Canada much of the ground remains permanently frozen a short distance under the surface. This permanently frozen layer, known as **permafrost**, gives much of the structure and characteristics to the land over these vast northern regions. Rapid warming in Alaska and northern Canada is already beginning to destroy much of the permafrost, resulting in increased erosion, landslides, and sinking of the ground. Additionally, there is considerable evidence that the ice sheet over Greenland is thinning and retreating.

A review of the maps for long-term projections (2090 to 2099) shows very much the same uneven changes, except that global warming begins to overwhelm the climate everywhere. The Northern Hemisphere and the Arctic still warm faster than the Southern Hemisphere, excluding Antarctica. Sometime toward the end of the 21st century, Antarctica will also begin to warm at a more rapid rate.

Temperature changes also affect wind patterns, which in turn will cause the weather to change. Rainfall will increase in some places but decrease in others. This would have a tremendous impact on agriculture. The IPCC has used computers to project precipitation changes for the period from 2090 to 2099 relative to 1980 to 1999. The main generalizations that can be drawn from these computer projections are that the major increases in precipitation will occur at the high latitudes and that significant drying will occur in the tropics and subtropics. Major drought could become a problem, for example, for the southwestern United States, southern Europe, Middle East, and Western Australia.

Although some places might become better suited for growing crops, it would take time to make the transition from areas where farming becomes poorer to areas where farming improves. Famine and civil strife would almost certainly be outcomes of such a situation. Currently, we have starvation in some areas of the world, but this is principally a problem of food distribution. Consider how much worse the situation would be if there really was not enough food.

In addition to shifts in the weather, the intensity of meteorological phenomena will generally increase. Heat waves will be hotter and longer, and the increased precipitation will tend to come in more intense rain events. Hurricanes will likely become more intense, although the effect of global warming on the number of hurricanes is unclear. The 2005 hurricane season in the North Atlantic was unusual both in the numbers of hurricanes (15) and intensity of hurricanes. Four hurricanes reached category five status on the Saffir–Simpson scale (on which five is the most intense). Since records have been kept on the intensity of hurricanes, this was the first time that there had ever been four category-five hurricanes in one season.

As if these effects are not bad enough, global warming could cause sea levels to rise. The reasons for this phenomenon are simple. First, water is a fluid that like any other fluid expands when it is warmed. The relationship between the volume of a fluid and its temperature is the principle on which an ordinary liquid thermometer operates. If the water in the ocean is warmed a little bit, then the sea level rises. Furthermore, warming encourages melting of glaciers and the polar ice caps. These two effects can significantly raise sea levels. Sea levels rose worldwide by 5 to 9 in. (12 to 22 cm) during the 20th century. From 1993 to 2003, the sea level was rising at a rate of 0.10 to 0.16 in. per year (0.24 to 0.38 cm per year) of which it is estimated that 0.05 to 0.09 in. per year (0.11 to 0.21 cm per year) was due to thermal expansion of ocean waters and 0.03 to 0.07 in. per year (0.07 to 0.17 cm per year) to melting of glacial and polar ice.

The actual rise in sea level may vary from place to place, as illustrated in Figure 11.7. The reason for the variation is that both the height of the land as well as the sea level can change. Note, for example, the rapid rise in sea level at Galveston, Texas. This effect is partly caused by sinking of the land, which resulted from excessive removal of groundwater (see Chapter 12). Conversely, in Sitka, Alaska, the sea level is actually falling. During the last ice age, the weight of the ice was so great that it compressed the ground underneath. After the ice melted, the ground began to expand upward and is still expanding, which has resulted in the falling sea level at Sitka and other far northern cities.

Estimates by the IPCC are that the global average sea levels should rise by an amount between 7 in. (18 cm) and 23 in. (59 cm) by the year 2100. One great uncertainty that hangs over this entire situation is the stability of the Greenland ice sheet. The paleontological record indicates that about 125,000 years ago the sea rose from 13 ft (4 m) to 20 ft (6 m) in response to temperatures similar to the highest temperatures projected for the 21st century. It is suspected that much of the water came from Greenland, the Arctic, and perhaps some parts of the Antarctic. What are the consequences of sea level changes anywhere between 7 in. (18 cm) and 20 ft (6 m)?

To consider this question, let us look at the coast of Louisiana and Texas (see Figure 11.8). All of the areas in black are less than 5.2 ft (1.5 m) above sea level. The entire area in black contains a large number of cities, including the cities of Galveston and New Orleans. Of course, even an increase of 2 ft (0.6 m), which is roughly the maximum projected by the IPCC, would not flood this entire area. Some parts of the area are more than 2 ft above sea level, and other parts that are not that far above sea level are protected by dikes and levies. The use of dikes and levies is particularly common in highly populated areas. In the aftermath of Hurricane Katrina, it is now well known that most of New Orleans is already below sea level and is kept dry only by levies.

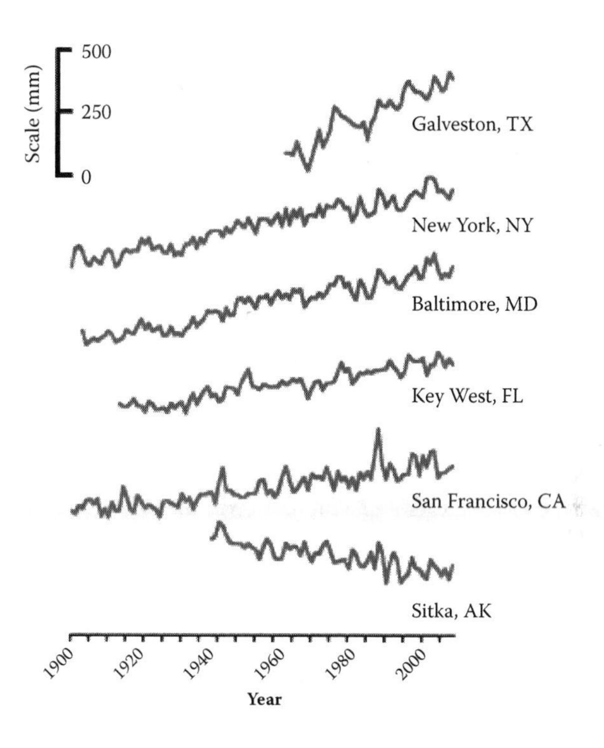

Figure 11.7 Sea level changes from 1900 to 2003. (From USEPA, *Climate Change Indicators in the United States*, U.S. Environmental Protection Agency, Washington, DC, 2006, http://epa.gov/climatechange/science/recentslc.html.)

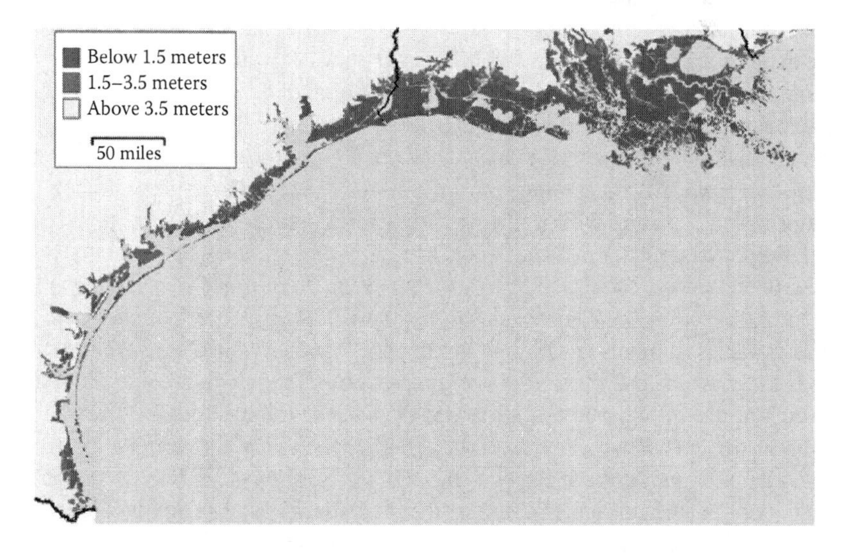

Figure 11.8 Sea level rise potential on the Gulf. (From USEPA, *Maps of Lands Vulnerable to Sea Level Rise: Gulf Coast*, Risingsea.net, http://papers.risingsea.net/coarse-sea-level-rise-maps-gulf.html.)

The sea level values that we have been referring to are averages; therefore, the actual sea level at any given time can vary with the tides and other meteorological factors. The effect of storms is probably the most significant issue. As the elevation of a coastal location above sea level decreases, the likelihood of serious flooding increases. Of course, if the Greenland ice sheet begins to destabilize, then all of the black and gray areas shown in the figure might be flooded. If rises in sea level become more serious than predicted, some interesting and tragic problems could be created. The nation of Kiribati in the middle of the Pacific Ocean is an island nation of 33 islands, most of which are coral atolls. Except for one island, the highest elevation is about 6.5 ft (2.0 m). Most of this nation could simply be wiped out by a significant rise in sea level. Even a modest sea level rise of 1 ft (0.3 m) would cover much of the land in such low-elevation islands.

We finally come to the big question. What can we do about global warming? It is at this point that things get really difficult. One of the biggest reasons for the difficulty is that we are now talking about *money, large amounts of money*. Many people worry about what will happen to the world economy if great sums of money are sucked up by the fight against global warming. Many Americans worry about what will happen to their standard of living if too many resources are directed toward efforts to combat global warming. Yet, one has to question whether doing nothing actually costs nothing.

The evidence strongly suggests that global warming will occur and that it will be significant. If that is the case, then money will have to be spent to counter the effects or will be lost to the economic disruption that would follow. As noted earlier, the disruption to agriculture could lead to mass starvation at the worst or to higher food prices at the best. Even in the best case, we lose money, and if things are really bad we lose human lives.

Rises in sea level that would lead to coastal flooding would have to be dealt with, for example. Do we build dikes as was done in Holland, or do we allow the flooding and relocate whole cities or parts of cities? We could cite more examples, but it would add little to the argument. The point is that we can pay now or we can pay later, but pay we will, unless we just happen to luck out on some obscure chance that no significant global warming will occur. Are we willing to take that risk?

If we decide to do something, then what is it that we should do? Many people hope that addressing the problem will involve very little disruption in the way they live and do business. This approach can be defined as *tinkering*. The idea here is to counteract the symptoms rather than getting to the causes of global warming. Some suggestions have included such notions as spreading dust into the stratosphere to cut out some of the sunlight before it reaches the surface or spreading 300,000 tons of iron in the oceans around Antarctica to encourage the growth of phytoplankton. The latter solution is based on the notion that rapid growth of these microscopic plants would remove carbon dioxide from the air. These ideas are just a couple of the many that have been suggested.

Without going into any of these suggestions in detail, there is one rather universal criticism that can be made of all of them. They all are attempts to counterbalance one effect with another—that is, to create a global cooling effect. The odds are that we will either underdo it or overdo it. We may end up with global cooling, or we may accomplish nothing and continue moving toward global warming. This approach seems risky at best. This leaves us with addressing the root cause of the problem—the production of greenhouse gases. The rate of production of many of these greenhouse gases is firmly linked to the functioning of our economy, and to change their production rate we must change how we do things. But, we would rather not change how we do things; we like our lifestyle, and we like the cheap energy that we get from burning things such as coal and oil.

We have two ways to cut down on carbon dioxide production. We can consume less energy, or we can find alternative ways to produce energy. In the end, doing both would be ideal, but that would involve making some real changes in how we do things. Both options require a real commitment from all of us, both individually and politically.

INTERNATIONAL AGREEMENTS ON GLOBAL WARMING

The beginnings of the current serious efforts to control global warming can be traced back to the Third World Climate Conference, which was held in Rio de Janeiro in 1992. The Rio conference was widely referred to as the "Earth Summit," as the conference attracted many world leaders and produced the first comprehensive approach to the control of climate change. The United Nations, however, had already taken two significant steps before this. In 1988, the IPCC was founded. This panel, comprised of thousands of scientists with expertise in the area of global climate assessment, has been able to develop the scientific foundations necessary for international discussions on global climate variability. Also, the United Nations Intergovernmental Negotiating Committee (INC) had developed a draft convention on climate change. It was this draft convention that was being considered at the Rio conference.

This convention, known today as the **Framework Convention on Climate Change (FCCC)**, was put into its final form and made available for signature at the Rio conference. To date, the convention has been ratified by 195 countries, including the United States. The principles on which the FCCC rests are as follows:

1. Developed nations and developing nations have common responsibilities but not always to the same degree.
2. Developed nations should take the lead to show others how to put into place methods to support international environmental needs.

A fundamental assumption within the framework is that wealthier developed countries (Annex II countries as specified by the FCCC) should assist other countries in making the changes necessary to combat global warming.

The FCCC calls for an annual meeting of the parties to the convention. This meeting, which is called the **Conference of Parties (COP)**, has been held every year since 1995. At the third COP, held in 1997 in Kyoto, Japan, the **Kyoto Protocol** was developed. This protocol called for mandatory restrictions on the emissions of six greenhouse gases—carbon dioxide (CO_2), methane (CH_4), nitrous oxide (N_2O), hydrofluorocarbons (HFCs), perfluorocarbons (PFCs), and sulfur hexafluoride (SF_6)—by about 40 industrialized countries (Annex I countries as specified by the FCCC). These restrictions required these countries on average to reduce their greenhouse gas emissions by 5.2% below the 1990 levels by 2008 to 2012. The actual reduction target varied from country to country, with some European nations expected to achieve a reduction of 8% (United States, 7%), whereas at the other extreme Iceland was allowed up to a 10% increase. Other means have been provided for a country to meet the emission target. Emissions trading is allowed so that industrialized countries can buy and sell emission credits among themselves. Also countries can receive credit for financing emission reduction projects in other countries.

For the protocol to take effect it had to be ratified by 55 parties and by parties whose total emissions accounted for 55% of the emissions of the Annex I countries. The protocol went into force on February 16, 2005, after it was ratified by the Russian Federation. For years it was believed that the Kyoto Protocol could never come into force without ratification by the United States, which contributes about 30% of the greenhouse gas emission from Annex I countries. To everyone's surprise, the rest of the world came together and put the Kyoto Protocol into force without the cooperation of the United States. As of now, the only Annex I country that has not ratified the protocol is the United States.

As much as the Kyoto Protocol represented somewhat of a victory in the international restraint on the ever-increasing emission of greenhouse gases, the agreement was never seen as the final agreement to halt global warming. Major players, such as China, who were not included in the Annex I countries were not part of the protocol. In addition, the initial commitments of the protocol were slated to expire in 2012. Hence, there has been considerable activity over the last several years to hammer out an agreement that would replace the Kyoto Protocol and be more comprehensive.

After a number of preliminary conferences and meetings, delegates, including many heads of state, from around the world met for COP 15 in Copenhagen, Denmark, in late 2009. The expectations for this meeting were very high, and when a short and not very detailed agreement was produced there was widespread disappointment. In many ways, however, the Copenhagen Accord was a beginning point that led to development of the more detailed Cancun Agreement at COP 16. This process of refinement is ongoing and has the potential to lead to significant control of greenhouse emissions.

The key provisions of the Cancun Agreement so far differ in three distinct ways from the Kyoto approach. First, each country is to propose internal controls to limit its own production of greenhouse gases at such a level that, with the participation of all countries, the global community might meet the goal of limiting future temperature rise to 2°C or less. All of the programs would then be analyzed by international bodies to see if collectively they would likely meet the goal. If programs are insufficient, countries would be encouraged to make deeper cuts. The United States has pledged to decrease emissions by 17% relative to 2005 by 2020.

Another key change from the Kyoto Protocol is that all countries, whether developed or developing, are expected to be involved in mitigation and control of greenhouse emissions, including for the first time major emitters such as China, India, Brazil, and South Africa. It was determined that financial help would be necessary if poorer countries were to be involved in greenhouse gas mitigation and control. At Copenhagen, it was agreed that developed countries would immediately create a fund of $30 billion in "fast start" money for assistance to the least developed countries to make the changes necessary to decrease their greenhouse gas emissions. An additional commitment was to create a Green Climate Fund of $100 billion by 2020.

The current climate negotiations are still a work in progress; therefore, it is difficult to say how successful they will be in the end. There seems to be very little question at this point of the need to take decisive action if really unpleasant consequences from climate change are to be avoided. All of these changes in our energy policy make good sense anyway for the reasons that we covered in great detail in Chapter 7. Why not change? In the long run, what do we have to lose?

CHEMISTRY OF THE STRATOSPHERE: OZONE LAYER

Another issue of global importance is stratospheric ozone depletion. Although ozone depletion is not directly related to global warming, the two issues are often confused, for two reasons. First, both phenomena involve some of the same gases (CFCs), and, second, they both have global atmospheric effects. It is important to stress, however, that they are *independent phenomena.*

The **ozone layer** is a region in the stratosphere in which there is a relatively high concentration of ozone. To understand the importance of this part of the atmosphere we need to consider the radiation that comes to us from the sun. This radiation is of many different types, including visible radiation, heat radiation, and **ultraviolet (UV) radiation**. Although these types of radiation are very different in their effect on us, they are really just different manifestations of the same phenomenon—electromagnetic radiation. Each type differs in its destructive potential. Heat and visible radiation have few if any destructive effects on human beings. UV radiation, however, ranges from mildly dangerous to extremely dangerous, depending on the type to which one is exposed.

Ozone in the stratosphere, as opposed to that found in the troposphere, is not a pollutant but occurs naturally and has a function. When ozone is struck by UV radiation it absorbs the radiation and yields an oxygen molecule and a free oxygen atom:

$$O_3 + UV \rightarrow O_2 + O$$

The UV radiation is gone, and its energy is contained in the oxygen atom and oxygen molecule. As discussed in Chapter 9, a free oxygen atom is very reactive and short lived. The oxygen atom will collide with a large number of molecules in a very short period of time. Other than a nitrogen molecule that is unreactive, the oxygen atom is mostly likely to collide with an oxygen molecule and then react with it. This reaction regenerates the ozone molecule, but what happens to all of that energy contained in the UV radiation? It appears as heat:

$$O + O_2 \rightarrow O_3 + \text{Heat}$$

It is this heat that warms the stratosphere and creates the temperature inversion that defines the stratosphere. The net result of these two reactions is that UV radiation is consumed and heat is produced. As a result, most UV radiation fails to reach the Earth's surface.

Stratospheric ozone is in a continual state of formation and destruction. Until recently the fact that the ozone was undergoing some destruction was no cause for alarm because it was balanced by the rate of formation. The process here is very much like pouring water into a bucket with holes at the bottom. If one can pour it in at the same rate at which it runs out, the water level in the bucket will remain constant.

The reaction that creates the ozone layer in the first place involves high-energy ultraviolet (HEUV) radiation, which is only found in the higher part of the atmosphere. This type of radiation interacts with most of the gases in the atmosphere, which explains why it is unlikely to reach the Earth's surface with or without an ozone layer. One thing that this type of radiation does is to cleave an oxygen molecule into two oxygen atoms:

$$O_2 + \text{HEUV} \rightarrow 2O$$

These oxygen atoms can then react with oxygen molecules as seen earlier, giving rise to ozone:

$$O + O_2 \rightarrow O_3 + \text{Heat}$$

The destruction of ozone occurs because the improbable happens once in awhile. As noted earlier, ozone absorbs UV light and produces an oxygen atom:

$$O_3 + \text{UV} \rightarrow O_2 + O$$

This oxygen atom has the greatest likelihood of reacting with an oxygen molecule, but occasionally one will react with an ozone molecule, producing two oxygen molecules instead of regenerating ozone:

$$O + O_3 \rightarrow 2O_2 + \text{Heat}$$

To the extent this reaction occurs, ozone will be destroyed.

IMPORTANCE OF OZONE LAYER

As long as human beings have been on the Earth, the ozone layer has been in place, blocking UV radiation. This layer developed along with early life forms that put oxygen into the atmosphere, and it remained there as we, *Homo sapiens sapiens*, evolved. Because humans and many other organisms have never experienced high levels of UV radiation, we are not equipped to handle it. Some UV radiation does pass through the ozone layer, but as long as our exposure to the sun is restricted to a moderate level this radiation does little harm to us. However, people who are "sun worshippers" or who are particularly sensitive to the sun may experience problems from UV radiation. The common garden-variety sunburn is a consequence of UV radiation. Whatever the health risks are from UV radiation, they would be much worse if there were no ozone layer.

Ultraviolet radiation is associated with several health problems, including sunburn, skin cancer, premature aging of the skin, cataracts, and immune suppression. Suppression of the immune system can in turn lead to, for example, the eruption of fever blisters. For elderly people, the formation of cataracts, which are estimated to blind or dim the sight of some 30 million people worldwide, could be encouraged by UV radiation.

Increased UV radiation could lead to additional cases of skin cancer each year. According to the National Aeronautics and Space Administration (NASA), it is estimated that a 1% decrease in the ozone concentration in the stratosphere will cause a 2% increase in the type of UV radiation that is usually associated with health problems, and this increase can result in a 4 to 6% increase in certain types of skin cancer. Luckily, most of these types of skin cancer are easily curable and are not generally fatal; however, there seems to be a correlation between UV radiation and the more serious, and often fatal, cancer known as melanoma. This cancer may not appear for 10 to 20 years after exposure. Finally, UV radiation affects many living things, not just humans. In a NASA report, it was noted that half of the plant species tested were sensitive to UV light of the type filtered by the ozone layer. Many marine species are also sensitive to UV radiation. These species are a large portion of our food supply.

OZONE DEPLETION AND CHLOROFLUOROCARBONS

In October of 1982, some British scientists were taking routine scientific measurements in Antarctica. Much to their surprise and disbelief, they found that most of the stratospheric ozone above them had vanished. After a few years of further study and measurement, it was announced that a "hole" in the ozone layer was appearing over Antarctica in late September and October every year. This is the now the infamous "ozone hole."

Figure 11.9 shows satellite images of the ozone levels over the Antarctica for September 30 of various years. In 1978, NASA began measuring global ozone level through satellite. The amount of ozone in the atmosphere is measured in Dobson units (DU), which were developed by G.M.B. Dobson, a pioneer in the collection of ozone data. A Dobson value of 300 units is roughly a normal reading. By 1982, the satellite readings showed the "ozone hole" to be only slightly apparent. Levels had dropped down to approximately 200 DU as compared to approximately 250 DU in 1979. However, over the ensuing years, the ozone levels continued to decline such that on September 30, 2011 (Figure 11.10), the ozone level dropped down to approximately 100 DU, or about one-third the nominal normal value in the Antarctic region.

Nimbus-7/TOMS Version 8 Total Ozone
for Sep 30, 1979

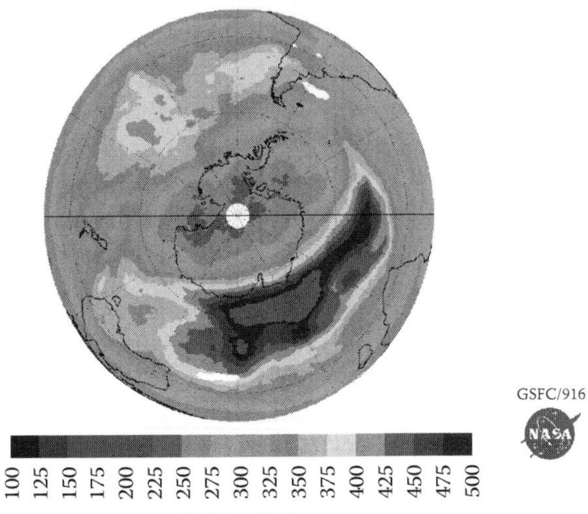

Dark Gray < 100 and > 500 DU

Nimbus-7/TOMS Version 8 Total Ozone
for Sep 30, 1982

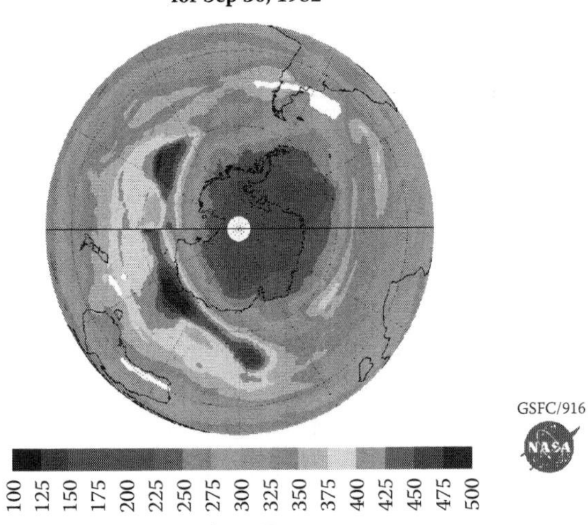

Dark Gray < 100 and > 500 DU

Figure 11.9 **(See color insert.)** Satellite image of total ozone over Antarctica in late September of 1979 (top) and late September of 1982 (bottom). (From *Total Ozone Mapping Spectrometer*, Goddard Institute for Space Studies, National Aeronautics and Space Administration, Greenbelt, MD, http://toms.gsfc.nasa.gov/ozone/ozone_v8.html.)

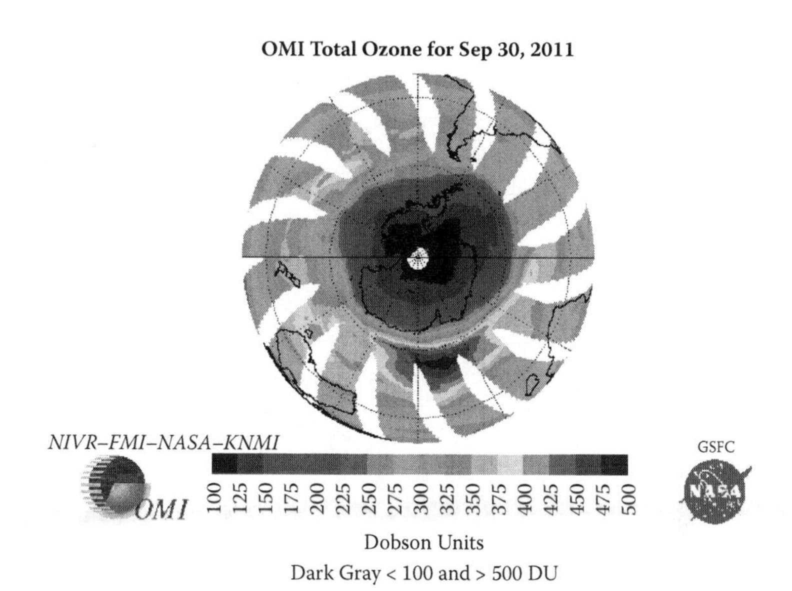

Figure 11.10 **(See color insert.)** Satellite image of total ozone over Antarctica in late September of 2011. (From *Total Ozone Mapping Spectrometer*, Goddard Institute for Space Studies, National Aeronautics and Space Administration, Greenbelt, MD, http://toms.gsfc.nasa.gov/ozone/ozone_v8.html.)

If the effect were limited to Antarctica, the problem would be serious but would not receive the kind of worldwide attention it has received. The concern is that, if there is almost no ozone over Antarctica, then what does that imply about the ozone over the more populous parts of the world both now and in the future? Data collected from 1967 to 2002 suggest that the stratospheric ozone levels over most of the northern mid-latitudes have decreased by about 5%. In 1999, the peak summertime levels of UV radiation over New Zealand had increased by about 12% over a period of 10 years. As the amount of UV radiation striking the Earth increases, the associated health risks also increase.

How did we get ourselves into this mess? To answer this question we need to look more closely at CFCs. Although CFCs have many different uses, they were originally developed as refrigerants. Refrigerating devices were originally developed in the first half of the 19th century. Mechanical cooling required the use of a gas that would liquefy under pressure. The gas most commonly used before 1850 was sulfur dioxide, which was replaced by ammonia (NH_3) after 1850. Both of these gases are corrosive and toxic. Any leak released toxic and irritating gas into the air. In the early part of the 20th century, the need for a better refrigerant led to a search for a replacement. In 1930, scientists at duPont struck pay dirt. They developed a group of compounds to which they gave the trade name of Freon. These compounds were stable, nonflammable, nontoxic, nonirritating, and noncorrosive and turned out to be useful for many other applications as well. Many of these materials found uses as propellants for aerosol spray cans, sterilizing agents, foaming agents in the production of foam packaging, solvents in degreasing and cleaning metal and sensitive electronic components, and ingredients in aircraft fire extinguishers. CFCs are probably a very good example of the *law of unintended consequences*. What possibly could go wrong here? The chemists had created chemicals that were absolutely harmless to anyone and anything. (Experiments have shown that animals can survive breathing an atmosphere of oxygen and CFCs.)

Oddly, part of the problem with these molecules is their nonreactivity in the lower atmosphere. Most chemicals, even fairly nonreactive ones, will react and break down in the atmosphere after some period of time, but CFCs do not. Because these molecules do not degrade, eventually they will find their way into the stratosphere. Despite the fact that CFCs are nonreactive in the troposphere, they are reactive in the stratosphere.

These molecules are made up of only carbon, chlorine, and fluorine, hence the name CFCs. The problem arises from the fact that UV radiation can break the carbon–chlorine bond (C–Cl). This produces a free chlorine atom which, like a free oxygen atom, is very reactive:

$$\text{CFCs} + \text{UV} \rightarrow \text{Cl} + \text{Other products}$$

Because the free chlorine atom is very reactive and there is quite a bit of fairly reactive ozone around, the chlorine atom reacts with the ozone:

$$Cl + O_3 \rightarrow ClO + O_2$$

The chlorine monoxide (ClO) molecule that is produced will react with any free oxygen atoms that it happens to encounter. But, where might these free oxygen atoms come from? They come from the cleavage of ozone by UV radiation. This reaction is one of the two reactions fundamental to functioning of the ozone layer:

$$O_3 + UV \rightarrow O_2 + O$$

$$ClO + O \rightarrow Cl + O_2$$

The result of this reaction is that free chlorine atoms are produced which can react with ozone to produce more chlorine monoxide. To make an important point about the chemistry of chlorine in the stratosphere, we will rewrite the chemical reactions next to each other:

$$Cl + O_3 \rightarrow ClO + O_2$$

$$ClO + O \rightarrow Cl + O_2$$

$$O_3 + O \rightarrow 2O_2$$

By careful examination of these reactions it should be apparent that chlorine atoms are both consumed and produced. For each chlorine consumed in the first reaction, one chlorine is produced in the second reaction. Clearly, the chlorine atoms are acting as a catalyst for the overall reaction. The chlorine atoms are not consumed, yet they provide a new way for ozone to be destroyed. Experiments show that a single chlorine atom can lead to the destruction of 100,000 ozone molecules. As noted earlier, the amount of ozone in the stratosphere depends on the rate of destruction vs. the rate of formation. The chlorine atoms provide yet another way for ozone to be destroyed. This situation is analogous to drilling more holes at the bottom of the bucket we talked about earlier, which means the water will run out faster. Because we cannot change the rate at which we put water in, the bucket will slowly begin to empty. In the same way, the level of stratospheric ozone will slowly decrease. This process has been most pronounced in the coldest regions of the Earth. The reasons for this are too complex to discuss in this chapter, but they do explain why the problem was first noticed in the Antarctic.

NATIONAL AND INTERNATIONAL RESPONSE TO OZONE DEPLETION

The solution to this problem again provides an example of international cooperation to solve an environmental problem. This issue, like global warming, is a problem for the entire planet and requires international cooperation to solve it. Scientists in the 1970s had already begun to suspect that CFCs were a danger to the ozone layer. At that time, the United States was responsible for about half of the CFCs released worldwide. In 1979, the U.S. government banned the use of CFCs as aerosol propellants, and the worldwide emission of CFCs dropped temporarily; however, this drop was short lived as CFC emissions grew in the rest of the world.

The worldwide perspective shifted with the discovery of the ozone hole in the early 1980s, which created a sense of urgency. We all became aware that *we* had a problem and that *we* were going to have to solve it. Representatives of most of the countries across the world met in Vienna in 1985 and then again in Montreal in 1987. These meetings led to the **Montreal Protocol**, which was signed by 57 countries. In this agreement, the parties agreed to cut their use of most CFCs by 50% by 1998. These countries represented most of the developed countries of the world, as the poorer, less-developed countries refused to sign the protocol. These countries argued that they would not be able to afford to use the more expensive CFC substitutes.

As seen from Figure 11.11, the limits of the Montreal Protocol were not going to do the trick in terms of reducing the levels of CFCs in the stratosphere. In 1990, another meeting was held in London. Two important things happened at this meeting. First, the signatories agreed to phase out all CFCs completely by the year 2000. Second, the developed countries of the world agreed to pay into a temporary fund to help less-developed countries make the transition away from CFCs. With this understanding, countries such as India and China agreed to sign the treaty. In all, 93 countries signed what has come to be known as the London Agreement.

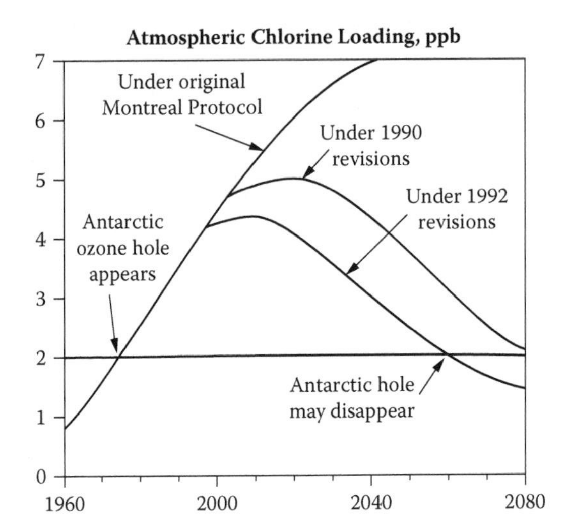

Figure 11.11 Measured (1960–1990) and projected (from 1990 onward) abundances of chlorine in the atmosphere under various agreements. (From Zurer, P.S., *Chemical & Engineering News*, 71(21), 9, 1993. With permission.)

Although Figure 11.11 makes it clear that the atmospheric chlorine levels should decrease under the London Agreement, many felt that the decline would take too long and by then too much damage would be done to the ozone layer. In 1992, representatives of many of these same countries met in Copenhagen. The agreement was made to move up the target date to phase out all CFCs to 1996. In addition, the fund to help less-developed countries make the transition away from CFCs was made permanent. Because of the long life of the materials, lowering levels of chlorine will still take some time, but "preozone hole" conditions could be achieved by the middle of the 21st century.

Although many environmental problems resist international solutions, it is good to know that international cooperation can be successfully applied in some situations. Let us hope that this type of cooperation will also work well for global warming.

DISCUSSION QUESTIONS

1. Do you think that life would have evolved if there were no greenhouse effect? Why? If life had evolved, how do you think it would have been different?
2. The average temperature of the Earth is about 60°F (15°C). Venus would be expected to be a little warmer because it is closer to the sun and might be expected to have an average temperature of perhaps 100 to 200°F (40 to 95°C). The atmosphere of Venus is almost completely composed of carbon dioxide at about 100 times Earth's atmospheric pressure. How does this information relate to the fact that the average temperature on Venus is roughly 840°F (450°C).

3. Many gases emitted into the atmosphere are stronger greenhouse gases than carbon dioxide, yet most of the focus on control of global warming is related to restrictions on the emissions of carbon dioxide. Why is this?
4. Given that historically there have been much greater temperature changes, up to 14°F (8°C), why are we now concerned about temperature rises of up to 11°F (6°C)?
5. Given that the 5-year running mean global temperature in 1975 was about 0°C on the scale used for the data in Figure 11.3, determine the temperature increase from 1975 to 2010. Using Figure 11.3, speculate on what you think the temperature increase might be by 2030. Explain your reasoning.
6. For a given amount of greenhouse gases in the atmosphere, does the presence of huge amounts of water in the oceans have a stronger or weaker effect on climate change? Explain.
7. Melting of ice due to global warming is one way to cause a rise in sea level. Explain how the effect of the melting of sea ice is different from the effect of the melting of land ice with regard to sea level rise. Describe another mechanism leading to sea level rise as global temperatures increase.
8. What do you think should be done about global warming and why? Are international treaties necessary to have an effect on climate change?
9. How does the ozone layer function to protect us from harmful radiation? Why doesn't the ozone get consumed in the process? What happens to the energy contained in the ultraviolet radiation after the radiation interacts with the ozone?
10. How do some chlorine-containing compounds function to destroy the ozone layer?
11. What are some of the uses of CFCs? If they pose such a problem for the environment, why were they developed and used?
12. What is the Montreal Protocol and why is it important?

BIBLIOGRAPHY

Adger, N. et al. and Members of Working Group II of the IPCC, *Climate Change 2007: Climate Change Impacts, Adaptation and Vulnerability—Summary for Policymakers*, Fourth Assessment Report, Intergovernmental Panel on Climate Change, Geneva, Switzerland, 2007.

Alley, R. et al. and Members of Working Group I of the IPCC, *Climate Change 2007: The Physical Science Basis—Summary for Policymakers*, Fourth Assessment Report, Intergovernmental Panel on Climate Change, Geneva, Switzerland, 2007.

Butler, J.H., *Radiative Climate Forcing by Long-Lived Greenhouse Gases: The NOAA Annual Greenhouse Gas Index (AGGI)*, Earth System Research Laboratory, National Oceanic and Atmospheric Administration, Boulder, CO, 2012, http://www.cmdl.noaa.gov/aggi/.

CCSP, *Methane as a Greenhouse Gas*, CCSP-H1, U.S. Climate Change Science Program, Washington, DC, 2006 (http://www.climatescience.gov/infosheets/highlight1/CCSP-H1-methane18jan2006.pdf).

Dlugokencky, E.J., Houweling, S., Bruhwiler, L., Masarie, K.A., Lang, P.M., Miller, J.B., and Tans, P.P., Atmospheric methane levels off: temporary pause or a new steady-state? *Geophysical Research Letters*, 30(19), 5-1–5-4, 1992.

King, D., Richards, K., and Tyldesey, S., *International Climate Change Negotiations: Key Lessons and Next Steps*, Smith School of Enterprise and the Environment, University of Oxford, Oxford, UK, 2011 (http://www.smithschool.ox.ac.uk/wp-content/uploads/2011/03/Climate-Negotiations-report_Final.pdf).

NASA, *Total Ozone Mapping Spectrometer*, Goddard Space Flight Center, National Aeronautics and Space Administration, Greenbelt, MD, 2012.

NOAA, *Ozone Depletion*, Earth System Research Laboratory, National Oceanic and Atmospheric Administration, Boulder, CO, http://www.esrl.noaa.gov/gmd/about/ozone.html.

NOAA, *Trends in Atmospheric Carbon Dioxide*, Global Monitoring Division, Earth System Research Laboratory, National Oceanic and Atmospheric Administration, Boulder, CO, 2013, http://www.esrl.noaa.gov/gmd/ccgg/trends/co2_data_mlo.html.

Schneider, S.H., *Global Warming*, Sierra Club Books, San Francisco, CA, 1989, p. 100.

Solomon, S. et al., Eds., *Climate Change 2007: The Physical Science Basis*, Intergovernmental Panel on Climate Change, Geneva, Switzerland, 2007.

Todaro, R.M., Ed., *Stratospheric Ozone: An Electronic Textbook*, Atmospheric Chemistry and Dynamics Branch, Goddard Space Flight Center, National Aeronautics and Space Administration, Greenbelt, MD, 2003, http://www.ccpo.odu.edu/~lizsmith/SEES/ozone/oz_class.htm.

UNFCCC, *Cancun Agreements: Main Objectives of the Agreement*, United Nations Framework Convention on Climate Change, Bonn, Germany, 2010, http://cancun.unfccc.int/cancun-agreements/main-objectives-of-the-agreements/.

UNFCCC, *Cancun Agreements: Significance of the Key Agreements Reached at Cancun*, United Nations Framework Convention on Climate Change, Bonn, Germany, 2010, http://cancun.unfccc.int/significance-of-the-key-agreements-reached-at-cancun/.

UNFCCC, *Report of the Conference of the Parties on Its Fifteenth Session, held in Copenhagen from 7 to 19 December 2009*, United Nations Framework Convention on Climate Change, Bonn, Germany, 2010 (http://unfccc.int/resource/docs/2009/cop15/eng/11a01.pdf).

USEPA, *Climate Change Indicators in the United States*, U.S. Environmental Protection Agency, Washington, DC, 2006, http://epa.gov/climatechange/science/recentslc.html.

USEPA, *Maps of Lands Vulnerable to Sea Level Rise: Gulf Coast*, Risingsea.net, http://papers.risingsea.net/coarse-sea-level-rise-maps-gulf.html.

Watson, R.T. et al., *Climate Change 2001: Synthesis Report—Summary for Policymakers*, Intergovernmental Panel on Climate Change, Geneva, Switzerland, 2001.

WHO, *Global Solar UV Index, A Practical Guide*, World Health Organization, Geneva, Switzerland, 2002.

Zurer, P.S., Ozone depletion's recurring surprises challenge atmospheric scientists, *Chemical & Engineering News*, 71(21), 9, 1993.

CHAPTER 12

Water

There is probably no single compound with which we have more intimate and continuous contact than water. Water is nearly everywhere and is used in nearly everything. We drink it, eat it, bathe in it, swim in it, and navigate through it. When we get into situations with no water, we could be in a great deal of trouble. Water is essential for life, and all living things contain water. Species that thrive in deserts have efficient ways of conserving water to ensure their survival. Only air is of more immediate necessity for life, but since it is nearly always present we are not as aware of air as we are of water. What makes water so special? Why is life so dependent on it? To understand the importance of water, we must first consider the properties of water.

PHYSICAL PROPERTIES OF WATER

1. *Water is a liquid at room temperature.* Life evolved in liquid water and could not exist without it. Interestingly, if one considers the small mass of a water molecule, one would not expect it to be a liquid at room temperature. Materials such as ammonia (NH_3), hydrogen fluoride (HF), and hydrogen sulfide (H_2S), which have similar or greater molecular masses, are gases at room temperature. There are some good reasons for this unusual behavior, but these reasons are beyond the scope of this chapter. The fact remains that the first important physical property of water is that it is a liquid.
2. *Water is an excellent solvent.* Water is not the universal solvent, but it probably comes as close as any "ordinary" liquid. A large variety of compounds dissolve in water, including many ionic salts, ionic bases, and covalent compounds. The processes of life take place in water, and many of the chemical species involved must be dissolved in water. The fact that water is a powerful solvent means that the water found in nature nearly always has other substances dissolved in it. Rainwater dissolves nitrogen, oxygen, and carbon dioxide, whereas seawater dissolves salts and a number of other things.

3. *Ice is less dense than liquid water.* One may not find it peculiar that ice floats on water, but in the world of physical reality it is an unusual event. There is no other commonly encountered substance where the solid phase of the compound floats on the liquid phase. The fact that ice floats may not seem very important, but the very existence of various life forms, and probably entire ecosystems in colder parts of the Earth, depends on ice that remains on top of the water. If water had properties like other liquids, then when it began to freeze the solid material would sink to the bottom and a partially frozen lake would have ice at the bottom. Normally, fish and other aquatic organisms live under a layer of ice in winter, which protects them from the extremely cold air temperatures. If the ice sank to the bottom, the upper exposed water would be much more likely to continue to freeze until the lake froze solid. In spring, a partially frozen lake would thaw much more slowly because it would be difficult to transfer heat from the warm air to the ice at the bottom. The result would be that lakes, oceans, and other bodies of water in the Arctic, Antarctic, and temperate regions of the Earth would contain a higher percentage of ice and maintain the ice for a greater part of the year.
4. *It takes a relatively large amount of heat to warm or evaporate water.* It takes more heat to warm 1 g of water by 1°C than any other common substance. The consequences of this fact are enormous. Large bodies of water warm very slowly in spring and cool very slowly during fall. These warming and cooling rates have a moderating effect on weather and climate of nearby landmasses. Additionally, because about 72% of the Earth's surface is covered by water, the high heat capacity of water has a significant effect on the climate of the entire planet.

Water also requires a tremendous amount of energy to convert a given quantity into vapor, whether by boiling or evaporation. This chapter focuses on the process of evaporation, as it is more important environmentally. *Evaporation* is the slow conversion of a liquid into its vapor without reaching the boiling point. We are all familiar with this process, as the water in clothes hung up to dry, water left in a glass, or water of the morning dew will eventually disappear. Whenever water is converted into water vapor without the addition of heat, cooling takes place. Heat is removed from the surroundings to provide the energy needed to turn the water into water vapor. A very efficient use of this property by nature is sweating. When one's body senses that it is getting overheated, sweat is produced. The moisture in the sweat evaporates, absorbs heat, and in the process cools the body.

WATER AND LIFE

Water is the solvent of life—all living cells contain water. All of the biochemical reactions of life occur either in water or at the interface between water and some cell structure. Because all living things are made up of one or more cells, all living things depend on water for their survival. The water content varies from species to species, but all species contain water. An adult human, for example, is made up of about 70% water. The presence of reasonably clean water, therefore, is necessary for the existence and maintenance of life.

LOCATIONS OF WATER

A short review of some of the material in Chapter 8, which described the water cycle, may be in order here. Water is constantly being evaporated from the oceans and other bodies of water, land, and vegetation into the atmosphere. After some time this water returns to the Earth as rain or in some cases snow. This water soaks into the ground, runs into the ocean, or evaporates again. A supply of water of some type would not be a problem if water was supplied to Earth's surface uniformly, but rainfall (or snowfall) is distributed unevenly over the Earth. Two key factors give rise to some of the dry areas of the Earth, one of which was discussed in Chapter 8. Warm, moist air rising at the equator tends to sink at 30°N and 30°S latitude as dry air. Many of the world's deserts lie at these latitudes.

The other factor that relates to the amount of rainfall in a particular area is the location of mountains. If air tends to flow in one direction over a range of mountains, a rain pattern will develop. As the air rises and moves up one side of the mountain it cools, forcing the moisture in the air to separate out of the air as clouds or fog. As the air moves further up the mountain, the clouds thicken, and eventually it will rain (or snow). After it rains, the air passes over the top of the mountain and begins its journey down the other side of the mountain. This air contains very little moisture, so the likelihood of rain is very small. The area on the other side of the mountain is said to be the mountain's **rain shadow**. The more predictable the winds and the higher the mountains, the stronger the effect. Much of Nevada is in the rain shadow of various mountain ranges. The rain shadow of the Andes Mountains in northern Chile is so strong that some towns in the Atacama Desert have had no rain recorded in history. According to the location of water, our planet can be divided into nine *water compartments*: oceans; glaciers, ice, and snow; **groundwater**; lakes and ponds; **soil moisture**; rivers and streams; the atmosphere; biological moisture; and **wetlands**.

Oceans

Oceans represent by far the largest compartment. The oceans contain about 97% of all the water on the planet. Although the oceans contain this incredibly large amount of water, it is not useful for some purposes because it is saltwater, which is not generally suitable for supporting human life. One can boat on or swim in it, and fish and other forms of marine life can live in it, but one cannot drink it, and it cannot be used to irrigate crops. The residence times for water in the oceans are extremely long. The water near the surface has the shortest residence time, about 3000 years, whereas deeper waters may have residence times as high as 30,000 years. Despite the fact that water stays in the ocean for a very long time, oceans still account for 88% of the total evaporation of water into the air. When water evaporates, the salt dissolved in it is left behind. Because the salt does not evaporate, the water vapor in the air contains no salt. The oceans provide most of the moisture to the atmosphere from which the world's rain is produced. The oceans, therefore, are indirectly the source of most of the world's freshwater.

Ice Caps, Glaciers, and Snow

With the exception of a few salt lakes, the rest of the world's water is freshwater. The catch is that three-fourths of this water is frozen. Incredible amounts of water are tied up in the polar ice caps. Additional water is frozen in mountain glaciers and in varying amounts of snow. Some of this water remains in the frozen state for extremely long periods of time, whereas some snow may not last long at all. The residence times vary all the way from less than 1 year to 16,000 years. The ice in some glaciers is much older than that. All of this frozen water represents about 2% of all the world's water, which, when added to the 97% found in the oceans, leaves only 1% remaining for human and other land-based uses.

Groundwater

All but a small percentage of the remaining water is groundwater. The oceans, frozen water, and groundwater account for 99.95% of the world's water. Groundwater has been important throughout human history. There are many biblical references to important wells in the Old Testament. At a fairly early point in human evolution, people figured out that if one dug a hole deep enough this hole would partly fill up with water. The well became an important source of freshwater, surpassed only by the more convenient stream or lake. To understand why one can dig a hole in the ground and have it partly fill up with water, one needs to look at where water is located underground.

Most soils contain a significant amount of air spaces between the soil particles. When rain falls on the ground, much of it soaks in and occupies the air spaces in the soil. Gravity acts on the water and pulls it further and further down into the ground. This process will continue until the water reaches some type of impermeable layer such as rock or clay. As more and more water infiltrates, a layer containing no air (only water mixed with soil) will develop above the layer of clay or rock. This layer is known as the **zone of saturation**. The soil above the zone of saturation that contains soil, air, and some residual water is known as the **zone of aeration**. The dividing line between these two zones, which is the top of the zone of saturation, is known as the **water table**. The water table is the depth below which one must dig for a well.

Soil and water positioned above a layer of rock or clay is known as an **aquifer**. What I have presented so far is a fairly simple, straightforward picture of an aquifer, but the situation is not always that simple. Consider, for example, the situation shown in Figure 12.1. Two aquifers are shown with one lying directly over the other, except in one small area. When drilling a well it would be possible to tap into either aquifer, depending on how deep the drilling goes. The big difference between these two aquifers is the ease with which each one is replenished with water. The upper aquifer, which is known as an **unconfined aquifer**, can receive any of the rain that falls above it. This aquifer is said to have a large **recharge zone**, which is the area where water infiltrates into the aquifer. The lower aquifer can only receive water at the small exposed area at one end. This type of aquifer is known as a **confined aquifer**. Such aquifers generally have small recharge zones, but under natural processes

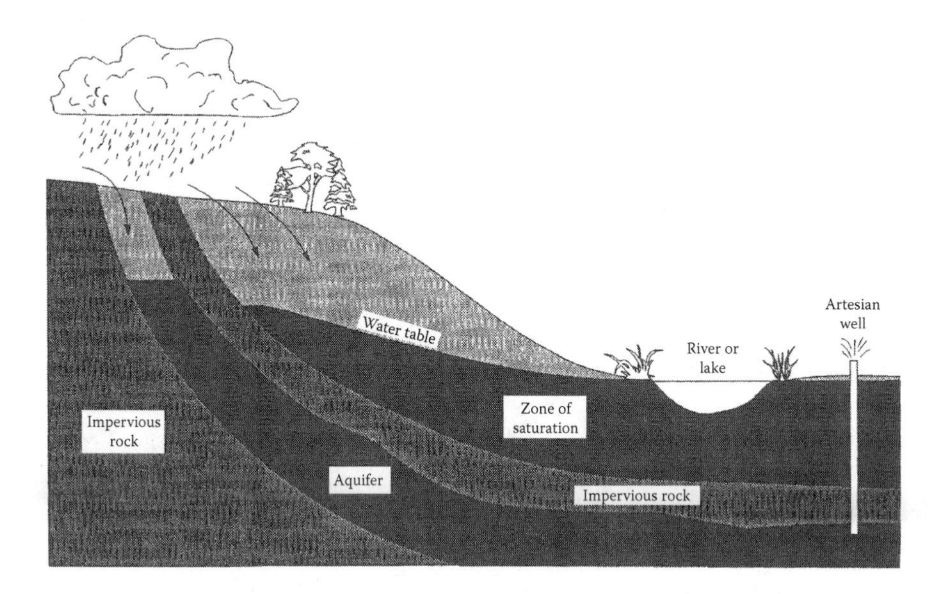

Figure 12.1 Confined and unconfined aquifer. (From Turk, J. and Turk, T., *Environmental Science*, 4th ed., Saunders, New York, 1988.)

the water is generally not removed from them at any significant rate. These differences between aquifers explain the large variation in residence times for groundwater. Water near the surface may have residence times of a few days, whereas water in some deep confined aquifers may have residence times of tens of thousands of years.

Aquifers can be a stable source of freshwater, if there are not too many wells. If the number of wells is too high or the recharge rate is too low, the aquifer can become depleted. Aquifer depletion can be a problem for unconfined aquifers in dry areas or for confined aquifers almost anywhere. The problem for confined aquifers relates to their small recharge zones relative to the volume of the aquifer. If one pumps water out of an aquifer faster than it is replaced, the volume of water in the aquifer will decrease. Sooner or later, the water will run out; hence, the water in the aquifer can be viewed as a nonrenewable resource.

Lakes and Ponds

Lakes and ponds are inland depressions that hold water. There is no strict difference between a lake and a pond; lakes are just bigger. In comparison to groundwater, this water compartment is fairly small, but in many places in the world lakes have been an important source of freshwater. The reason is simple: convenience. Early in human history many villages sprang up by lakes. When water was needed, digging a well was unnecessary. All one needed to do was get water from the lake. Lakes and ponds are temporary features of the geological landscape. All freshwater lakes have inlets and outlets. As time passes, silt is brought in the inlet and the ridge that forms the outlet is eroded away. The lake will over time either fill in with silt or empty when the outlet ridge is eroded away. Lakes vary considerably in size, as does the residence time of the water in them. The residence time of water in some deep lakes may be as much as 100 years, whereas the water in some very shallow lakes may have a residence time of only about 1 year. Lakes and ponds are the largest surface freshwater compartment and contain over 100 times the amount of water in rivers and streams. The Great Lakes of Canada and the United States alone contain about 20% of all the surface freshwater in the world.

Soil Moisture

This is the water found underground, in the zone of aeration. It is not counted with groundwater because one cannot access it by digging a well. It is instead the dampness in the soil that one finds when digging a shallow hole. This moisture is important for the growth of plants, but it is not accessible for human consumption. The amount of water in the soil moisture compartment is about half that in lakes and ponds. Residence times are generally short, about 2 weeks to a year.

Rivers and Streams

The remaining compartments are extremely small, but not unimportant. The rivers and streams compartment has been very important to the history of mankind. Historically, cities and entire civilizations grew up along rivers. Rivers provide drinking water, transportation, nutrients for the soil, and a way of removing unwanted materials, such as human wastes. Much of the attractiveness of rivers and streams comes from the fact that they generally contain reasonably fast-moving water. Such water can arrive fresh and drinkable, and when it leaves it takes with it unwanted materials. As expected, the residence times are short, about 30 days or less.

Atmosphere

Another very important and very small water compartment is the atmosphere. Although the amount of water in the atmosphere at any one given time is very small, all water that ends up in rivers, lakes, soil, and groundwater comes from the atmosphere as rain or snow. The trip of water through the atmosphere is fairly short, with about a 10-day residence time.

The two remaining small compartments, *biological moisture* and *wetlands*, are not discussed in this chapter. These compartments are important, but they are more biological in nature.

TYPES OF WATER USE

When we talk about the use of freshwater by humans, it is necessary to define what we mean by the use of water. There are many ways in which people use water. They may take the water out of a lake or stream to use it and then return it to the

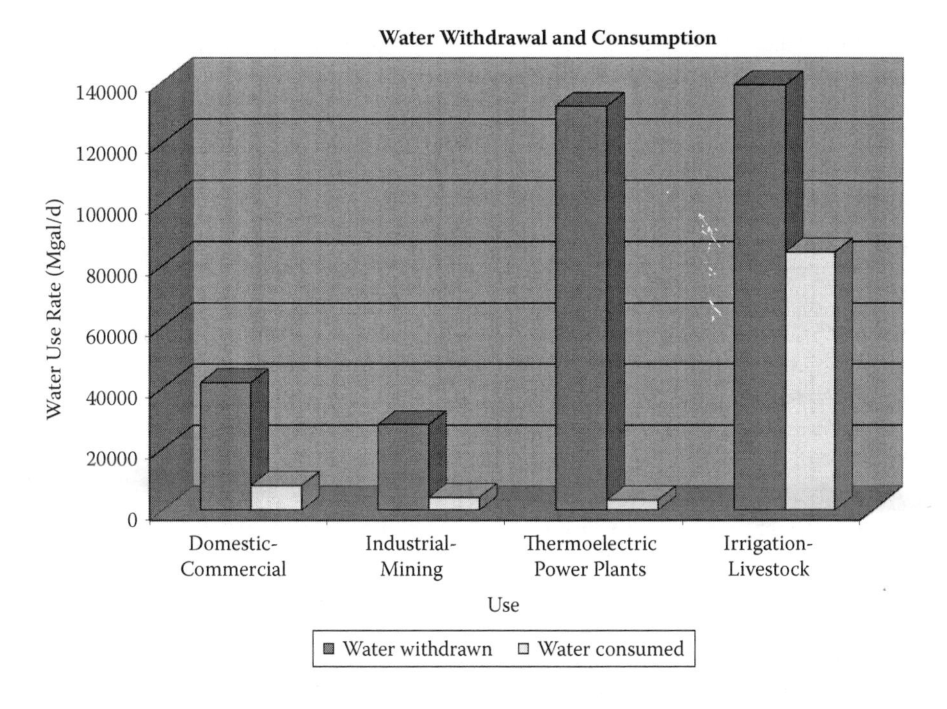

Figure 12.2 Water withdrawal and water consumption in the United States for 1995. (Based on data from Solley, W.B. et al., *Estimated Use of Water in the United States in 1995*, USGS Circular 1200, U.S. Geological Survey, Denver, CO, 1998.)

same body of water with very little change. Or, the water removed may not be returned to the lake or stream from which it was removed. Water that is removed from a body of water or groundwater and not returned as liquid water to the general area from where it was removed is said to be *consumed*. Water may be lost in any number of ways. It may evaporate into the atmosphere, be incorporated into biological material or into soil moisture, or undergo chemical transformation. Some water may be returned to its source but at a lower quality than when it was removed. This contaminated water is said to be *degraded*. All of the water that is taken for any purpose, whether returned (degraded or not) or consumed, is said to be water that is *withdrawn*.

The amount of water used for various purposes depends a great deal on whether one is discussing water consumption or water withdrawal. As seen from Figure 12.2, a very large water user in the United States is the thermoelectric power industry. Power companies use the water to cool spent steam in their electric turbine generators. This water is largely returned to the lake or stream from which it was removed very little changed, except for the fact that it is a little warmer; hence, the water is withdrawn but not consumed.

Agricultural irrigation is also a major withdrawer of water, but there is a major difference when compared to the use of water by power plants. As irrigation water flows out into the fields, a large amount of the water evaporates and that which does not soaks into the soil, as it should. Once in the soil, the water is taken up by the plants, becomes part of the soil moisture, or evaporates. This water will not be returned to the water source from which it was taken and is, therefore, said to be consumed. Irrigation is the largest consumer of water in the United States, as seen from Figure 12.2.

Although domestic and commercial use of water is dwarfed by irrigation and power plant use, it is certainly important. As seen from Figure 12.2, only a small percentage of the water withdrawn is consumed. Much of the water we use in the home is for flushing toilets or for various types of washing. This water, although contaminated, is not lost but is returned through the sewer system to a river or lake. In modern industrial societies, this water goes through a sewage treatment plant before being returned to a body of water. Very few domestic activities actually consume water. Watering houseplants, the lawn, and the garden along with cooking would be some examples of domestic water consumption.

FRESHWATER SHORTAGES

A shortage of freshwater is not a new problem. In certain areas of the world where water has always been scarce, humans have been struggling with a lack of water for centuries; however, other regions of the world have not historically had a freshwater supply problem. Our current water shortage difficulties have been aggravated by three related demons: population, development, and pollution. As the world's population has increased, areas with marginal water supplies are straining an already limited resource. Furthermore, many parts of the world desire a higher standard of living. As these societies undergo industrial development, water consumption goes up. A case in point is the North China Plain, which is a semiarid region supporting a population of more than 200 million people that includes the major cities of Beijing and Tianjin. Water for irrigation and other purposes is being pumped out of the underlying aquifers at a rate that cannot be sustained. The nearby Yellow River has had water withdrawn in such quantities that at times the river does not flow into the sea.

Population and development can add to the water problem in another way. In more mobile societies of the developed world, people like to move to warmer, sunnier locations, but sunnier locations are often also drier locations. Many of these mass migrations aggravate already serious water problems. The increase in population and the drive for industrial development can lead to greater pollution, thus reducing the supply of usable freshwater.

A lack of proper wastewater treatment can become a serious issue. In the developed world, we tend to expect that raw sewage will not be dumped into streams, but sewage treatment is the exception, not the rule, for most of the world. Many areas of the world are faced with the unpleasant choice of either finding alternative water sources or drinking polluted water.

PROBLEMS FROM OVERUSE OF GROUNDWATER

A population that wishes to use more water than is available in rivers, streams, and lakes may look to groundwater to make up for the difference. If the need for water is moderate, no adverse effects may be seen; however, excessive withdrawal of groundwater can cause some serious problems.

Aquifer Depletion

When wells are drilled in an aquifer, the aquifer can provide water to the wells indefinitely, as long as the rate of water withdrawal does not exceed the rate of water gain. In many semiarid regions of the world, wells are a common and usually dependable source of water, but during a prolonged drought the well may dry up. The reason for this is that no water is being added to the aquifer while more water is being taken out of it to irrigate crops and to water livestock.

When the population is small, aquifer depletion is usually a local, and perhaps temporary, event; however, aquifer depletion becomes a more significant problem as populations grow or when the per capita consumption of water becomes excessive. Both of these factors are at work in today's world. Since 1950, the population of the United States has roughly doubled; yet, the U.S. withdrawal from the public supply of water has gone up by more than a factor of three. Obviously, there is more at work here than simple population pressure. In absolute terms, the greatest increase in water withdrawal since 1950 is for irrigation. Globally, the withdrawal of water for agricultural purpose increased by 136% between 1950 and 2000. As the various countries of the world struggle to maintain or expand their food supply, farmers press into production many areas receiving marginal rainfall.

Water contained in some large confined aquifers is water that has been there for many thousands of years. Removing water from these aquifers faster than it can be replaced amounts to mining "fossil water." This water represents a nonrenewable resource. An example of such a reservoir of underground water is the Ogallala Aquifer of the central United States. Located under Nebraska and seven other states, the Ogallala Aquifer is the largest known underground freshwater reservoir in the world. The aquifer contained at one time about 16 times the amount of water in all of the surface freshwater bodies put together. This water was first exploited by settlers in the latter part of the 19th century. They used windmills to pump water for livestock and for personal use. The small amounts they used had little effect on the vast amounts of water in the underground reservoir. For many decades, the farmers in the area carried out what is called "dry land" farming (i.e., farming without irrigation). Crop yields were modest and an occasional drought year caused some complete crop failures. The Dust Bowl drought of the early 1930s nearly devastated the entire region. Then, in the 1960s, mechanized irrigation equipment became available that could irrigate large fields using water from the Ogallala Aquifer. Soon crop yields shot up as huge quantities of water were drawn from the aquifer. At the height of irrigation, more water was being pumped from the aquifer each year than the entire annual flow of the Colorado River. As one may have already guessed, this rate of flow cannot be maintained. The water table has dropped as much as 150 ft (45 m) in some places, and the aquifer has dried up completely at other locations. The aquifer still has lots of water at some locations, but clearly unless there is some change it will give out at some point.

Subsidence

Groundwater filters into the soil and fills in the air spaces; however, it has more of an effect than simply filling air spaces. Just as a ball filled with water is much less compressible than one filled with air, soil saturated with water is much less compressible than soil containing many air spaces. If we pump large amounts of water out of an aquifer, the soil begins to compress and the surface begins to sink. This process is called **subsidence**. In the earlier mention of the water problem in the Beijing–Tianjin area of China, we noted the unsustainable withdrawal of water from the underlying aquifer. Because Beijing has a shortfall in water supply, and the difference is made up by drawing on groundwater, the water table has dropped about 250 ft (75 m) in some areas, and as a result the ground level of Beijing has sunk about 1.5 ft (0.5 m).

One of the more striking examples of subsidence is the San Joaquin Valley of central California. This heavily irrigated agricultural valley has sunk about 30 ft (9 m) in the past 75 years. Many cities of the world are experiencing severe subsidence; Table 12.1 lists some of the more extreme examples, and a striking illustration of subsidence can be seen in Figure 12.3. The most graphic result of subsidence is a phenomenon known as a *sinkhole*. Although sinkholes can and have occurred in other places, they have most commonly occurred in Florida. The soil in Florida is very sandy, and when aquifer depletion leads to subsidence a portion of the ground may simply collapse, leading to a large hole in the ground which may grow over time. As shown in Figure 12.4, entire houses can be swallowed up in these sinkholes. These holes can form without warning in a matter of minutes.

Saltwater Intrusion

Saltwater intrusion into wells is a problem in coastal areas experiencing aquifer overdraft. In coastal areas, saltwater from the ocean seeps in underground until it meets underground freshwater. An interface develops between them, with the

Table 12.1 Subsidence of Some of the World's Cities

Subsiding Cities	Maximum Subsidence
Tokyo	14.8 ft (4.5 m)
San Jose	12.0 ft (3.7 m)
Los Angeles	29.5 ft (9.0 m)
Mexico City	30.0 ft (9.0 m)
San Joaquin Valley	29.0 ft (8.8 m)
New Orleans	6.5 ft (2.0 m)

Figure 12.3 A photograph of the San Joaquin Valley that strikingly illustrates subsidence. (From Galloway, D.L. et al., *Land Subsidence in the United States*, FS-165-00, U.S. Geological Survey, Denver, CO, 2000.)

saltwater remaining under the freshwater because it has a higher density. As seen from Figure 12.5, coastal wells draw from the freshwater above the underground seawater, but if too much freshwater is removed then saltwater will intrude into the freshwater well.

WATER SHORTAGE SOLUTIONS

Generally, there are only three solutions to water supply problems: (1) clean up water that has been already used, (2) bring in water from somewhere else, or (3) use less.

Desalination

Obviously, the largest water compartment in the world is the oceans. If one is in a location near the ocean that is experiencing a water shortage, it would probably occur to that person that it would be nice to use the water in the ocean. But, how can

Figure 12.4 Sinkhole in Winter Park, Florida, covering a major portion of a city block. (From Galloway, D.L. et al., *Land Subsidence in the United States*, FS-165-00, U.S. Geological Survey, Denver, CO, 2000.)

that be done? Saltwater is not directly useful to support most forms of nonmarine life, but there are a couple of ways to make seawater useful for human consumption or agricultural purposes. These processes are known are **desalination**.

Anyone who has put saltwater in a pan, put a lid on the pan, and then heated the pan on a stove has seen how to get freshwater from saltwater. When the lid is removed, water will be found on the underside of it. This water is nearly pure (assuming no saltwater splashed up on the lid). The water in this case came from water vapor (steam) condensing back into liquid water on the cooler surface of the

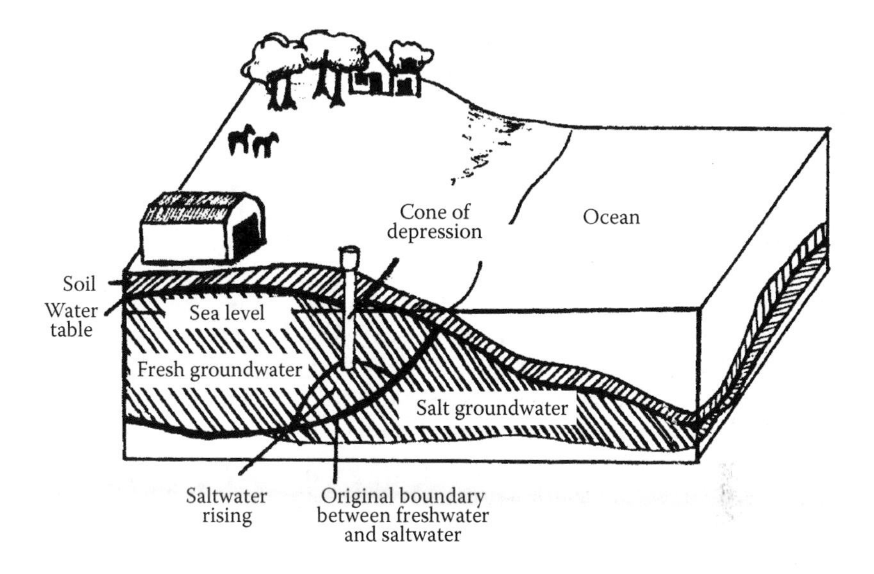

Figure 12.5 Illustration of saltwater intrusion into a well. (Courtesy of Amy Beard.)

lid. When water is heated, water vapor can be driven off, but the salt does not vaporize. If we boil seawater and then cool and collect the steam, we can get freshwater. This process is known as **distillation**.

The other process for obtaining freshwater from seawater is known as **reverse osmosis**. The process involves the use of a membrane that allows water molecules to pass through it but does not allow salt to pass through. Such a membrane is said to be **semipermeable**. If one puts seawater on one side of this membrane and applies pressure, the water passes through the membrane but the salt does not. The water that appears on the other side is freshwater.*

The biggest problem with either of these technologies is their cost. The energy and equipment costs are so high that only more developed countries can afford them. As one might guess, the biggest application of these processes is in the Middle Eastern oil-producing countries, which are probably the only countries in the world with more oil than freshwater. If there is no other way to get water and one has the money, these methods become attractive. Currently, the Middle East has about 47% of the global capacity. The major countries involved include Saudi Arabia, Kuwait, and the United Arab Emirates. Desalination is also attractive on islands that have limited rainfall, few if any flowing steams, and little groundwater. Many islands of this type in the Caribbean, for example, must either ship in freshwater or carry out desalination. The island of Curaçao in the Netherlands Antilles has been desalinizing water continuously since 1928.

* This process is referred to as *reverse osmosis* because without the application of pressure the water would flow the other way—that is, from the freshwater to the seawater.

Over time, the process is becoming less expensive and the cost of other water sources is becoming more expensive. From December 1986 to December 1989, the number of desalinization plants more than doubled. There are plants in more than 120 countries, and the United States is now second in the world, with most of the U.S. plants being found in South Florida.

Moving Water: Canals and Aqueducts

The idea of moving water through aqueducts or canals is not new. Many cultures of the ancient world had extensive irrigation canals. The Romans built aqueducts that are still in use today. The key difference between then and now is the sheer scale of water being moved. Of course, any large-scale water-moving project is going to be expensive, and, as is the case with desalination, it can only be carried out by the more developed countries. The reason why moving water is attractive is because water consumption is largest in the driest areas, where people have to irrigate their crops, water their lawns, and so on. In contrast, water consumption is less in areas where there is water. The western part of the United States is generally dry, and historically water rights have been a significant issue in the development of this region. The eastern United States, however, has generally had sufficient water, with the possible exception of Florida..The western U.S. water situation, which was already difficult, was complicated by the California and Arizona population explosions, which added millions of people to an already dry area. This situation was driven to crisis levels with a drought in the late 1980s and early 1990s.

The story of California water diversion begins in the early 1900s. The city of San Francisco proposed to build a dam in the Hetch Hetchy Valley in the Yosemite National Park to provide water and electricity to the city. It was decided that the water would be brought to San Francisco via an aqueduct. John Muir, who is best known as the founder of the Sierra Club, was outraged that such a reservoir would be placed in the midst of the beautiful Yosemite National Park. Muir and others put up a long, hard fight against the project, but they lost and the dam was built. This dam marked the beginning of the era of diverting water to the urban and agricultural areas of California.

At about the same time Los Angeles began to look for water. In 1913, Los Angeles built the Los Angeles Aqueduct to divert water from Owens Valley on the eastern slope of the Sierra Nevada Mountains. Owens Valley, near Yosemite, is a landlocked valley with no water outlet. The water from the streams in the valley collects at the valley floor, forming a salt lake called Mono Lake. As Los Angeles grew, so did the amount of water diverted from Owens Valley. By the 1950s, the Los Angeles Department of Water and Power (LADWP) was taking water from four of the five streams that fed Mono Lake. Mono Lake began to dry up and its salinity rose. Many birds were threatened when their source of food, freshwater shrimp, died due to the increased salinity. Stream trout also disappeared as some streams dried up completely.

The years of 1981 and 1982 were wet years in California. These years were so wet that in fact there was more water than the LADWP could use. The streams were refilled with water and trout returned. With the expectation that dry years would return, the National Audubon Society brought suit against the LADWP to prevent them from withdrawing water from the largest stream because under California law it is illegal to endanger an existing fishery. Out of this lawsuit came the concept of the **public trust**. The California Supreme Court ruled that government has some obligation to maintain the natural state of public lands or, in other words, to protect the public trust. The decision was not clear cut, however, as the court also ruled that LADWP has certain rights to water to protect the population and economic status of Los Angeles. In 1994, the California State Water Resources Control Board required that the water to the lake be sufficient to keep Mono Lake at least 6392 ft above sea level. Any water beyond that could be diverted to Los Angeles.

The Los Angeles Aqueduct was also the beginning of what came to be known as the California Water Plan. This system involves utilizing a massive collection of aqueducts to bring water from the northern mountains of California, where two-thirds of the rain and snow in California falls, to central and southern California, where 80% of the water is used. These water transfers have not been without controversy. Northern California does not give up its water easily, and the aqueduct systems used to transfer the water are expensive. In addition, in a hot, dry climate, much of the water evaporates before it reaches its destination. Because all natural waters contain some salts, evaporation leads to an increase in salinity of the water, which can lead to problems when the water is used for irrigation.

Not all of the California water comes from the northern mountains. The other main source is the Colorado River. This river is a particularly important source for San Diego and the Imperial Valley of extreme southern California. The Colorado River is one of the important rivers of the American West. It is so important that the water in the river has been carefully divided up legally between the several states bordering the river and Mexico, where the river empties into the Gulf of California. Arizona was entitled to one-fifth of the water flowing in the river but had no way to transport that amount of water to where it was needed. In 1985, the Central Arizona Project was completed which pumps water uphill to Phoenix and Tucson. With the completion of this project, Arizona was able to retrieve its portion of the Colorado flow at the expense of California, which had been using some of Arizona's portion.

Given the desperate need of the American West for water, even more ambitious projects have been proposed. Western North America does have water but it is further north in Washington State, in the Canadian Province of British Columbia, and in Alaska. For example, the Fraser River of British Columbia has an annual flow equal to almost twice the annual run-off of the entire state of California. Given the legal and practical complications of current projects, it is not difficult to imagine the problems to be encountered with an international long-distance water-transfer project. The obvious benefit of these water projects is that the water gets to where it is needed, but there is a price to be paid. Again, there is the problem of increased salinity from evaporation, and the consequences of building dams, which were discussed in Chapter 7. (Nearly all water diversion projects start by damming up the source river.)

Water diversion can have other profound effects on the source river. When large amounts of water are removed from a river, the water flow obviously slows. The flow rate of a river when it empties into the sea determines how far the river will push freshwater out into the ocean. If the river slows too much, seawater may back up into the river channel. Clearly, large-scale water diversion projects can dramatically affect the salinity in areas around the mouth of a river.

One of the issues in environmental studies is that of energy consumption (see Chapter 7). Long-range water diversion usually involves moving water over hills or through mountain passes. Energy must be expended to lift the water to the top of a hill or a pass. Some of this energy can be recovered as the water flows down the other side, but not all of it. There is always frictional resistance to the flow of water which converts kinetic energy into heat energy. As a result, most water-transfer projects consume huge amounts of energy; for example, the estimated energy consumption required to move the irrigation water in California is more than 10 billion kWh of electricity annually. This amount of electricity could power about 2 million homes.

Clearly, the large populations in some arid areas of the world are not going to suddenly move elsewhere. Because water transfer or desalination is difficult and expensive, what are the other options? The answer is water conservation: Use less water. This option is the least expensive and in the long run the least complicated option.

Water Conservation

It is estimated that 30 to 50% of the water used in the United States is wasted. For those living in a part of the world that has a plentiful water supply, such wastage is probably not a big issue. In the arid regions of the world, however, wastage can mean the difference between having to build a new water-transfer project or having sufficient water under current circumstances. Water wastage can be reduced; for example, in Israel, in the years between 1950 and 1980, water wastage was reduced from 83% of total consumption to 5%.

One reason for wasting water is having cheap water. For example, consider the case of Vancouver, British Columbia, in comparison to Seattle, Washington. When domestic water was not being metered in Vancouver, the average daily per capita consumption was 198 gal (751 L), whereas just a little further south in Seattle, where consumption was metered, the average daily per capita consumption was 127 gal (480 L). Under really severe conditions, even bigger savings can be noted. During a California drought, the average water bill in Santa Barbara, California, went up from $22 to $135 per month and water usage dropped by 64%.

The fact is that water in most arid areas of the United States is artificially cheap. Local government officials try to keep water inexpensive to encourage development and to keep voters happy. Until the mid-1970s, most giant water projects were heavily subsidized by the federal government. As federal, state, and local subsidies continue to help water bills stay low, wastage continues, and everyone pays for it in taxes (or increased government deficit).

Traditional irrigation is wasteful, as 63% of the water is lost, mostly by evaporation. Despite the wastefulness of traditional irrigation, it will continue to be used as long as water is inexpensive, so there needs to be some incentive to conserve the water. The typical method of irrigation involves running water out into the field either in ditches or as a general flood over the entire field. Unfortunately, much of the water, rather than going to the plants, either evaporates or simply runs off of the field and is wasted. In certain places, such as the Great Plains and in the Middle East, sprinkler systems have been used. If calibrated properly, such systems can cut down on runoff, but evaporation is still a problem because the water is sprayed into the air. Such systems are fairly capital intensive and tend to be preferred when groundwater is used rather than surface water.

The best way water can be conserved is by using drip irrigation. The water is run through a pipe with holes in it near the base of the plants. These systems can be expensive and sophisticated or can utilize simple pipes with holes drilled into them and attached to an elevated hand-filled reservoir. The less expensive systems make it possible to use drip irrigation in places such as Nepal.

Water can also be conserved on the domestic and industrial fronts. One of the major ways in which water is lost in urban areas is through leakage. Many of the water systems in urban areas are old and are not well maintained. In Mexico City, for example, the water system loses enough water each year to supply the needs of the city of Rome. Leak detection can be made an urban priority. In addition to the municipal water system being inspected for leaks, property owners can be encouraged to find leaks by charging them for the water they actually withdraw and by providing assistance in searching for leaks. This approach was followed by New York City after a drought in the early 1990s left their supply about 90 million gal of water per day short of their needs. By being diligent about stopping water leaks and taking other measures, New York avoided the need to find new water sources.

One of the other measures that New York City took to save water was the conversion to low-flow toilets. Over a 3-year period from 1994 to 1997 the city provided free low-flow toilets to anyone who would use them to replace older, higher flow toilets. This issue is not small in terms of water conservation, as toilet flushing accounts for 38% of the water used in the typical U.S. household. Most of the older standard toilets use 5 gal of water or more per flush, whereas the low-flow models usually use about 1.6 gal per flush. The water saving here is obvious, and the low-flow models seem to work just as well.

The reuse of gray water is another approach to water conservation. **Gray water** is water resulting from, for example, bathing, dishwashing, and hand washing. This water can be reused for watering gardens or flushing toilets. One difficulty in setting up a system to recycle gray water is that these arrangements sometimes violate existing health codes.

The fact is that any water, no matter what it was used for, can be cleaned up and reused for any purpose. Most of us simply have an aversion to drinking water that was purified from the effluent of a sewage treatment plant, although most water is recycled from somewhere else. If city A has an intake for the municipal water supply on the local river and city B is upstream from city A, then some of the sewage treatment effluent from city B is going to end up in city A's drinking water supply. Water is recycled many times through the natural water cycle. One city that was literally forced to recycle sewage water is Windhoek, Namibia, in southern Africa. Southwest Africa is a desert region, and Windhoek as the capital city and major commercial center grew from a population of 61,000 to 230,000 in about 30 years. The options for water were simply too limited. The nearby rivers only had water in them for part of the year and the draw on the groundwater supply was more than could be sustained. The nearest major, continuously flowing river was about 400 miles away, and transporting the water that far was simply too expensive in this case. Despite having access to very clean water produced from the sewage effluent, many people in Windhoek are still resistant to drinking it.

A lot of water conservation is about common sense and simple, practical steps. Some ideas include taking short showers rather than baths, using low-flow showerheads, and not letting the water run while brushing your teeth or shaving. Many more ideas can be found on the Internet at various sites, including a New York City site, http://www.nyc.gov/html/dep/html/ways_to_save_water/index.shtml, and the California Earth911 site at http://california.earth911.org.

Domestic conservation programs can have lasting effects. In 1977, California experienced a drought and Marin County put in place a water conservation program to encourage residents to conserve water. Water consumption fell by 65%, and after the drought consumption was still down by 45%.

DISCUSSION QUESTIONS

1. What would be the consequences to life in the temperate zone and the subarctic regions if ice were more dense than liquid water?
2. Explain why weather is influenced significantly by oceans.
3. Why is water called the solvent of life?
4. Explain why there is usually little rainfall on the leeward side of mountains. What would be your expectation for rain on the windward side of mountains?
5. Oceans, frozen water, and groundwater account for 99.95% of the world's water. Indicate the problems related to using each of these water compartments for human activities. Explain any procedure that would allow the use of each.
6. Both thermoelectric power plants and irrigation use very large amounts of water; however, the impact on a water system is very different between the two. Explain.
7. Why do the driest areas of the United States have the greatest consumption of water?
8. In some cases, water removed from an aquifer is considered a renewable resource and in some cases it is considered a nonrenewable resource. What factors account for this difference?
9. Referring back to Figure 11.7 on sea level rise, why is the sea level rising faster in Galveston than in other U.S. coastal cities?

10. Some islands in certain parts of the world get very little rainfall. Some of these islands with significant populations often either import water or desalinize seawater to obtain water for domestic purposes. Another approach would be to drill a well. What might be some difficulties in using well water in these situations?
11. Explain how reverse osmosis works to purify seawater. How does reverse osmosis differ from osmosis? Is there any method for desalination of seawater that does not require significant input of energy? Explain.
12. What is the public trust doctrine and how does it relate to environmental issues? Does this doctrine have application to environmental issues other than water? Give some examples.
13. What are some of the major problems with long-distance water transfer to solve water shortages in certain locations?
14. How do you think California should solve its water problems?
15. What are the major domestic indoor uses of water, and for each use how can the water be conserved?
16. What are some of the advantages and disadvantages of the reuse of sewage effluent water? Would you drink the water?
17. Should population be controlled in areas with very limited water resources?

BIBLIOGRAPHY

Arms, K., *Environmental Science*, Saunders, New York, 1990, p. 217.

Buros, O.K., *The ABCs of Desalting*, 2nd ed., International Desalination Association, Topsfield, MA, 2000.

Cech, T.V., *Principles of Water Resources: History, Development, Management, and Policy*, John Wiley & Sons, New York, 2003.

Committee on Advancing Desalination Technology, *Desalination: A National Perspective*, National Academies Press, Washington, DC, 2008.

CUWCC, *Save Water, Money, Energy Now! Top 5 Actions*, California Urban Water Conservation Council, 2006, http://www.h2ouse.org/action/top5.cfm.

Department of Community Affairs, Office of Environmental Management, State of Georgia, *Every Drop Counts*, Earth911.com, http://earth911.com/location/georgia/water/every-drop-counts/.

Dolan, R. and Goodell, H.G., Sinking cities, *American Scientist*, 74, 38–47, 1986.

Galloway, D.L., Jones, D.R., and Ingebritsen, S.E., *Land Subsidence in the United States*, FS-165-00, U.S. Geological Survey, Denver, CO, 2000.

Gaughen, B., Making every drop count: water conservation in Santa Barbara, *Western City*, September, 14–17, 1990.

Gleick, P.H., Making every drop count, *Scientific American*, 284(2), 40–45, 2001.

Henthorne, L., *The Current State of Desalination*, International Desalination Association, Topsfield, MA, 2011.

Kenny, J.F., Barber, N.L., Hutson, S.S., Linsey, K.S., Lovelace, J.K., and Maupin, M.A., *Estimated Use of Water in the United States in 2005*, Circular 1344, U.S. Geological Survey, Denver, CO, 2009 (http://pubs.usgs.gov/circ/1344/).

Klenin, G., *California's Water–Energy Relationship, Final Staff Report*, CEC-700-2005-011-SF, California Energy Commission, Sacramento, 2005.

Kusky, T.M., *Geological Hazard: A Sourcebook*, Greenwood Press, Westport, CT, 2003.

Larmer, B., Bitter waters: can China save the Yellow—its Mother River?, *National Geographic Magazine*, May, 146–169, 2008 (http://ngm.nationalgeographic.com/2008/05/china/yellow-river/larmer-text).

Leake, S.A., *Land Subsidence from Ground-Water Pumping*, U.S. Geological Survey, Denver, CO, 2004 (http://geochange.er.usgs.gov/sw/changes/anthropogenic/subside/).

Martindale, D. and Gleick, P.H., How we can do it, *Scientific American*, 284(2), 38–41, 2001.

Milstein, M., Water woes, *National Parks*, May–June, 39–45, 1992.

Mono Lake Committee, *The Public Trust Doctrine at Mono Lake*, Mono Basin Clearinghouse, January 10, 2007, http://www.monobasinresearch.org/timelines/publictrust.htm.

Polak, P., The big potential of small farms, *Scientific American*, 293(3), 84–91, 2005.

Postel, S., Growing more food with less water, *Scientific American*, 284(2), 46–51, 2001.

Reisner, M., *Cadillac Desert: The American West and Its Disappearing Water*, Penguin Books, New York, 1993.

Rogers, P., Facing the freshwater crisis, *Scientific American*, 299(2), 46–53, 2008.

Rudolph, M., Urban hazards: sinking of a titanic city, *Geotimes*, July 2001, http://www.geotimes.org/july01/sinking_titanic_city.html.

Sachs, J.D., The challenge of sustainable water, *Scientific American*, 295(6), 48, 2006.

Shiklomanov, I.A., *Evolution of Global Water Use: Withdrawal and Consumption by Sector*, State Hydrological Institute, St. Petersburg, and United Nations Educational Scientific and Cultural Organisation, UNESCO, Paris, 1999.

Solley, W.B., Pierce, R.R., and Perlman, H.A., *Estimated Use of Water in the United States in 1995*, USGS Circular 1200, U.S. Geological Survey, Denver, CO, 1998.

Turk, J. and Turk, T., *Environmental Science*, 4th ed., Saunders, New York, 1988, p. 494.

Tyler Miller, Jr., G., *Living in the Environment: An Introduction to Environmental Science*, 6th ed., Wadsworth, Belmont, CA, 1990, p. 250.

UNEP, *Vital Water Graphics: An Overview of the State of the World's Fresh and Marine Waters*, 2nd ed., United Nations Environment Programme, Nairobi, Kenya, 2008, http://www.unep.org/dewa/vitalwater/index.html.

USEPA, *Cases in Water Conservation: How Efficiency Programs Help Water Utilities Save Water and Avoid Costs*, EPA 832-B-02-003, U.S. Environmental Protection Agency, Washington, DC, 2002.

Ward, D.R., *Water Wars: Drought, Flood, Folly, and the Politics of Thirst*, Riverhead Books, New York, 2002.

WWAP, *UN Water Development Report (WWDR4)*, Vol. 1, *Managing Water under Uncertainty and Risk*, United Nations World Water Assessment Programme, Paris, 2012.

CHAPTER 13

Water Pollution

Cities from ancient to modern times have grown up along rivers or in coastal areas, which is certainly not an accident. All urban societies produce waste materials, and rivers or oceans provide a convenient means to remove these wastes. Until modern times there was very little concern about what ultimately happened to the wastes. As populations of urban areas began to increase, the ability of rivers to cleanse themselves was soon pushed to the limit. We were faced with the situation where one city's sewage became another city's drinking water.

In Köhln, a town of monks and bones
And pavements fang'd with murderous stones
And rags, and hags, and hideous wenches;
I counted two and seventy stenches,
All well defined, and several stinks!
Ye Nymphs that reign o'er sewers and sinks,
The river Rhine, it is well known,
Doth wash your city of Cologne;
But tell me, Nymphs! what power divine
Shall henceforth wash the river Rhine?

Samuel Taylor Coleridge, 1828

From this poem it is clear that even long ago there were some individuals who began to question whether waterways could continue to absorb wastes without limit; however, it was a long time before anyone really considered doing anything about it. In fact, matters got worse before they got any better. By the mid-19th century, industrialization was in full swing in Britain, Germany, and the United States. The amount of waste materials discharged into rivers and streams began to increase drastically. Yet, little action was taken in response, and large quantities of pollutants were added to many waterways.

Probably no river better illustrates the deteriorating situation of waterways in industrial locations than the Cuyahoga River near Akron and Cleveland, Ohio. During the 1950s, the river was being polluted by 155 tons of toxic chemicals per day. Nearly all life forms in the river were destroyed, and the stench of rotting refuse from the river was overwhelming. Large parts of the river were covered with a floating slick of organic chemicals. Then, in 1959, the river caught fire and burned for 8 days. The sight of a burning river certainly caught the attention of the public. Rivers are not supposed to burn. Unfortunately, even the burning river did not lead to any action. The foul river was just accepted as part of life in modern industrial America. Ten years later the river caught fire again, but by then the environmental consciousness of society was beginning to let itself be known, and the country was ready to act. In 1972, an act for controlling water pollution, later known as the Clean Water Act, was passed by the U.S. Congress.

NATURE AND SOURCES OF WATER POLLUTION

Water pollution is more difficult to define than one might expect, as all natural waters have various substances dissolved in them. Furthermore, the types and concentrations of dissolved materials vary from location to location, making a universal definition of water pollution difficult to come by. Whether water is polluted or not depends to a large extent on what one wants to do with the water. A different level of water quality is required for drinking water than for water used to wash a car or flush a toilet. Generally, we would wish that natural waters are not toxic to the many varieties of aquatic organisms we expect to find in them. **Water pollution** can be defined as any chemical or biological change that adversely affects the aquatic life normally found in water or that makes water unfit for a desired use.

How does water get polluted? The view that most people probably have is of a pipe from some industrial plant discharging toxic stuff into the river. Although this view of the source of water pollution is not unfounded, it is not the entire picture. A significant portion of water pollution comes from rainwater washing across contaminated soil on its way to a river or stream. The source of pollution can be feedlots covered with manure, heavily fertilized fields, or construction sites. Pollution sources are generally separated into point sources and nonpoint sources. A **point source** is the discharge of a water pollutant or pollutants from a specific location. A **nonpoint source** is the generalized discharge of a water pollutant or pollutants from runoff over a large area.

TYPES OF WATER POLLUTANTS

Disease-Causing Agents

In most of the less-developed countries of the world the spread of biological agents associated with waterborne diseases is a major problem. The problem stems from inadequate wastewater treatment in these areas, an issue that is discussed later in this chapter. The individual disease-producing agents, such as bacteria and parasites, are not discussed here, as the focus is on chemistry.

Organic Materials

This category is a very large and diverse collection of substances, including human and animal wastes, pesticides, and herbicides. Some of these substances are directly toxic, but all of them have one thing in common: They are all organic compounds. Most of the compounds that are found in wastewater are broken down in the same way; that is, they are consumed by microorganisms. If there is enough oxygen, the principal microorganisms will be oxygen-consuming microorganisms, or **aerobes**. These microorganisms use the organic material as food and break the materials down into carbon dioxide, water, nitrates, phosphates, and sulfates. A key feature of the process is that it requires oxygen, and the amount of oxygen required to decompose a given quantity of organic material is known as the **biochemical oxygen demand (BOD)**. The oxygen that is used in this process comes from that which is dissolved in the water. This is the same oxygen that fish and other aquatic organisms need to survive.

Oxygen found in water is a result of underwater photosynthesis by aquatic plants and also comes from the air. The oxygen is removed by aquatic organisms of various kinds, including fish and microorganisms, to support life. An adequate level of oxygen in the water can be maintained as long as the rate of removal of oxygen does not exceed the rate at which oxygen is added. If the removal rate is too high, the level of **dissolved oxygen (DO)** will fall. If the level falls below 8 ppm, certain species of fish will be put in jeopardy, and as the level drops further more and more aquatic organisms will be affected. When the level drops below 2 ppm, very few forms of life can survive, and those that do are not the forms of life we usually like to have around us. These survivors include worms, fungi, and non-oxygen-consuming microorganisms known as **anaerobes**. It is the anaerobes that are generally responsible for the unpleasant odors associated with oxygen-depleted water. These microorganisms decompose organic material without the use of oxygen and produce such products as methane, ammonia, hydrogen sulfide, and various small organic sulfur- and nitrogen-containing compounds. With the exception of methane, nearly all of these compounds have horrendous odors.

Bodies of water that are contaminated by large amounts of organic wastes do have the ability to clean themselves, if the amount of organic material is not too large. The natural state of a river or other body of water is known as the **clean zone**. A clean zone of a body of water is characterized by a high DO level and a low BOD. In other words, there is plenty of oxygen for the fish, and very little organic material in the stream is using up the oxygen. In such an aquatic environment, a wide variety of fish, such as trout, perch, and bass, is usually found.

A **decomposition zone** is found immediately downstream from any major waste discharge area. In this area, the effects of organic wastes are first felt. The BOD is very high, and as oxygen is consumed to break down wastes the levels of DO begin to fall rapidly. This rapid fall of the DO is known as **oxygen sag**. Many aquatic species begin to disappear. Only fairly hardy fish, such as catfish and carp, can survive in these waters. As the DO continues to fall, the situation gets even worse. At very low DO levels, no fish can survive, and the only life forms other than anaerobic microorganisms are sludge worms, midge larvae, and mosquito larvae. This is known as the **septic zone**, and it continues downstream as long as the BOD remains high enough to require large amounts of oxygen.

As this mass of organic waste material flows further downstream it is gradually decomposed. Eventually, the process gets to a point where most of the material has been decomposed. This is where the **recovery zone** begins. The recovery zone is marked by continued decline in the BOD, continued increase in the DO, and the return of some species such as catfish and carp. Eventually, when the BOD and the DO have returned to normal, the stream returns to its natural state, or clean zone. The general assortment of species usually found in a healthy stream returns, and the stream remains in this state until another insult of organic wastes occurs.

Some organic wastes have impacts beyond their effect on DO levels and are directly toxic to either humans or certain species. One group of such compounds is pesticides. The water becomes polluted when these chemicals are washed from their points of application into nearby streams. Other toxic organic chemicals often have industrial sources and usually get into the environment by accident or negligence. These chemicals include solvents, chemical intermediates, and textile dyes. The improper disposal of household chemicals is another source of organic chemicals in water. Many items in the household, from bug spray to paint to lubricants, can contribute to water pollution if not disposed of properly.

Plant Nutrients

One of the most interesting and complicated issues in water pollution is the problem of plant nutrients. Green plants, unlike animal species, manufacture their own food via photosynthesis; hence, the raw materials or nutrients that plants require are much simpler than the complex foods needed by animals. One key element, carbon, is present in the air as carbon dioxide in an almost limitless supply. Another key ingredient is water, which can at times be in short supply; however, in our discussion of water pollution, we will assume it is plentiful. The other key elements that plants require for growth are nitrogen, sulfur, and phosphorus. Plants grow by drawing water containing these elements through their roots. These elements are usually present as inorganic salts such as nitrates, sulfates, and phosphates.

Nutrients get into the water in a number of different ways. One way is through the application of commercial fertilizers. As one would expect, commercial fertilizers are made up of nutrient salts, which are generally applied to fields somewhat in excess. Farmers apply an excess of fertilizer to be sure that the crops will have as much of the nutrients as they can possibly need. The excess fertilizers, being water soluble, will dissolve in rainwater and are washed away into nearby lakes or streams. Fertilizer is also used on golf courses, lawns, flower gardens, and so on. These nutrients are also found to a varying extent in natural soils. Some nutrient runoff from soil would be considered normal, but under poor farming practices or poor forest management excessive leaching of water-soluble nutrients and soil erosion can occur. The eroded soil will carry with it a full complement of essential plant nutrients.

In the earlier discussion of organic material, we noted that organic wastes discharged into a stream would eventually decompose; however, if the law of conservation of mass means anything, the wastes must be decomposed into something. The organic material is broken down into carbon dioxide, water, nitrates, sulfates, and phosphates. The carbon dioxide is lost to the air, but the other products remain. Hence, inorganic plant nutrients can come from sewage, animal wastes, or other organic materials such as crop residue. Standard sewage treatment does not help with this problem because sewage treatment breaks the sewage down into inorganic nutrients unless supplementary processes are added.

To understand why these plant nutrients are such a problem, we must consider the process known as **eutrophication**. Lakes and streams can be classified by their nutrient level. Cold, clear lakes or streams with few nutrients and a relatively low abundance of life are referred to as **oligotrophic**. Lakes or streams with a large supply of nutrients and an abundance of life are **eutrophic**, and lakes or streams of intermediate nutrient level are **mesotrophic**. Lakes generally evolve naturally from oligotrophic to mesotrophic to eutrophic over time. Over a long period of time nutrients gradually wash in from the surrounding terrain, changing the nutrient level of the lake. This is the process known as eutrophication, during which the amount and varieties of life in the lake increase.

When humans add large amounts of nutrients to a lake or stream, the process of eutrophication speeds up. The process of rapid eutrophication due to the addition of nutrients produced by humans is called **cultural eutrophication**. The growth of plants generally is limited by the nutrient that is in shortest supply. Once this point is reached, the plant must slow down its growth until more of the nutrient is available. With algae, for example, the limiting nutrient is usually phosphate. If there is a large excess of phosphate in a water body, algae can grow wildly and cover the entire surface of the water. Although there is much life in the lake, the water is certainly much less appealing for many human uses. In extreme cases, a septic condition can result. Excessive algae growth can produce multiple layers of algae, not all of which receive light. The algae at the bottom die and become organic matter to be decomposed. This situation leads to an increase in the BOD and a decrease of the DO. If the situation is severe enough, some parts of the lake will have little oxygen and the smell of anaerobic decomposers will become evident.

Ordinary Salt (Salinity)

In Chapter 12, water was described as being nearly the "universal solvent." A large variety of different substances will dissolve quickly into water; therefore, it should be no surprise that natural waters have a large collection of substances in them. As water percolates through the ground or runs over the soil, it will pick up and dissolve many things. Some of the ions that will normally be found in natural water include sodium, chloride, calcium, bicarbonate, and magnesium.

Of the ions found in natural water, sodium and chloride are the most persistent ones. Sodium chloride, or table salt, is very water soluble and once dissolved in water cannot be removed easily. As water dissolves increasing amounts of sodium chloride, its salinity increases. In those regions of the world where rain is plentiful, salinity is not generally a problem. When the amounts of rainwater are large, the concentration of salt in the rainwater will be small. If large amounts of water continually run over the soil and rocks, most of the available salt will have already been dissolved. It is in the more arid parts of the world that salinity is a problem.

All water flowing over the land or percolating through the soil will pick up some salt in the process because the weathering of soils and rocks will slowly release salt. If water is diverted to flow over land that has historically had little water flow, then additional salt will be dissolved because this soil would still contain much of the salt load ever associated with it. This problem is common in irrigation, when the water flowing over relatively dry soil picks up salt. As the water flows to, through, and from the fields, the sun's heat evaporates some of the water and the remaining water becomes even saltier.

Is this salty water a problem? The answer is sometimes yes and sometimes no. The salt generally poses a problem for terrestrial plants because crops require freshwater to grow. As water becomes saltier, it must be kept away from the crops. One of two requirements must be met to avoid salinity problems: Either the saltier water must move down away from the surface or it must flow out of the area. For example, salinity is usually not a problem in west Texas because the underlying aquifer is so deep that the water percolates to very deep regions underground, well away from crops. In coastal areas, excess underground water and excess surface water flow slowly into the ocean, which is already salty. Salinity is a problem for irrigation in California's San Joaquin Valley because the underground aquifer is very shallow and the water does not flow away very well. Water is brought in for irrigation, but little of the water leaves except by evaporation, which leaves the salt behind.

Another difficult salinity problem exists on the lower Colorado River. In 1947, the U.S. Congress authorized the Wellton–Mohawk Project near Yuma, Arizona. This project, which was completed in 1952, diverted water from the Colorado River to irrigate the Gila River Valley. This area, like the San Joaquin Valley, has a shallow and poorly drained aquifer. To prevent salinity problems from becoming more severe, the Wellton–Mohawk irrigation district put into operation a system of drainage wells to pump excess groundwater back to the Colorado River. The Colorado River had a salinity at that time of about 800 ppm, whereas the excess irrigation water being pumped back into the river was at about 6000 ppm. The addition of this water to the Colorado River increased the salinity of the river significantly. The river flows from there into Mexico, where some of the water is used to irrigate the Mexicali Valley. Needless to say, the Mexicans objected, and after 11 years of negotiation the United States guaranteed that the river water entering Mexico would increase by no more than 145 ppm from the salinity at the point of removal of water for the Wellton–Mohawk project (the Imperial Dam).

One approach to preventing salinity levels from becoming too high in the river is to not return the salty water to it. Instead, the water can be sent through concrete-lined canals directly into the Gulf of California. This tactic, of course, decreases the flow of the Colorado River into Mexico, but, by treaty, the United States has guaranteed that about 10% of the average river flow of the Colorado will enter Mexico. In some years, it may be necessary to use the irrigation runoff water to ensure this 10% volume. To do so, to meet both the volume and the salinity requirements, a desalinization plant was built near Yuma, Arizona, to remove salt from irrigation runoff and return the water to the river. The plant, one of the largest in the world, has only been operated for less than a year, because the natural water flow has been sufficient. In recent years, though, drought has returned to the west and use of this plant may be required in the near future. A recent test run of the plant was successful, but funding for the plant continues to be a major issue.

Soil and rocks in the desert are not the only sources of salt and related materials in our water supply. In the northern parts of the United States and in some parts of Canada millions of tons of sodium chloride and calcium chloride are spread on roads to melt ice. Because these salts are water soluble they find their way into surface water and groundwater. Certain types of oil wells produce considerable amounts of saltwater which, if not handled properly, can pollute the ground and surface water. These wells, which are known as injection wells, are discussed further in Chapter 7.

Heavy Metals

Mercury

Metallic mercury is one of those fascinating substances that nearly everyone has seen and perhaps played with. Part of its fascination is that it is one of the few liquid metals. Mercury balls up and scatters all over when poured out on a surface. It does interesting or disastrous things to jewelry, depending on one's perspective. It is commonly found in thermometers and other types of measuring devices, and although it is a rare element in the Earth's crust we all encounter it enough to feel comfortable with it. As was discussed in Chapter 6, mercury is fairly harmless as a liquid if its vapors can be avoided. These observations led to rather lax standards for handling elemental mercury in the past. Mercury was often dumped into waterways on the assumption that it would sink to the bottom and remain there in a harmless elemental state.

In the 1950s, in the small fishing village of Minamata, Japan, people began to fall ill. The effects were traced to a nearby chemical plant that was dumping mercury wastes into the Minamata Bay. Before the episode was over, 43 people had died out of 111 reported cases. Sadly, 19 babies were born with congenital defects from the mercury. Despite the publicity from this event, mercury contamination of water was not taken very seriously. A U.S. survey of trace metals in surface water between 1962 and 1967 did not even measure mercury. This state of affairs quickly changed in 1970 when mercury was discovered in fish in Lake Saint Claire between Michigan and Ontario. Furthermore, new surveys found that other U.S. waters were also contaminated with mercury. Some chemical plants were found to be dumping 31 lb (14 kg) or more of mercury into the water each day. How was mercury sitting on the bottom of the lake getting into the fish?

It turns out that anaerobic bacteria are the culprits. These bacteria convert mercury to monomethyl mercury chloride and dimethyl mercury. These mercury compounds are removed from water by fish, which store the compounds in their fatty tissue. Fish concentrate the mercury, resulting in an approximately 1000-fold biomagnification. The consumption of sufficient quantities of these fish leads to mercury poisoning in humans.

Lead

An extensive discussion on lead is found in Chapter 6. Lead enters drinking water principally from the lead solder used in plumbing connections. The amount of lead in drinking water seems to be declining; however, if one lives in a house with fairly old plumbing, one will almost surely find considerable amounts of lead in the water. When returning after a long period of time, let the water run awhile before drinking it. The standing water may leach lead out of the pipes.

Cadmium

Cadmium generally occurs in water as a result of improper wastewater handling by industry or mining operations. Cadmium has found many uses in the metal plating industry. Because metal plating involves large quantities of solutions containing cadmium, industrial plants must take great care to remove cadmium from the solutions before returning the water to the environment.

Acids

Acids get into natural waters from industry, acid deposition from the atmosphere (covered in Chapter 9), or mining. The industrial sources are many and varied, but because it is not difficult to adjust the pH of water by using bases, industries simply should not be discharging acidic waters into the environment. Classically, one of the worst offenders in the area of acid discharge has been coal mines. The mine drainage that is pumped out of coal mines contains sulfides from deep underground sources. These sulfides are oxidized by bacterial action and converted into sulfuric acid. The acid drainage is often dumped into mountain streams, which become exceedingly acidic. All life in the stream may be killed for several miles downstream.

Sediment

As water flows into lakes and streams, it runs over land and picks up soil particles. These particles wash into the lake or stream and cause the water to appear murky or muddy. The appearance of sediments in a river or lake is natural. These sediments wash downstream, and as the river flow slows down they begin to settle out of the river. The fertility of many lower river valleys is dependent on the settling out of these sediments to form new land. The histories of the Nile River Valley, the Tigris–Euphrates Valley, the Ganges, and the Mississippi are all closely connected to the deposition of sediments in the delta regions near the mouth of a river. Whole civilizations thrived on the fertile soil deposited in these regions.

Sediment deposition, however, can be overdone. Many human activities disturb the soil and remove the protective layer that holds it in place. It is often principally vegetation that holds the soil. Deforestation and improper farming practices expose the soil and allow it to be carried off at an excessive rate by rainwater. Large amounts of sediment in water can cause real problems; for example, sediments may fill in shipping channels and reservoirs, and the cost of producing drinking water may rise because of the expense of removing large amounts of sediment. High levels of sediments can also interfere with living organisms. Sediments can clog the gills of fish, reduce the survival of eggs and young ones, cover the living areas of organisms, and block the light used by plants for photosynthesis. The reduction in photosynthesis will in turn lower the DO level.

POLLUTION OF SURFACE WATER

Industrial and Mining Sources

The types of pollutants discharged from industrial and mining operations can be some of the most insidious ones imaginable; however, such pollution is among the most controllable types because it is nearly all point source pollution. If national laws are strong enough, the polluter can be identified and forced to cease or reduce the discharge of the offending pollutant. Industrial and mining pollution is still a more serious problem in the developed world than in the developing world because of the sheer numbers of plant sites. Regulation is often difficult and expensive; as a result, some polluters may be missed. Levels of industrial and mining-related pollution are on the rise in the developing world. To encourage industrialization, laws regarding the operation of plants or mining sites tend to be weak; hence, industries in these countries can pollute waterways without significant concern for economic or legal consequences.

Domestic Sources

From a global perspective, domestic wastes are the most important sources of surface water pollution. In the world as a whole, sewage treatment is not the norm in many countries. Even in the United States, widespread sewage treatment is a rather recent development. Before passage of the Clean Water Act of 1972, many communities dumped raw sewage directly into rivers and streams. For example, in 1948 only one-third of U.S. cities had municipal sewage treatment. By 1986, 70% of U.S. cities had municipal sewage treatment, and 80% of these had secondary treatment.

Such standards are only maintained in highly developed countries. Table 13.1 indicates that even in Europe there is considerable variation in treatment levels among countries. Sweden, Spain, and Germany treat nearly all of their sewage, whereas in Latvia and Romania sewage treatment is spotty. Whatever the situation is for Europe, the rest of the world outside of the United States, Canada, Australia,

Table 13.1 Percentage of People with Sewage Treated in Various Countries in 2004

Country	Percent of People with Public Sewage Treatment (%)	Percent of Those People with Adequate Sewage Treatment (%)
Canada	78	100
China	22	44
Congo	2	27
Fiji	35	72
Germany	93	100
Haiti	0	30
India	9	33
Jamaica	17	80
Latvia	67	78
Romania	49	—
Spain	97	100
Sweden	99	100
United States	83	100

Japan, and New Zealand is considerably worse. The fact is that most of the world does not treat a large percentage of its sewage, and as a result waterborne diseases are rampant. One of the more advanced developing countries is Thailand. Its capital city, Bangkok, has 10 million inhabitants; yet, before 1994, Bangkok not only lacked a sewage treatment plant but also did not even have a sewer system. Such a system was constructed in the 1990s, which finally stopped the dumping of 10,000 tons of raw sewage into a series of canals and four rivers each day.

Agricultural Sources

The most significant nonpoint source of pollutants is agriculture. In a world that is trying to keep food supplies ahead of a rapidly growing population, intensive agriculture has become an absolute must. This intensive agriculture has involved large applications of fertilizers and pesticides, both of which can wash off into rivers and streams. As more and more land (some of which should not be cultivated) is used for agriculture, a massive amount of soil erosion occurs. This eroded soil ends up in the water as sediment. The search for more land also leads to the use of arid areas and all of the salinization problems associated with irrigation. Another aspect of intensive agriculture is the livestock feedlot. One cow, for example, averages about 14 lb (30 kg) of manure per day (the equivalent of that produced by 10 people). Multiply this by the large number of cattle in a feedlot and then consider the rainwater that runs off from the lot. This rainwater will flow into the nearest stream and pollute it with essentially untreated animal sewage.

POLLUTION OF THE OCEANS

The oceans become polluted in many of the same ways as surface freshwater. Coastal regions of the oceans receive sewage discharge, industrial discharge, and agricultural runoff along the coastal watershed. Human perceptions of the seriousness of ocean pollution vary considerably. People have generally viewed the oceans as being so big as to be able to absorb any insult and remain relatively unchanged by it. After all, the oceans contain 97% of all of the water and cover 71% of the Earth's surface! Anything this big ought to be able to absorb tremendous quantities of pollutants and not be affected very much, or at least this is the popular understanding. The fact is that this is one of those beliefs that is almost true, but the little bit that makes it not true is important.

One important thing to understand about the ocean is that it is not homogeneous although it may look as though it is. The ocean can be viewed as being made up of four zones, which can be arrived at by first dividing the ocean into the **coastal zone** and the **open sea**. The coastal zone starts at the high-tide mark and extends out through the relatively shallow waters near the shore to the edge of the continental shelf. These waters, in addition to being shallow, are fairly warm and nutrient rich. The coastal zone contains only about 10% of the ocean's waters but is home to about 90% of all ocean life.

The remaining three zones come from dividing up the open sea according to depth. The divisions are made roughly according to the amount of sunlight received. At increasing depth less and less light penetrates, which affects the amount of photosynthesis possible and, therefore, the type of life found there. The top layer of the ocean, called the **euphotic zone**, receives plenty of light and contains most of the life of the open seas. The next layer is in a condition of constant twilight during the day. Called the **bathyal zone**, it is colder and darker and contains somewhat less life. Below this zone is found the **abyssal zone**, which is in total darkness all of the time and contains only a few specialized life forms.

At this point, let us return to the issue of the human impact on the ocean. Although it is difficult for us to have a significant effect on the ocean overall, we can have a devastating effect on the coastal zone where most of the life is. The coastal zone does not exist in isolation, but the exchange of water with the greater ocean is a relatively slow process. About one-third of the Earth's urban population lives within 35 miles (60 km) of the world's oceans. Not only do these populations produce pollution, which often finds its way into the coastal waters of the oceans, but the drive for coastal development also leads to the clearing or modification of wetlands, mangroves, salt marshes, and seagrasses. The destruction of these areas alters the habitat for various species, contributes to their decline, and aggravates problems created by pollution. Pollution of coastal waters includes most of the well-known types of water pollution, such as nutrient overloading, oxygen depletion from organic pollution, toxic chemicals, heavy metals, disease-causing agents, oil, sediment, and plastic refuse. In some major cities, the harbors are so totally polluted that they contain no life to speak of.

For many years, certain materials were knowingly dumped at sea by responsible governmental agencies. From 1986 to 1992, about 42 million tons of wet sewage sludge were dumped into the ocean off the Atlantic coast. (The nature of sewage sludge is discussed later in this chapter.) This area was known as the 106-mile dumpsite, because the location was 106 nautical miles southeast of New York Harbor. Actually, the general area had been used for the dumping of sewage sludge since 1924. The practice was finally halted when the general public became concerned about the effects of this practice on the ecology of the ocean off the Atlantic coast.

The restriction of ocean dumping is controlled internationally by the Convention on the Prevention of Marine Pollution by Dumping of Wastes and Other Matter, which was adopted in 1972. All dumping is forbidden except for certain items specifically listed in the treaty. The dumping of sewage sludge is permitted by the treaty, but many countries, including the United States, no longer dump sewage sludge into the oceans.

In the United States, the terms of the treaty are enforced through the Marine Protection, Research, and Sanctuaries Act of 1972 and the Clean Water Act. The most common discharge from the United States into the oceans is dredged material (sediment) removed from waterways to keep the channels deep enough for navigation. This dredged material can only be dumped by permit from the Environmental Protection Agency (EPA) and must be tested to guarantee that toxic materials are not contained in the sediment.

The beach-going public became aware of an ocean pollution problem in the summer of 1988 when they found garbage washing up on beaches, as well as sewage, drug paraphernalia, and medical wastes. As disgusting as some of this material was, it probably did not represent as serious a problem as some other pollution situations we have. This type of material has been dumped at sea for many decades and probably ended up on the beaches because of some unusual weather conditions. The dumping of some materials at sea is very difficult to control. Ships, for example, dump garbage and trash in the ocean all the time. Trying to ban such activity would be nearly impossible to enforce. The ocean is, after all, a very big place.

To the public, the most spectacular ocean pollution issue is oil spills. These spills receive considerable press attention and cause severe environmental problems in local areas. The spills are usually the result of accidents involving tankers or offshore oil rigs. Despite the tremendous amounts of oil that oil spills can release, as much as half or maybe even 90% of the oil released into the world's oceans comes from runoff from land.

Of the oil spills that have become newsworthy, the recent major oil spill in the Gulf of Mexico had a major impact on the public and caused considerable anxiety during the aftermath. The spill was caused by an explosion on an oil rig in the Gulf known as the Deepwater Horizon. The cleanup efforts were difficult and slow, and it took four months to cap the well. By the time the well was capped, the spill had become the second largest oil spill in history.

Another Gulf of Mexico blowout occurred in 1979 at the Ixtoc oil rig in the southern Gulf of Mexico. Although the amount of oil released was slightly smaller than in the Deepwater Horizon incident, it took eight months to cap that well. The worst oil release was not a spill at all, but a deliberate act of terrorism or war, depending on one's perspective. On January 19, 1991, Iraqi soldiers opened valves at Kuwait's Sea Island oil terminal and released millions of gallons of oil into the Persian Gulf. The impact of this release was worsened by the fact that the Gulf is shallow and the water takes about three years to exchange with water in the Indian Ocean through the very narrow Straits of Hormuz.

Another type of oil spill that usually makes the news is tanker accidents. One of the most well-known tanker oil spills in the United States was the spill from the *Exxon Valdez*, which occurred when the tanker hit a submerged reef in Prince William Sound in Alaska in March of 1989. The spill was remembered for a number of reasons. The accident was the only tanker accident to have occurred in U.S. waters, it occurred near an economically important fishing area, and it occurred in the midst of some of the most beautiful scenery in the world. This accident resulted in more restrictive oil spill legislation in the U.S. Congress. In spite of the fact that the *Exxon Valdez* had a major impact in the United States, it was small as far as spills are concerned. The *Castillow de Bellver*, a tanker that caught fire off the coast of Capetown, South Africa, released about seven times more oil than the *Valdez*. Oil rig blowouts tend to release considerably more oil than tanker accidents. A comparison of the oil releases we have discussed here are shown in Figure 13.1.

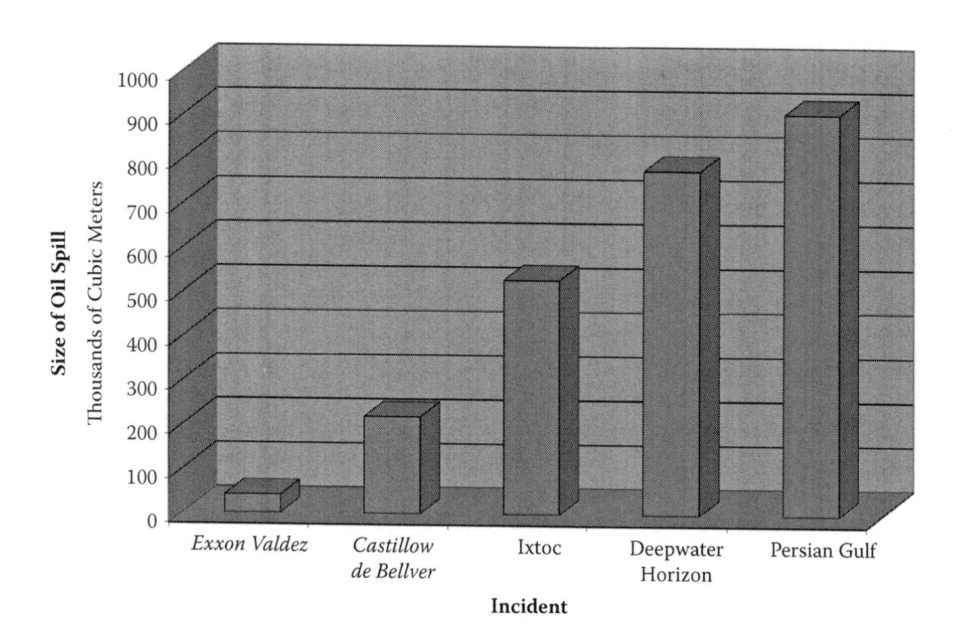

Figure 13.1 The amount of oil spilled in incidents: *Exxon Valdez*, tanker, Alaska (1989); *Castillow de Bellver*, tanker, South Atlantic Ocean (1983); Ixtoc, offshore oil rig, Gulf of Mexico (1979); Deepwater Horizon, offshore oil rig, Gulf of Mexico (2010); and Persian Gulf, intentional dumping (1991).

The environmental impact of oil spills varies greatly from one spill to another. Petroleum is a complicated mixture of compounds, as discussed in Chapter 7. Some compounds have low boiling points and evaporate rapidly, whereas others evaporate more slowly. Some of the compounds with very high boiling points are viscous and gooey. It is the sticky material that does the greatest damage to marine life. Obviously, the amount of environmental damage done will be related to the size of the spill and the percentage of the oil that is of the sticky, gooey variety.

Oil spills are also affected by their location. Those occurring far from land are more likely to break up before they reach the coastal waters, where the greatest potential for harm exists. If an oil slick does reach coastal waters, it will be slowly broken up by the forces of nature. This process seems to occur much faster in warm tropical waters than in the more frigid waters of the far north. The colder water temperature was a major factor in the recovery time for the *Exxon Valdez* accident.

POLLUTION OF GROUNDWATER

Many of us have an innate belief that water pumped out of the ground in a well will be safe to drink. But, just because water is underground does not mean that it cannot be contaminated. The reason why groundwater contamination is more difficult to deal with than surface water contamination is that it occurs out of sight. When surface waters are contaminated, we can observe the results. The water has an odor, a cloudy appearance, a funny color, or dead fish in it. Any of these can be signs of contaminated surface water, yet they are rare in contaminated groundwater. Generally, there are three ways by which groundwater can be contaminated: natural processes, wastes, and human activities unrelated to waste disposal.

Natural Processes

This type of groundwater contamination is most prevalent in fairly arid areas of the world. We discussed the salinization of water earlier in this chapter. The soil in any arid region is much more likely to contain many substances that are water soluble. When water is present, it will dissolve some of these soil chemicals and wash them into the groundwater. In addition to ordinary salt, the contaminants usually include, sulfates, nitrates, fluorides, and iron, as well as arsenic or uranium in some locations. Human activity can aggravate this natural process. In the process of agricultural irrigation, additional water is added to that which is naturally percolating through the soil. This water often has washed over great expanses of soil before soaking in. Irrigation can bring in large quantities of naturally occurring contaminants.

Waste Disposal

In our attempts to dispose of wastes, we often will contaminate groundwater. Liquid wastes can present some of the most difficult disposal problems. In many cases, it is very easy to just "pour it down the drain," but, as we learned in earlier chapters of this book, "There is no 'away.'" Even if we dump the waste out of sight, it is still somewhere. Liquid wastes can be broadly divided into three classes: saline wastes (saltwater), human and animal wastes, and various industrial and commercial wastes.

Human and animal wastes have in some sense been seeping into groundwater ever since humans and animals evolved on this planet. The soil and microorganisms in the soil have the ability to purify sewage-contaminated water that enters the ground. As a result, groundwater from preindustrial times was rarely contaminated; however, the ability of the soil to purify water has its limits. As population density increases, so does the strain on the ability of natural systems to purify sewage. Admittedly, dumping raw sewage into a river or stream is not a good idea, but holding the sewage on land may simply transfer the sewage contamination problem from the river to the groundwater. The results depend on the population density and on whether any attempt is made to treat the sewage. Large amounts of untreated sewage or animal wastes in a holding pond (cesspool) will contaminate the groundwater to some extent. Even various methods of sewage treatment may, via unintended leaks, lead to groundwater contamination.

Saline wastes can be very difficult to deal with. They come from irrigation, steam injection into oil wells, and various industrial processes. Because salt is not toxic except in high concentrations, one solution is simply to dilute it. However, this solution is only practical for those who live in areas with a high level of rainfall. In areas with limited rainfall, such as irrigated agricultural regions, freshwater is precious and could not be used to dilute saltwater wastes. Saltwater would cause no problem if discharged into the ocean or if discharged into a salt lake basin, but not everyone is located within easy access of the ocean or a salt lake basin.

Another solution to this problem is to intentionally pollute groundwater with saline wastes. This idea is not quite as bizarre as it seems. Underlying much of the land area of Earth is not just one aquifer or two, but often three or more. Some of the deepest aquifers are rarely tapped for any use because shallower aquifers can be tapped more easily. The theory is that, because these aquifers are deep and not likely to be used, there is little danger in pumping liquid wastes into them. The pumping of the wastes into the aquifer is done by use of what is known as an **injection well**. Liquid wastes are pumped down these wells under pressure into the deep aquifer. These wastes may include not only saltwater but also treated sewage and chemical plant wastes, as well.

This process may or may not be safe. There are some real concerns about our level of knowledge of the connections between various aquifers. One of the key assumptions on which the safety of this procedure is built would be false if there is an easy connection between aquifers. The presumption is that whatever is put down there stays there, but how sure are we about that?

The process of injecting wastes into deep aquifers was not very tightly controlled until recently. This situation changed markedly in the United States with passage of the 1986 amendments to the U.S. Safe Water Act and the 1984 amendments to the U.S. Resource Conservation and Recovery Act (RCRA). Restrictions are imposed on what substances can be injected and where the wells can be located.

Specific requirements are imposed on the construction and operation of the wells, and periodic testing of surrounding aquifers is required. Similar measures have come into law in other nations.

Miscellaneous Contamination

Solid Waste Disposal Sites

These sites include landfills and dumps, both legal and illegal. The groundwater contamination arises from water percolating down through the solid wastes and leaching out various potentially toxic substances. These materials then mix in with the groundwater. We will not discuss this in detail here, as this topic is covered in Chapter 14.

Underground Storage Tanks

There are probably millions of underground tanks buried throughout the world, most of which are at the site of gasoline filling stations, both currently operating and abandoned. Most of these tanks are made out of steel and corrode over time; as the tanks get older, the likelihood of a leak increases dramatically. A small amount of leaked gasoline can have devastating effects on an aquifer. As little as 1 gal leaked per day can render the water in an entire aquifer unsuitable for drinking. Even abandoned tanks can be a hazard because they are seldom pumped completely dry. The EPA has estimated that there have been over 400,000 leaks of petroleum from underground storage tanks. One situation that tends to aggravate the leaking tank problem is the fact that many of the tanks are owned by small filling station operators. These operators are not inclined to report leaks, because if they do they will be required to dig up the tank and repair the leak. Such an expense might run a small operator out of business. Despite the risk to the financial stability of some small operators, federal and some state laws were tightened up. Under regulations published in 1988 as part of the Resource Conservation and Recovery Act as amended in 1984, all new tanks installed after 1993 must have leak detection systems and must be repaired immediately if they leak. After 1998, all tanks were required to be double-walled or to be placed in a concrete vault; in addition, they must have leak detection systems, overfill prevention devices, and spill prevention devices. Furthermore, corrosion protection devices are also required.

Wells

Wells drilled into confined aquifers can be a problem if the unconfined aquifer above it is more polluted than the confined aquifer below. Because the well cuts through the clay or rock divider between the two aquifers, it provides a path for the pollutants to pass from the surface or the unconfined aquifer to the deeper, confined aquifer. Particularly bad are abandoned wells that are often forgotten but still provide various types of direct access to a deep aquifer. These wells should be capped, but the caps often corrode or crack and fail in their purpose.

Agricultural Chemicals

Each year in the United States about 22 million tons of fertilizer are spread on the soil along with over 600,000 tons of pesticides. To be effective, both of these materials are applied to the soil in excess. Because fertilizers are a collection of water-soluble materials containing nitrogen, phosphorus, and potassium, these materials will very quickly dissolve in any rainwater or irrigation water reaching the field. The nutrient-rich water will either run off and contaminate surface water or soak in and contaminate the groundwater. Pesticides are not on the soil in quite as high a level as fertilizer nor are they generally as water soluble; however, they are usually more toxic than fertilizers. All of these materials end up in the groundwater as well as the surface waters. In heavily agricultural states, contamination of wells can be a real problem. In 1986, a survey found that half of the wells in Iowa were contaminated with pesticides or other synthetic chemicals. In addition, one-fifth of the wells had high nitrate levels from fertilizer contamination. In that same year, Florida shut down 1000 drinking water wells due to toxic chemical contamination, primarily pesticides.

Mining

Deep subsurface mining represents a very deliberate disruption of underground aquifers. Although mining operations do not intentionally contaminate groundwater, it is very difficult to operate underground with heavy equipment and not contaminate the groundwater. Additionally, the presence of air, certain bacteria, and underground sulfides often leads to the production of sulfuric acid, which is a serious environmental problem for mining.

Highway Salt

As noted earlier, spreading salt on the highway may be a convenient way to melt ice and snow from roadways, but what most people fail to consider is the fact that this salt, like everything else in the environment, has to go somewhere. Highway salt dissolves in the water from the melting snow and ice; in fact, the very nature of the process by which salt causes ice to melt requires that the salt dissolve in the water produced. This salty water ends up in the surface runoff or in the groundwater.

Saltwater Intrusion

A problem in the contamination of groundwater in coastal areas is saltwater intrusion. Ocean saltwater can contaminate an aquifer if freshwater is removed at too high a rate. This problem is discussed in Chapter 12.

Accidental Spills

This category includes any method that would cause anything to seep into the ground so that it might soak down to an aquifer. The category includes spilling, pumping, and leaking. There are many ways for polluting chemicals to get into the environment, such as tanker truck accidents, train wrecks, and chemical plant accidents. Such spills are covered in the United States by the Comprehensive Environmental Response, Compensation, and Liability Act and many state statutes. Generators and transporters of the chemicals are responsible for the cleanup of any spill.

WATER POLLUTION CONTROL

Human Wastes

Human wastes, one example of animal wastes, are quite natural and in and of themselves do not necessarily represent pollution. Such wastes have been scattered around the landscape ever since higher species of animals began to evolve. There are plenty of natural processes to decompose waste materials. Problems begin to develop when we put large numbers of people or animals in a relatively small area. Large amounts of waste in a small place simply overwhelm the natural processes that break down the material.

As cities began to develop, an early solution to this problem was the municipal sewer. Sewers were built as long as 5000 years ago in the Indus Valley of India. For thousands of years after this, sewers had but one purpose and that was to remove sewage from a town. In most cities, the sewage was simply discharged into the nearest river or the sea. As long as cities were scattered and their populations were not excessive, no problems were encountered for the most part. The natural factors in the river could break up the sewage and return the river to a clean state before it reached another city.

Eventually, however, population density and industrialization outstripped the ability of rivers to clean themselves, and it became clear that something more than simply channeling the sewage away from the town into the river was needed. In the 19th century, many of the cities of the industrialized world underwent successive cholera epidemics, primarily because they were drawing their drinking water from the same place where sewage was dumped. In Chicago, for example, Lake Michigan was the source of drinking water and also the dumping place for sewage. When cholera struck they simply extended the sewage discharge pipes farther out into the lake. When this solution failed to work, the city reversed the flow of the Chicago River and dumped the sewage into it. The river was diverted to flow into the Illinois River and eventually to the Mississippi.

Onsite Disposal

In many places where the population density is lower, sewage can be handled quite well at the site of its generation. In the middle of the 20th century, the most common type of such onsite disposal was the pit privy or outhouse. It consisted of a small building set over a pit, into which the excrement was deposited. When the pit filled up, a new pit was dug, the building was moved to the new site, and the old pit was covered over with soil. Of the outhouse much has been written. Consider, for example, the poem, "The Passing of the Backhouse" by James Whitcomb Riley:

> But when the crust was on the snow
> and the sullen skies were grey,
> In sooth the building was no place where
> one could wish to stay.
> We did our duties promptly,
> there one purpose swayed the mind.
> We tarried not nor lingered long
> on what we left behind,
> The torture of that icy seat
> would make a Spartan sob ...

Outhouses were, of course, smelly, outdoors, and not very pleasant to care for. They were not unsafe for the most part, if the population density was low and the groundwater did not move directly from the outhouse pit toward the drinking water well. An improvement on the outhouse that can be connected directly to modern plumbing is the **septic tank** and **tile field**. As seen from Figure 13.2, the septic tank is

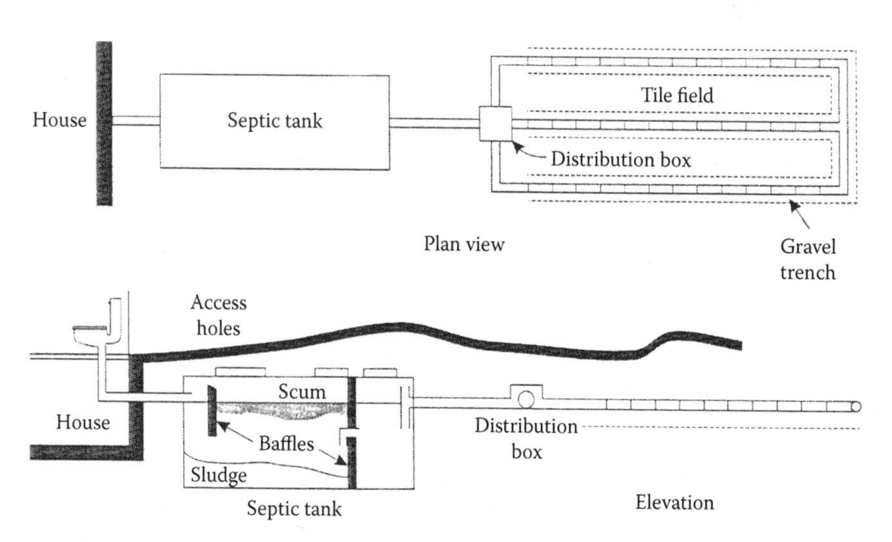

Figure 13.2 Diagram of a septic tank and tile field system. (From Vesilind, P.A. and Peirce, J.J., *Environmental Pollution and Control*, Butterworth, Boston, MA, 1983. With permission.)

a concrete tank into which human wastes are emptied. The tank is constructed in such a way as to encourage the settling or flotation of solid material. The liquid effluent is then spread out through the tile system into a gravel drainage bed to percolate down into the soil. The natural microorganisms in the soil are used to break down the organic material dissolved in the water. Some of the solid material in the septic tank will also break down due to bacterial action. Septic tanks need to be opened and cleaned after a number of years to remove accumulated undecomposed solid material.

Septic tank systems work well provided that the soil drainage is adequate and that there are not too many systems in a small area, as septic systems depend on natural processes in the soil to purify the water. A large number of systems too close together may overwhelm the ability of the soil to handle the effluent load.

Centralized Treatment

When the population density of an area is sufficiently high, it may become desirable to treat all of the sewage in a central facility. The general purpose of such treatment is to return the water to a river or other water body in such a state that it does not cause any damage to the environment. The aim, however, is not necessarily to produce water identical to the water already in the river. According to the Clean Water Act of 1977, discharge water must meet certain standards with regard to BOD, suspended solids, fecal bacteria, and pH. In other words, water discharged into surface waters of the United States must have a low BOD, little suspended material, and a nearly neutral pH and must not be infectious.

Primary Treatment

The first stage in sewage treatment is **primary treatment**, which is the simplest and oldest of the treatment processes. The purpose of primary treatment is to remove solid material from the sewage stream. The first step is referred to as **screening**. The intention here is to remove fairly large objects from the water. The screen used is very coarse. Next, the flow rate of the water is slowed down a little as it passes through a tank. Larger particles, such as sand and coffee grounds, in the wastewater will settle down in the tank. This step is called **grit removal**. From here the water is moved to a very large tank, where the water moves very slowly. This step is called **primary sedimentation**. Very small solid particles either sink to the bottom or float on the top. The floating solids are skimmed off and heavy solids are removed from the bottom of the tank. The water that flows from the tank still has some solids in it because extremely fine particle will not settle down even in perfectly still water. This water also contains a considerable load of dissolved organic material. Still, the water is much improved in looks, odor, and actual content. About one-fourth of the BOD is removed as a result of removing about 70% of the suspended solids.

Secondary Treatment

We have not yet met the goals of wastewater treatment, though. About three-fourths of the BOD remains, and the water still contains some infectious organisms. **Secondary treatment** is aimed principally at BOD reduction. The main reason for a high BOD in the water at this point is the dissolved organic material, which is relatively high energy and as such can serve as a food source for the proper organisms. The useful species are a whole collection of the naturally occurring microorganisms. These organisms consume the dissolved organic material and either incorporate the organic matter in the biomass or metabolize the material to produce energy.

These microorganisms extract energy from the organic material by using the oxygen present in air to convert the carbon, hydrogen, nitrogen, phosphorus, and sulfur into carbon dioxide, water, nitrates, phosphates, and sulfates:

$$\text{Organics} + O_2 \rightarrow CO_2 + H_2O + NO_3^- + H_2PO_4^- + SO_4^{2-}$$

The carbon dioxide is emitted into the air, and the water just becomes part of the water in the surroundings. The nitrates, phosphates, and sulfates do not contribute to the BOD.

It is important to keep oxygen levels high because lack of oxygen would cause different types of organisms to be active, thus converting the organic material into another class of products, most of which are toxic and foul smelling. These include hydrogen sulfide (rotten egg gas), ammonia, foul-smelling organic acids, and methane:

$$\text{Organics} \rightarrow CO_2 + CH_4 + NH_3 + H_2PO_4^- + H_2S + \text{Acids}$$

Not only are these products much less desirable than those produced in the presence of oxygen, but they still have a BOD. Because the goal is BOD reduction, the aerobic process is clearly better than the anaerobic process.

The other way that organisms use these organic compounds is to incorporate them into the cellular material as the cells reproduce and grow. As the growth process of the microorganisms continues, the cells usually clump together and form solid particles that can settle out. At this point, one may wonder if anything has been accomplished. In primary treatment, we tried to get rid of all of the solids and now we just made more. Remember, however, that the purpose is to remove organic material with its high BOD, and it is all right if some of the dissolved organic material becomes solid cellular material that can be settled out and removed.

For secondary treatment, it is clear that we need lots of the right kinds of microorganisms and plenty of oxygen. Under these conditions, the organic material will either become solid cellular material or be converted into carbon dioxide, water, and various inorganic ions that have no BOD. There are two approaches to secondary treatment, and some treatment plants use both of them. One is the **fixed-film biological process** and the other is known as the **activated sludge process**. Both of these processes use microorganisms in the presence of air to reduce the BOD of the water.

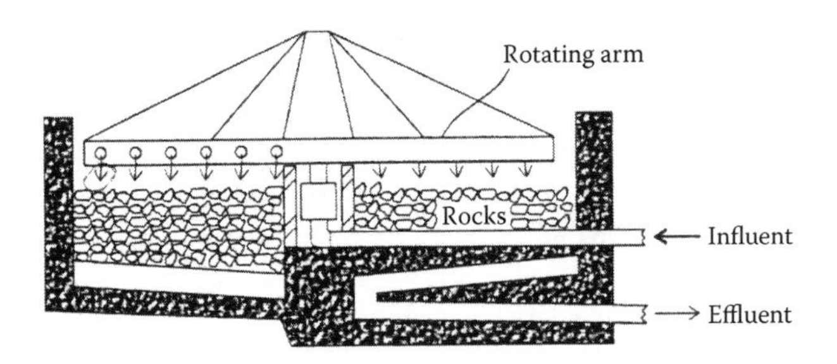

Figure 13.3 Diagram of a trickling filter. (From Vesilind, P.A. and Peirce, J.J., *Environmental Pollution and Control*, Butterworth, Boston, MA, 1983. With permission.)

The fixed-film biological process includes two distinctly different devices. One is a trickling filter, which allows water to trickle over rocks or other support material coated with microorganisms. As seen from Figure 13.3, the water falls from a rotating arm, which allows the water to be in good contact with the air. The microorganisms grow on the support, using the organic matter in the water as food. Excess microorganisms slough off from the rocks and are carried out with the effluent. The solid material in the effluent is removed in a later **secondary sedimentation** process. The trickling filter is the oldest of the secondary treatment devices. An experimental model was tried out by the Massachusetts State Board of Health in 1889, and a filter was put into operation in Baltimore in 1907.

Another version of the fixed-film process is to expose the film to the air rather than trickling the water through the air. In this process, large plastic disks are attached to a horizontal shaft. The microorganisms attach themselves to the rotating disks. The disks are roughly half-way immersed in the wastewater. As the disks rotate, the organisms on the disks are alternately exposed to the air and the wastewater. This approach, known as a *rotating biological reactor*, is illustrated in Figure 13.4. The microorganisms involved in secondary treatment come from the wastewater. Normally, the concentration of these organisms is too low to effectively treat the water, but the solid supports provide a place to collect large amounts of the organisms.

In another completely different approach, the activated sludge process, a high concentration of the appropriate organisms is provided by using recycled sewage sludge. Sewage sludge contains a wide variety of microorganisms, and by blowing air through the sludge the aerobic organisms thrive and become dominant. This material is known as *activated sludge*. Again, we appear to be undoing some of what we just did. We are adding back sludge just like that removed during primary treatment to help further reduce the BOD.

In the activated sludge process, air is forced into the water using large air pumps or mixers. The process starts with wastewater from primary treatment, activated sludge, and air from the pumps or mixers. After about 4 to 8 hours, the mixture is moved to a secondary sedimentation tank where the sludge settles out. Because the microorganisms in the sludge grow on the organic material, the amount of sludge has increased. Part of this sludge is recycled back to the aeration tank to be used with a new quantity of wastewater.

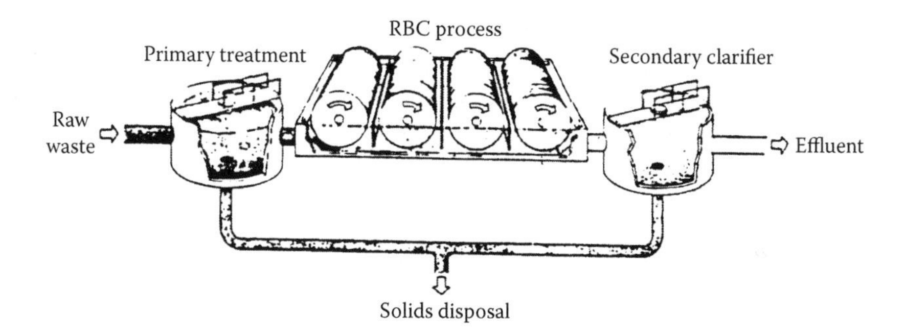

Figure 13.4 Diagram of a rotating biological reactor. (From Antonie, R.V., *Fixed Biological Surface—Wastewater Treatment*, CRC Press, Boca Raton, FL, 1976. With permission.)

Tertiary Treatment

When the sedimentation stage is completed, secondary treatment is finished. The water that comes from secondary treatment will have a fairly low BOD and few suspended solids; however, it may contain some disease-producing organisms and will certainly contain inorganic plant nutrients. In some cases, this water is discharged into a stream or lake, but usually some type of **tertiary treatment** is carried out.

Tertiary treatment is not nearly as well defined as the primary and secondary stages. At the very least, this stage would probably include disinfection to destroy any disease-producing organisms. Generally, disinfection is done either by adding chlorine or ozone or by radiation with ultraviolet light. Other operations that might be carried out include phosphate removal, nitrate removal, pH adjustment, further BOD reduction, and further removal of suspended solids.

What has been described here is the standard high-tech sewage treatment plant of the developed world. One problem with modern sewage treatment plants is that they are expensive. Most of the less-developed countries of the world simply cannot afford such plants, which explains why most of the world simply dumps their sewage directly into waterways. Low-tech alternatives are available that should be used wherever feasible. Westerners tend to be suspicious of such approaches. When describing a pebble filter for suspended solid removal, one text of several years ago noted that "this device will never succeed in the United States because the pebble filter is entirely too simple, with no moving parts. Nobody wants to manufacture it, nor will engineers design such a 'primitive' device" (Vesilind and Peirce, 1983).

Generally, low-tech approaches are inexpensive to build and operate but require the use of fairly large tracts of land. Variations of such systems have been built and are operating in Arcata, California; Lima, Peru; Bangladesh; China; Germany; Hungary; India; Indonesia; Israel; Malaysia; Taiwan; Thailand; and Vietnam. Such systems usually involve moving water through a series of holding ponds and wetlands. At first, solids are settled out and then later in the process algae and various plant life grow, followed often by the introduction of fish, which are harvested for food. The largest aquaculture system in the world is the Calcutta sewage system, which raises about 7000 tons of fish annually.

Industrial Wastes

Industrial wastes are a much more difficult problem than human wastes, principally because these wastes are not a standard collection of materials. Human wastes are roughly the same wherever they are found. The best approach to industrial waste is for each industry to treat the wastes generated from that particular industry. This may be expensive, but sewage treatment plants are not set up to remove unusual materials. Individual industries know what is in their wastewater and can determine how to remove specific pollutants. It is particularly troublesome when heavy metals are involved. Because of the toxicity of these metals, water that contains such metals will often kill the microorganisms on which the operation of the waste treatment plant depends.

Wastewater Treatment Sludge

What does one do with 1000 tons of sludge? This is one of the biggest problems facing modern wastewater treatment today. Initially, when sludge is drawn off it is a semi-liquid material, comprised of 96 to 99% water. Generally, to stabilize this material it is subjected to digestion, which involves heating the sludge to 95°F (35°C). Generally, there are two types of digestion: aerobic or anaerobic. If air is excluded, this process converts much of the organic material into water and methane. The methane can often be removed and used as fuel for heating. If air is bubbled through the sludge, carbon dioxide is formed. In either case, many of the disease-producing organisms are destroyed.

Following digestion, the main effort is to remove water from the sludge. When enough water is removed to decrease the volume to about 20% of its initial size, the sludge becomes a very thick solid/liquid material. Finally, with a further reduction to about 5% of the original volume, the material becomes a damp solid. We have now traded in a liquid waste problem for a solid waste problem. The three legal options for disposal of this sludge are incineration, land filling, and land spreading. Incineration and land filling are discussed in Chapter 14. Land spreading involves using the sludge as fertilizer because it has high plant nutrient value. The biggest problem is the disease-producing organisms. Some organisms survive digestion, and spreading potential disease producers on our crops is not a good idea. Sterilizing the material before spreading is one option. Another approach is to use the sludge on animal food crops or on nonfood crops.

Even if the sludge can be made sterile, another concern is the content of heavy metals. Sewage sludge is an effective absorber of heavy metals and is another reason why heavy metals need to be carefully restricted from the wastestream. Sludge contaminated with heavy metals becomes a hazardous waste and is very difficult to dispose of.

Source Reduction

In the end, the best approach to water pollution control, clearly, is to produce less waste in the first place. Many industries could profit in the long run by recovering chemicals from the wastestream and reusing them. Farmers could save money by utilizing techniques involving minimal amounts of fertilizers, irrigation water, and pesticides. Fertilizers and pesticides would not run off into streams, and less irrigation water would be used.

DISCUSSION QUESTIONS

1. How would the requirements differ for the purity of water taken from a lake or stream based on the following uses?
   a. Drinking water
   b. Cooling water from an electric plant
   c. Irrigation
2. Indicate whether each of the following represents point source or nonpoint source water pollution:
   a. Effluent from a sewage treatment plant
   b. Runoff from interstate highway construction
   c. Chemicals from farming in a rural stream
   d. Pesticide from a golf course
   e. Pesticide byproducts from an industrial plant
3. What happens to the BOD and DO after a load of oxygen-demanding wastes is added to a stream?
4. Speculate on the effects on a stream or a lake when each of the following is added in excess:
   a. Untreated sewage from a town
   b. Phosphate fertilizer
   c. Animal waste
5. How can irrigation of desert areas lead to salinization of both land and streams?
6. Icy roads in the wintertime can be very dangerous. Suggest ways to improve the safety of such roads that do not threaten to pollute the groundwater or nearby streams.
7. How does elemental mercury that is dumped into streams get into the food chain and cause illness in humans?
8. Discuss how the likelihood of adequate sewage treatment relates to the country or geographical area in which one lives.
9. Suggest how a farmer could economically deal with feedlot runoff containing dissolved materials from raw animal sewage.
10. How would you treat water in a holding pond containing each of the following?
    a. Acid
    b. Sediment
    c. Salt
    d. Dissolved organic materials
11. Explain why the coastal zones of the ocean are so much more likely to be polluted than the open seas.

12. Why are oil rig blowouts generally worse than spills from oil tankers?
13. What factors seem to control the amount of damage that will be caused by an oil spill?
14. Much of human activity involves digging in the ground. Some of these activities can have an effect on groundwater. What are some of the ways in which human activities underground affect groundwater and why?
15. What is the logic behind the use of injection wells to dispose of liquid wastes? Do you feel that the process is safe or flawed? Why?
16. Discuss and define primary, secondary, and tertiary treatments as they apply to sewage treatment.
17. What are the similarities and differences in terms of the physical, biological, and chemical processes between standard sewage treatment in a treatment plant and the operation of a septic tank and tile field?
18. Why are heavy metals such a problem for sewage treatment plants, and how can industries with such wastes avoid problems?

BIBLIOGRAPHY

Anon., Plans announced to restart desalination plant near Yuma, *USWaterNews.com*, 2005, http://www.uswaternews.com/archives/arcsupply/5planannο5.html.

Anon., Preventing more BP-type oil disasters, *Earthjuctice.org*, 2013, http://earthjustice.org/features/preventing-more-bp-type-oil-disasters.

Antonie, R.V., *Fixed Biological Surfaces—Wastewater Treatment*, CRC Press, Boca Raton, FL, 1976, p. 10.

Chowdhury, A.M., Arsenic crisis in Bangladesh, *Scientific American*, 291(2), 87–91, 2004.

CLUI, *Yuma Desalinization Plant*, The Center for Land Use Interpretation, Culver City, CA, 2013, http://ludb.clui.org/ex/i/AZ3149/.

Collie, M. and Russo, J., *Deep-Sea Biodiversity and the Impacts of Ocean Dumping*, NOAA Research, National Oceanic and Atmospheric Administration, Silver Spring, MD, 2000, http://www.oar.noaa.gov/spotlite/archive/spot_oceandumping.html.

Curtius, M., A sewage plant tourists love, *SFGate.com*, December 18, 1998, http://www.sfgate.com/news/article/A-Sewage-Plant-Tourists-Love-Arcata-s-low-tech-2973312.php.

Davis, T., Test run on Yuma desalination plant a success, *Arizona Daily Star*, June 28, 2012, http://azstarnet.com/news/science/health-med-fit/test-run-on-yuma-desalination-plant-a-success/article_9b5ecc07-96da-57f5-a1b7-d2dd94726a00.html.

Doyle, R., Darkness on the water, *Scientific American*, 296(4), 26, 2007.

Embach, C., Oil spills: impact on the ocean, *WaterEncyclopedia.com*, 2013, http://www.waterencyclopedia.com/Oc-Po/Oil-Spills-Impact-on-the-Ocean.html.

IBWC, *Colorado River Boundary Section*, International Boundary & Water Commission, El Paso, TX, http://www.ibwc.gov/Water_Data/Colorado/Index.html.

JMP, *Joint Monitoring Programme (JMP) for Water Supply and Sanitation*, WHO-UNICEF, Geneva, Switzerland, 2013, http://www.wssinfo.org.

Mallin, M.A., Wading in waste, *Scientific American*, 294(6), 53–59, 2006.

Mason, W.R., Pollution of groundwater, *WaterEncyclopedia.com*, 2013, http://www.waterencyclopedia.com/Oc-Po/Pollution-of-Groundwater.html.

NOAA, *1998 Year of the Ocean: A Survey of International Agreements*, National Oceanic and Atmospheric Administration, Silver Spring, MD, 1998, http://www.yoto98.noaa.gov/yoto/meeting/intl_agr_316.html.

NOAA, *Gulf Oil Spill*, NOAA Education Resources, National Oceanic and Atmospheric Administration, Silver Spring, MD, 2011, http://www.education.noaa.gov/Ocean_and_Coasts/Oil_Spill.html.

Querna, E., A plan for water, *Scientific American*, 291(2), 25–26, 2004.

Senior, J., Yannawa wastewater project, Bangkok, Thailand, *ARUP Journal*, 3, 35–41, 2000.

USDA, *U.S. Fertilizer Use and Price*, Economic Research Service, U.S. Department of Agriculture, Washington, DC, 2012, http://www.ers.usda.gov/data-products/fertilizer-use-and-price.aspx.

USEPA, *Ocean Dumping and Dredged Material Management*, EPA-842-F-05-001d, U.S. Environmental Protection Agency, Washington, DC, 2005.

USEPA, *2000–2001 Pesticide Market Estimates: Usage*, U.S. Environmental Protection Agency, Washington, DC, 2006, http://www.epa.gov/pesticides/pestsales/01pestsales/usage2001.htm.

Vesilind, P.A. and Peirce, J.J., *Environmental Pollution and Control*, 2nd ed., Butterworth, Boston, MA, 1983, pp. 87, 95.

Wellton–Mohawk, *Challenges*, Wellton–Mohawk Irrigation and Drainage District, Wellton, AZ, 2004, http://www.wellton-mohawk.org/challenges.html.

CHAPTER 14

Solid Wastes

Trash! What to do with all that trash? One of the consistent byproducts of living is refuse. Ever since humans began inhabiting this planet, we have been producing garbage and trash. Of course, it is more of a problem for some societies than for others. The early hunter–gatherer societies struggled very little with the accumulation of refuse because they were mobile. When too much garbage or trash accumulated where they were living, or if they encountered any other problem with that location, then they would just pack up and move on. In addition, nearly all of their refuse was of natural origin; therefore, over time the material would biodegrade back into nature from which it had come.

With the development of agriculture the buildup of trash became somewhat more significant, but not severe. True, farmers did not pack up and move on so they had to deal with their refuse, but because they had plenty of land they usually found some place to hide it. It was not until humans began to live in settled cities that trash became a serious problem. Various solutions have been developed to handle solid waste problems. In ancient Troy, the solution was simply to throw the trash on the ground. As the trash built up to higher and higher levels on the ground, the occupants raised the doors and ceilings to make more headroom in the house. The ground level of Troy increased by about 5 in. (13 cm) per decade. In fact, the altitude of many ancient cities increased due to the accumulation of debris. From 9000 BCE to 1300 BCE the city of Jericho rose by nearly 70 ft (21 m).

Eventually, the citizens of urban centers began carting their trash outside the town to dumpsites. The success of these waste disposal sites depended on their location, the population of the town, and on how much trash each person produced. The huge piles of garbage outside the city gates of Paris about 1400 CE were so high that they blocked the view of potential enemies approaching the city and compromised the defense of the city.

However bad the situation with trash was, it was about to worsen with the Industrial Revolution. The accumulated discards of the new industries were added to the human refuse of the crowded, highly populated cities. Initially, little thought was given to the garbage and trash that were accumulating in the cities. Most of it was thrown into the streets or waterways. The result was that the streets of London reeked of garbage and the river became badly polluted.

The situation was not any better on the other side of the Atlantic. In 19th-century New York City, pigs roamed through the city rooting through the garbage. In fact, many cities valued the pigs for their garbage-eating abilities and had laws protecting them. In 1800, President John Adams hired a private citizen to cart the garbage away from the White House, but another 56 years were to pass before Washington had citywide garbage collection.

SOURCES OF SOLID WASTES

The materials that are released into the municipal wastestream are only a small percentage of the total solid wastes that are produced. Most solid wastes in the United States are, in fact, handled at the site where they are produced. The major contributors are agricultural and logging wastes, which account for about 48% of all solid wastes. These wastes include everything from crop residue to animal manure. Other major contributors are the mining industries and metal-producing plants, which account for about another 38% of solid wastes. These wastes include such items as mine tailings, strip mine overburden, and smelter slag. Although these two sources account for about 86% of the solid wastes, they are not the focus of this chapter. Most of these wastes are disposed of or processed at the site where they are produced. Sometimes the wastes are disposed of properly and sometimes they are not, but improper disposal leads to water and air pollution, not to solid waste problems.

Another 8% of solid wastes comes from industries. Although a small amount of this refuse may be added to the public burden, most of the material is handled by the industries themselves. Again, sometimes the refuse is handled well and sometimes it is not. The remainder represents about 6% of all solid wastes produced in the United States, but what a 6%! Referred to as **domestic waste**, the public sector is responsible for handling and disposing of this material. Domestic waste includes all of the materials discarded by individuals and commercial enterprises. This refuse is the kind we are all aware of. We throw the trash away and either pay a fee or are taxed for its disposal. Municipal trash and garbage are of public concern and an expense.

COMPOSITION OF DOMESTIC SOLID WASTE

Every year the United States disposes of over 250 million tons of trash and garbage. This amount is equivalent to over 1600 lb (720 kg) of trash for every man, woman, and child in the country. Figure 14.1 indicates that the United States is not the only country with so much trash to throw away; however, it is also apparent that decreasing levels of wealth are accompanied by decreasing levels of domestic trash. Less-developed countries simply cannot afford to throw away much material as waste because these countries need everything they have. In fact, many people in poorer countries make their living by picking through the dumpsites in larger cities, looking for items to salvage.

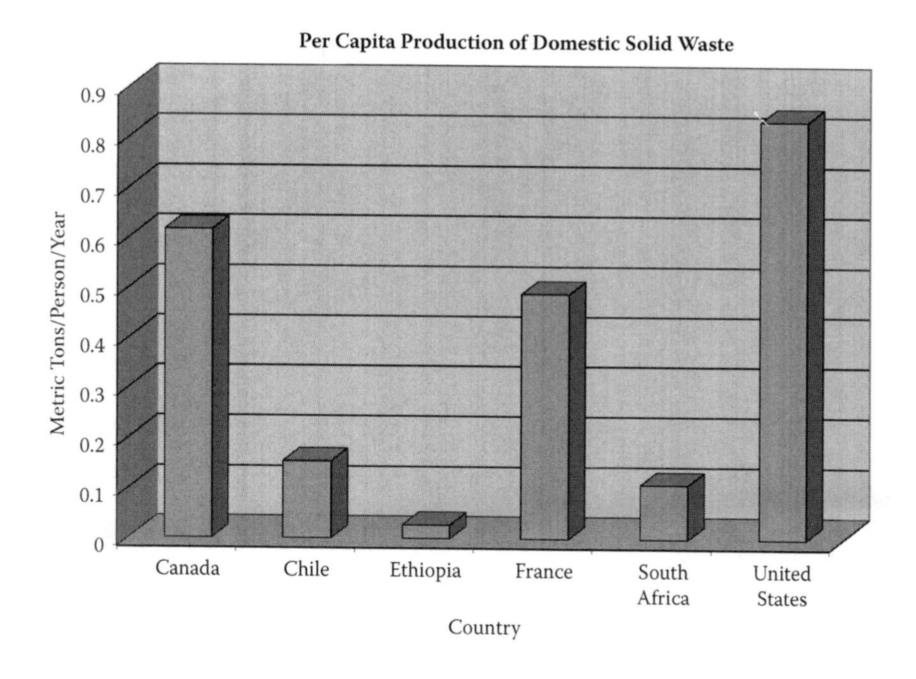

Figure 14.1 Production of domestic waste in selected countries on a per-person basis. (Adapted from Nshimirimana, J., Attitudes and Behaviour of Low-Income Households towards the Management of Domestic Solid Waste in Tafelsig, Mitchell's Plain, PhD thesis, University of the Western Cape, Capetown, South Africa, 2004.)

Solid wastes by their very nature are much more heterogeneous than pollutants found in the air or water. All kinds of things make up solid wastes: newspapers, magazines, cans, bottles, boxes, food, yard waste, writing paper, etc. To understand the ultimate fate of municipal wastes, we need to know something about the composition of this material. Most people would be surprised to know that the largest category of municipal refuse is paper and cardboard; in fact, in 2010, 28.5% of the domestic wastestream was paper or cardboard. The distribution of various types of municipal wastes in the United States is shown in Figure 14.2. The category titled "Other" in the figure includes yard waste, food waste, wood, textiles, rubber, leather, and numerous miscellaneous items. Another important thing to note about the composition of municipal waste is that most of the material is of natural origin. Natural materials include paper, cardboard, yard waste, food waste, wood, some textiles, and leather.

SOLID WASTE DISPOSITION

This section deals with the various ways of handling the solid wastestream. We are not going to discuss every possible way to get rid of trash, only the major ones. Generally, the solutions fall into one of two categories: either make the trash useful

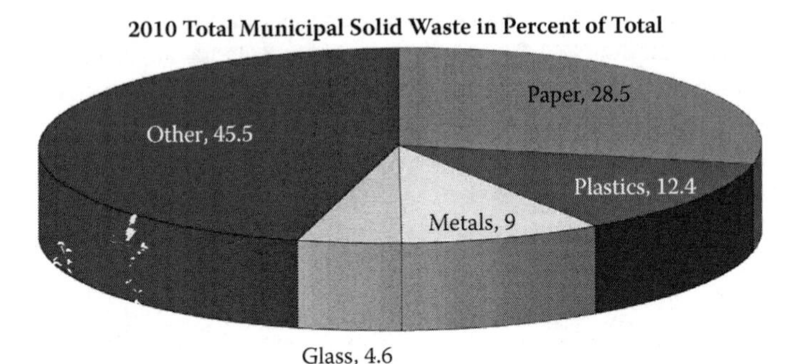

Figure 14.2 Percentage composition of U.S. domestic waste in 2010. (Based on USEPA, *Municipal Solid Waste*, U.S. Environmental Protection Agency, Washington, DC, 2012, http://www.epa.gov/epawaste/nonhaz/municipal/index.htm.)

again or get rid of it. When we try to make trash useful again, we can either use the items again as they are (that is, **reuse** them) or use trash items to make new products, which is called **recycling**. If the choice is to dispose of items, the most common methods are either *land disposal* or *incineration*. These are very old methods; however, in the modern age they have been improved upon significantly. The choice is largely dictated by the amount of land available for land disposal. As seen from Table 14.1, countries where land is precious, such as Japan, tend toward incineration as the method of choice, whereas countries such as Australia with lots of land are more likely to use land disposal.

Land Disposal

Land disposal includes any disposal method that dumps the material somewhere on the land. These methods run the gamut from high-tech to no-tech and legal to illegal. The open dump was the earliest practice of waste disposal on land. In early times, the main purpose was to get the garbage away from where the people lived.

Table 14.1 Landfilling vs. Incineration for Municipal Wastes in Various Countries

Country	Year	Landfill (Thousands of Metric Tons)	Incineration (Thousands of Metric Tons)
Australia	1980	9800 (98%)	200 (2%)
Canada	1989	13,448 (~90%)	1416 (~10%)
France	1989	7684 (52%)	6970 (48%)
Japan	1987	16,486 (34%)	32,616 (66%)
Switzerland	1989	460 (17%)	2270 (83%)
United Kingdom	1989	14,000 (85%)	2500 (15%)
United States	1986	138,705 (~90%)	15,000 (~10%)

With sparse populations and lots of open land, such an arrangement was usually adequate. In the Middle Ages, the practice of setting dumps afire began. This practice reduced the volume of wastes, cut down on the odor, and drove out some of the dumpsite pests. Other than the innovation of burning the dump, the operation of the dumpsite changed very little until the 20th century.

It became more and more difficult to keep dumpsites at a respectable distance from people and their homes. As a result, citizens began to object to the odor, to the rats and other pests, and to the air pollution resulting from burning the trash. In response to these concerns, the **sanitary landfill** was developed. The sanitary landfill is a process of land disposal where each day's refuse is spread out in a thin layer and covered with a thin layer of soil on the same day. After the landfill has been filled to a certain depth, the landfill is covered with a few feet of soil and compacted as much as possible, and the site is closed. The sites are considered sanitary because the refuse is exposed for only a short period of time. When the trash and garbage are covered, the rats, birds, insects, and other pests are denied access to it. In open dumps or improperly operated landfills, many of these pests can spread diseases picked up from the rotting refuse.

Because landfills have some clear advantages over open dumps, it does not mean that they are universally used. In the United States and most of the developed world, open dumps are illegal. Yet, open dumps exist and are in use. A good sanitary landfill is more expensive to operate than an open dump. In many areas, the use of a landfill requires payment of a fee, particularly for large items. Furthermore, in many cases the individual must transport the refuse to the landfill or a collection site. Some people cannot afford the landfill fees and others just choose not to bother. Open roadside dumping can become a real problem in areas of chronic poverty and weak law enforcement—for example, in the Appalachian region of eastern Kentucky, West Virginia, western Virginia, and eastern Tennessee. Many of the narrow valleys of this scenic, low-mountain region are cluttered by trash dumped there by the impoverished and poorly educated residents.

Less-developed countries of the world lacking the money and equipment to operate landfills often have no alternative to open dumping. In large cities, open dumps become huge unsanitary piles of trash and garbage covering large tracts of land. Literally thousands of people will try to survive by picking through mountains of trash, looking for anything of value. People searching through garbage for a living is very much a fact of life in Manila, which has a number of huge open dumpsites. One of them, Smokey Mountain, was about 90 ft (30 m) high, and about 30,000 people lived and worked around it. The rotting garbage often caught fire due to spontaneous combustion, and the smoke from the fires gave the dump its name. This dump was closed in 1990, and 21 medium-rise tenement buildings were built on the site. In 2000, at another larger dumpsite in Manila, more than 60 people were killed when the side of a mountain of garbage fell onto squatters' homes.

To discuss more completely the proper operation of a sanitary landfill, we need to consider what happens after the trash is put in the landfill. Do we expect that after a couple of years none of the trash will be there anymore? Such an expectation is unrealistic. Some things in a landfill biodegrade slowly, whereas some others do not degrade at all. Glass, aluminum, and some types of plastic simply do not biodegrade and may remain in the landfill indefinitely. Many other types of plastic will only degrade very slowly.

One of the widely held myths is that if one buried organic material, such as paper, for a long period of time it will automatically biodegrade. The reality is that the material may degrade and then again may not. Scientists in Arizona, who call themselves "garbage archeologists," regularly dig up newspapers, phone books, and other related items that are 30, 40, or 50 years old and are in excellent shape. Ancient manuscripts have been found in Egypt that date back thousands of years. The fact is that for biodegradation to occur water is required, but both of the regions just mentioned are very arid places.

Food, wood, yard waste, some textiles, paper, and cardboard can certainly be encouraged to biodegrade if water is present. For this reason, water should not be excluded from a landfill; however, water that soaks into the landfill will also move out of the landfill into the groundwater. This water, as it passes through the refuse, will dissolve any water-soluble materials with which it comes into contact. The dissolving of substances from a landfill into water is called *leaching*. The liquid that flows from the landfill is the **leachate**.

Leachate is an extremely complex mixture that is similar to sewage in many ways. The leachate becomes loaded with organic materials when it passes over food, contaminated boxes, dirty bottles, and messy containers. Disposable diapers may, in fact, contribute some real sewage components to the leachate. Other substances that may be present in the leachate include inorganic salts, various dyes from clothing and printed material, and residues of a large variety of products. This liquid leachate is free to soak down into the groundwater and contaminate it.

When wastewater treatment was discussed in Chapter 13, it was noted that the presence or absence of oxygen controls the type of products obtained due to the biodegradation of organic matter. The processes in a landfill are most likely to be anaerobic because the oxygen in a buried landfill would deplete very quickly. Two of the gases produced are methane and hydrogen sulfide. The first can be explosive and the second potentially toxic. These gases can also seep into basements or underground utility lines. The possibility of a methane explosion is a concern for the buildings that were built on the site of the former Smokey Mountain landfill in Manila. Before construction no consideration was given to venting the methane that would inevitably be emitted from the trash and garbage in the landfill.

Because of these problems, landfill operators in the United States are now under tighter rules of operation. Sanitary landfills are required to be more "secure,"* which means that it must be operated in such a way that protects the surrounding environment. A diagram of a secure landfill is shown in Figure 14.3. To prevent the leachate from contaminating the groundwater, the landfill is lined with a plastic liner. A drainage system with collection pipes is installed to remove the leachate. The leachate is considered to be contaminated wastewater and is either treated onsite or pumped into a sewage treatment plant. It is very important that toxic chemicals be

* A secure landfill is one that is operated under the rules for hazardous waste.

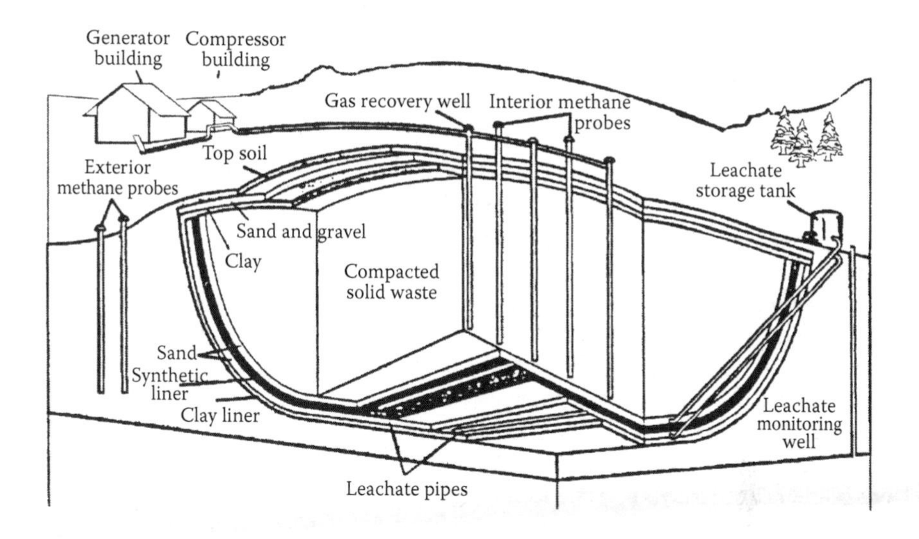

Figure 14.3 Diagram of a modern sanitary landfill. (From Arms, K., *Environmental Science*, Saunders, New York © 1990; Brooks/Cole, a part of Cengage Learning, Inc., www.cengage.com/permission. With permission.)

kept out of landfills because they will find their way into the leachate, and, as already noted, sewage treatment plants are not set up to remove unusual materials. As discussed in Chapter 15, an amazing number of household chemicals contain hazardous substances and should not be placed in a landfill.

As the landfill begins to reach its capacity in certain sectors, methane recovery probes are implanted to remove methane as it accumulates. This gas can be collected, purified, and sold as a substitute for natural gas for heating purposes. The Puente Hills landfill near Whittier, California, has one of the world's largest landfill gas recovery facilities. It produces about 14 billion ft^3 of gas per year, most of which is used to generate electricity.

It is difficult to find locations for landfills. In countries such as the United States there is no lack of land, particularly land that meets the technical requirements for a landfill, but there are other complicating factors. In large metropolitan areas where land is scarce, two facilities usually require a large amount of land: the airport and the landfill. Often land is acquired for both right next to each other. Airports attract airplanes and landfills attract birds; this is not a good combination. Airplanes have a fairly high probability of hitting the birds either during take-off or landing. It is particularly dangerous if birds are sucked into an engine during take-off, when the plane is at its peak power consumption level and anything that cuts engine power might well have disastrous consequences. A crash in 1960 at Boston's Logan Airport in which 62 passengers died occurred when the plane passed through a flock of starlings. Then there was the recent ill-fated US Air Flight 1549, which, on January 15, 2009, after a bird strike on both engines, was famously landed by Captain "Sully" Sullenberger in the Hudson River. It is not clear that birds from a landfill had anything to do with this airplane event, but it does make clear that flocks of birds and aircraft are not a good mix.

Another problem in finding sites for landfills is the **NIMBY syndrome**. NIMBY stands for "not in my backyard." When they have been covered over and closed, landfills are not the least bit unpleasant; however, while they are in operation there are issues of heavy truck traffic, litter scattered by the wind, and dust raised when the dirt is moved around. In addition, people just do not like the idea of living near the "dump." Often it is politically difficult to find a location for a landfill because of local opposition.

Once a landfill reaches its capacity and is closed, the resulting land area can be quite pleasant and useful. One of the more successful uses of a completed landfill can be found in Virginia Beach, Virginia, where trash was used to build a hill 68 ft high. This hill, called Mt. Trashmore, became the center of a recreation park, complete with a soap-box derby ramp, freshwater lakes, skateboard ramps, parking areas, concession stands, picnic areas, and playgrounds. This project was so successful that another Mt. Trashmore is in the process of being built. Scheduled to be complete in 2015, it will be a 573-acre recreational area with two artificial lakes and a 140-ft-high Mt. Trashmore.

The largest landfill in the world, until it closed in 2001, was the Fresh Kills Landfill on Staten Island in New York City. The site was so large that it could be seen with the naked eye from space, and the highest mound was higher than the height of the Washington Monument. The site was briefly reopened in 2002 to accommodate debris from the World Trade Center disaster. The City of New York is planning to convert the site into a 2200-acre park over a 30-year period.

Incineration

Setting trash on fire is an old and well-proven method of disposal. The practice of setting dumps on fire, as noted earlier, began in the Middle Ages. This method of trash disposal continues even today. As pointed out earlier, though, there are some problems associated with burning trash; for example, simply setting a pile of trash on fire produces massive air pollution from the smoke and incompletely burned trash. There are some obvious advantages to incineration, however. Incineration reduces the volume of trash and produces heat, which can be used as an energy source. To take full advantage of these benefits, modern incineration must be done carefully and under closely monitored conditions.

A drawing of a modern incinerator is shown in Figure 14.4. The incinerator must be operated to ensure that everything that will burn does burn, that everything that burns does so completely, and that only nontoxic gases are emitted from the stack. Any incineration facility must be built keeping in mind the complexities of the domestic wastestream. People throw all kinds of things into the trash, and not all of it will burn; therefore, such items must either be removed before burning or dealt with afterward. Whether the noncombustible material is removed before or after burning, the substances that enter the combustion chamber are a complex collection of items. To be sure that everything combustible burns completely, it is important that the fire be very hot. A good incinerator will bring all materials in contact with plenty of air and heat them to at least 1800°F (1000°C). The three T's for operation of an incinerator

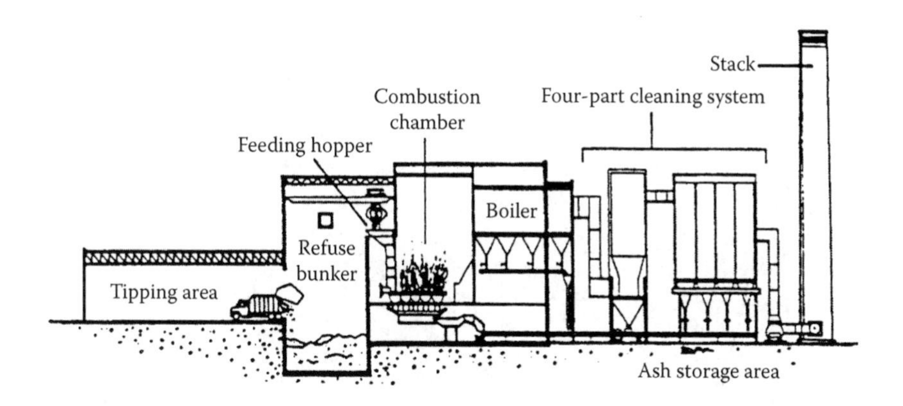

Figure 14.4 Diagram of a modern refuse incinerator. (From Cunningham, W.P. and Saigo, B.W., *Environmental Science, A Global Concern*, William C. Brown, Dubuque, IA, 1990. With permission of The McGraw-Hill Companies.)

are *time*, *temperature*, and *turbulence*. The material must be kept hot enough for long enough and undergo significant physical mixing. One is aware from personal experience that to burn a pile of solid material it must be stirred up from time to time.

From the law of conservation of matter we know that all the atoms we put into an incinerator have to go somewhere. They are either in the ash that falls to the bottom or in the effluent rising toward the stack. Most of what rises above the flames are gases, but this effluent also contains some solid fly ash. To avoid changing a solid waste problem into an air pollution problem, the material escaping from the stack must be controlled carefully. Devices are installed to remove all solid material from the effluent, and some types of toxic gases are removed from the stack gases as well. The easily removed gases include the reactive gases such as sulfur dioxide and hydrogen chloride.

If the wastestream is not sorted before incineration, then about 25% of the mass will be recovered as ash (both bottom ash and fly ash). Of this material, about 25% of it is iron metal, which can be separated by magnetic attraction. For the remaining material there are various processes that use the density of the ash to separate aluminum from other materials. Some of the other residues can be used to make items such as construction blocks. This leaves about 10% of the original ash that cannot be used and must be sent to a landfill.

Because incinerators release heat, these plants can be used to generate electricity. The fire could also be used to make steam or hot water for space heating. A good incineration plant could produce 500 kWh of electricity from 1 ton of municipal solid waste. This amount of electricity would run 5000 100-W bulbs for 1 hour or 500 hair dryers for 1 hour, or it could heat a typical apartment for 1 month.

Despite the definite advantages of incineration, there are also some drawbacks, the most persistent being the emission of dioxins and furans. These relatively toxic materials are nearly always produced in the incineration of complex organic materials such as municipal refuse. The seriousness of the threat from these chemicals is partly related to how well the incinerator is operated. The dioxins and furans are

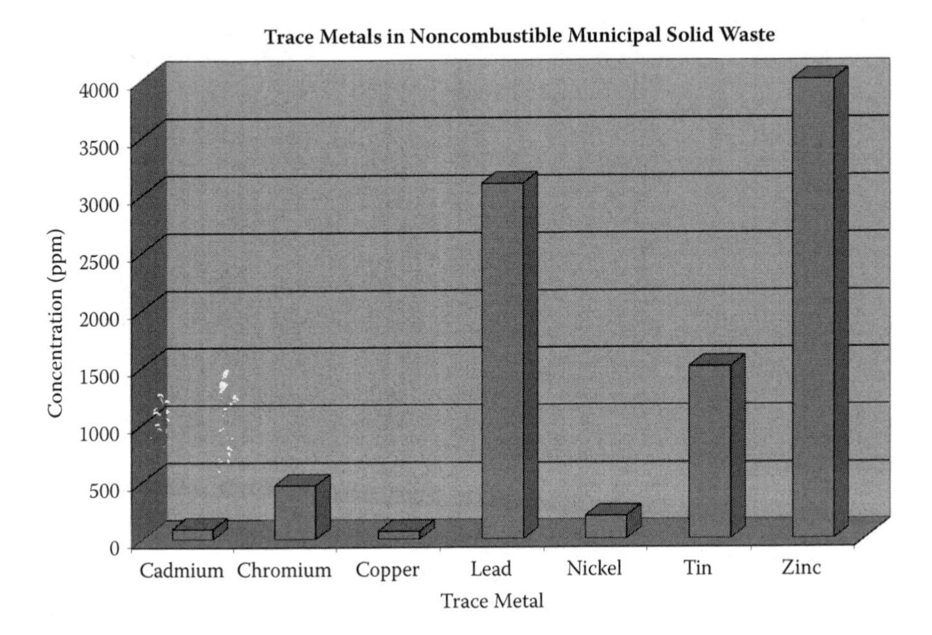

Figure 14.5 Typical concentration of trace metals in the noncombustible portion of U.S. urban refuse in 1990. (Adapted from Tchobanoglous, G. and Kreith, F., Eds., *Handbook of Solid Waste Management*, McGraw-Hill, New York, 2002.)

usually incorporated into the ash, so part of this material goes up the stack as fly ash. If the system is very efficient at removing fly ash, then very little of the toxic dioxins and furans make it into the atmosphere. The fact is, however, that no fly ash removal system removes all of the ash, and the removal efficiencies usually run about 95 to 99%.

The actual operating conditions of the plant can reduce the emission of dioxins and related materials. Dioxins will decompose in the incinerator, but not easily. If the incinerator is operated in such a way that everything in it reaches 1800°F (1000°C) for at least 1 second, the dioxins and furans would be destroyed.

Given the complex nature of the wastestream, the odds are fairly high that a lot of metals will be present in it. Many of these metals are not toxic in low concentration and do not pose a health hazard; however, some metals in the wastestream are known to be toxic. The relative concentrations of some of these trace metals in urban refuse are shown in Figure 14.5. Two of the most serious environmental heavy metal poisons are mercury and lead. Mercury does not appear in high enough concentration in the wastestream to appear in the graph and does not seem to pose a serious threat; however, lead is a more serious problem, as it is second only to zinc among the quantities of trace metals. Zinc, though, is not nearly as toxic as lead. Lead appears in many places in our environment and is difficult to control. Sources of lead in solid waste include automobile batteries, lead in old plumbing, old lead-based paint, automobile radiators, printing ink, glass, pottery, and electronic components. Upon incineration, these metals end up in the ash and, if the incinerator is not properly controlled, perhaps in the air. Metals are elements and therefore are not chemically destructible; hence, once they are put in the incinerator, they will either end up in the air or the ash. The only way to avoid this problem is to remove the metals before they reach the incinerator. Such procedures are equipment and manpower intensive and drive up the cost of incineration.

Finally, incinerators do not solve the solid waste problem but rather reduce it. Remember that, after incineration, separation of iron metal, and separation of material useful for construction, about 10% of the original mass still remains. Many of the toxic materials in the original trash, particularly the heavy metals, will now be more concentrated. Although not considered by the U.S. Environmental Protection Agency (EPA) to be a hazardous waste, many environmentalists feel that incinerator ash should be handled with greater care. Currently, it is generally sent to the landfill, but should special precautions be taken? It is possible that in incineration, a large volume of waste is being converted into a small volume of hazardous waste.

Recycling

Humans are the only animals who have a waste problem. For all other animals, everything is provided by nature and everything is returned to it. Everything is recycled indefinitely. The animal world is perhaps a paradigm for how we humans should handle the material of our world. To quote Arthur C. Clarke, "Solid wastes are the only raw materials we're too stupid to use." How much sense does it make to expend great energy, expense, and effort to mine some minerals from the ground only to later throw them into a landfill? Recycling deals with solid wastes by reducing the wastestream or, in other words, by redefining some waste materials into useful materials. Recycling requires more effort up front than does simple waste disposal, but recycling of certain materials is less expensive in the long run. The tendency of the universe toward disorder (higher entropy) is not a problem if we are going to dump all the material in a landfill, but if we want to recycle then the wastestream must be carefully controlled. Different types of materials must be separated from one another because each substance is recycled in a different way. For recycling to be successful, significant citizen involvement and commitment are required. If citizens separate their trash into various categories, then the expense of sorting need not be borne by the agency in charge of the recycling. This involvement by the citizen can often make a difference in whether or not recycling is economically feasible. The major categories of recyclable materials are paper, glass, metal, plastics, and compost materials.

Paper

Paper is an obvious material to begin our discussion with, as it is the single largest component of the wastestream. Paper and related products are used for many things. We write on it, print on it, package in it, bag in it, wipe with it, decorate with it, and read from it. We are constantly in contact with paper products. Any recycling

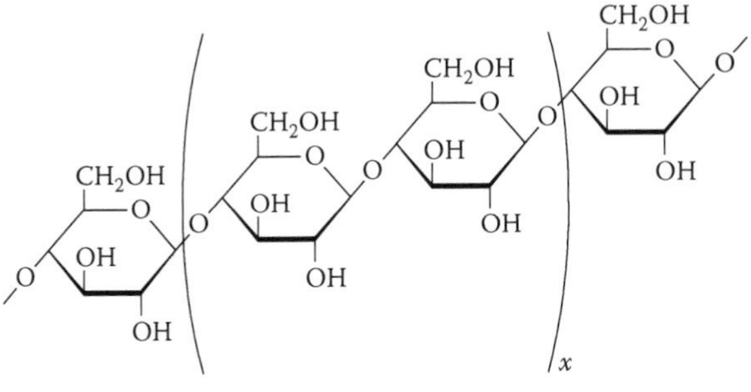

Cellulose

Figure 14.6 Structure of cellulose.

system that has any hope of having a significant impact on the solid waste crisis must be able to handle paper.

Paper is composed principally of cellulose from wood products. Cellulose is one of the most abundant compounds found in nature and is very common in plant material. The structure of cellulose is shown in Figure 14.6. Cellulose is a naturally occurring **polymer**. A polymer is a very large molecule made up of groups of atoms that are repeated over and over again. Cellulose molecules are very long, thin fiber-like molecules. The cellulose molecules in paper are held together by intermolecular forces to create thicker and longer fibers. A large collection of these cellulose molecules stick together to make the cellulose fibers of paper.

A number of processes must be carried out to make paper recyclable. The ink has to be removed, which destroys some of the paper fibers. After this, the paper is ground and treated to make new paper. This process breaks up and shortens the fibers, so recycled paper is inevitably of somewhat lower quality than paper from virgin sources; therefore, paper cannot be recycled indefinitely. Paper recycled once can be used for most purposes. If the paper is recycled more than once, the quality will deteriorate significantly, but the paper can still be utilized in a variety of ways. It is estimated that about 50% of the paper in use is of a quality that could be recycled.

Difficulties encountered from the very beginning with recycling paper include the fact that there are many different types of paper products and that to recycle paper we must somehow obtain "pure paper" to reprocess. Dirty paper contaminated with food products and the like cannot be recycled because these impurities would end up in the recycled paper, making it unfit for use. Even fairly clean paper usually has been written or printed on and the ink represents a contaminant. As a result, most paper must be de-inked before it can be recycled. Additionally, treated paper will carry their treatment chemicals into the recycling process. An example would be the slick paper from magazines and Sunday newspaper inserts which must be sorted out and handled separately. Historically, the first type of paper that was widely recycled was newspaper; therefore, the methods for recycling this type of paper are well established. Recycling newspaper is relatively straightforward because all of the paper is of similar quality, and there is a lot of it.

Recycling saves more than the trees that go into making the paper. Producing paper from recycled sources saves 30 to 55% of the energy compared to making paper from wood pulp. It also reduces air pollution by 74 to 95% and water pollution by 35%, and it saves water.

Glass

Glass is not a limited resource. The material is made from silicon dioxide, the principal component of sand. The sand is melted and small amounts of various chemicals are added, depending on the type of glass being made. Silicon dioxide is one of the most common compounds in the Earth's crust because silicon and oxygen are the two most abundant elements found in the Earth's crust. The problem with glass is that it does not degrade. Glass thrown into landfills will simply pile up. It melts but does not burn in incinerators. We keep melting sand, making glass, and piling it up while we make more. None of this is necessary because glass is one of the easiest materials to recycle. As long as it is sorted by type, the glass can simply be remelted and blown into new glass objects. Most glass used in standard packaging applications is the same type of glass. The only difference is in the color of the glass. If colorless glass and each of the other colors of glass are carefully separated, then glass recycling becomes pretty straightforward. Energy is saved when glass is recycled because old glass is easier to melt than sand, and the lower melting temperature leads to tremendous savings in energy consumption. The energy savings may be as much as 30%, which leads to savings in air pollution and water consumption for the production of that energy.

Metals

Without a doubt, the most successful recycling effort has been the recycling of aluminum cans. Aluminum is the third most plentiful element in the Earth's crust, but despite the large amounts present aluminum is very difficult to extract from its ore because it is a very reactive metal. (The history of the development of the aluminum industry was discussed in Chapter 1.) The problem with the process of extraction of aluminum from its ore is that it requires tremendous amounts of energy. The aluminum ore, bauxite, is mixed with a salt called cryolite and melted. The aluminum is then extracted by passing huge amounts of electricity through the molten mixture.

Recycling saves all of the electrical energy used to extract the aluminum from its ore. This saving amounts to about a 95% reduction in energy requirements. During recycling, old aluminum is simply melted and remolded into new aluminum products. Aluminum recycling has been very successful because of the economic incentives. Aluminum companies can easily afford to pay 25 to 30 cents/lb for scrap aluminum because of all the energy being saved in the process. At times, the price for aluminum has been as high as 80 cents/lb. Currently, about 58% of aluminum cans are recycled.

Recycling rates for other metals vary with the metal, from 24% for chromium to 74% for lead. Many factors control these recycling rates, including the price of the metal and the ease of extracting it in a pure form from the material. The so-called tin can is not a can made from pure tin, but is in fact made of steel with a thin layer of tin on each side of the steel. The tin and the steel must be separated before the steel is melted to be made into new products. This process is carried out chemically, and the tin can be recovered as well.

Plastics

Like paper, plastics are an example of polymers. The key difference is that cellulose in paper is a naturally occurring polymer, whereas plastics are a very large and diverse collection of man-made polymers. They are usually created by taking a small molecule and allowing it to react with itself over and over again to make an extremely large molecule. A very simple example would be the polymerization of ethylene to make polyethylene.

$$nCH_2=CH_2 \rightarrow \cdots CH_2CH_2CH_2CH_2CH_2CH_2 \cdots$$

The n is a very large number that corresponds to the number of ethylene molecules going into one polyethylene molecule.

A major problem in recycling plastics is the many different kinds of plastics that exist today. Plastics have been developed for many different uses, and for each application chemists have often developed a new plastic with unique properties. Not every type of plastic is recyclable. One important class of nonrecyclable plastics is **thermosetting plastics**, which have a three-dimensional cross-linked structure, such as Bakelite® (shown in Figure 14.7). Once such a plastic object is created in a specific shape, it cannot be heated and reshaped. In a sense, the object's shape is a result of

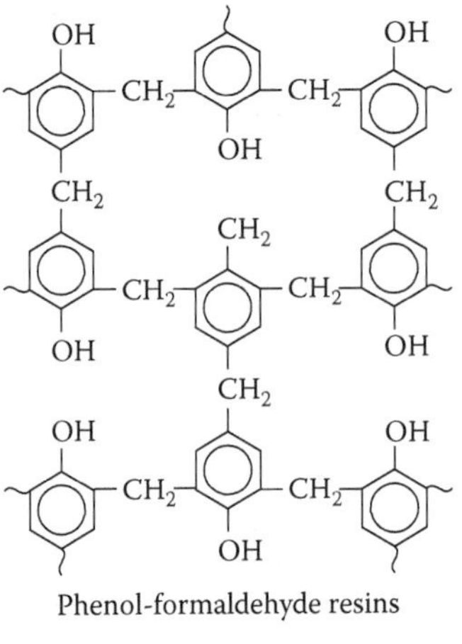

Phenol-formaldehyde resins

Figure 14.7 Structure of Bakelite®.

chemical reactions, and so the object is one giant molecule in that shape. Attempts at melting this type of plastic will usually result in charring and decomposition. Recyclable plastics are known as **thermoplastics**, which can be softened by heating and therefore can be remolded into different shapes over and over again.

Given the large number of different kinds of plastics that can be recycled, it is important to note that they cannot just be dumped into a large vat and melted together. Each polymer has specific properties that give the plastic its usefulness. Mixing them together would give one at worst a mess and at best an "average" plastic that would probably be of little value for anything. Luckily, about six different plastics predominate in the wastestream. Each of the classes of plastic is given a number. If an item is recyclable, the number can be found on the bottom of the item inside a special recycle symbol (Figure 14.8). For example, the bottom of a milk bottle has a "2" and the bottom of a soft drink bottle has a "1." Table 14.2 lists each code, the type of plastic, and the percentage of plastic bottles on which that code is found. If plastic recycling is to be successful, then a means must be found to sort and segregate the different types of plastic.

Figure 14.8 Recycle symbol.

The most common recycled plastics are from plastic bottles, most of which are beverage containers. These bottles may or may not be recycled back into the same kinds of bottles from which they came. Plastics recyclers often try to find ways to make other useful products out of the recycled plastic, such as fiber for carpet, fiber filling for pillows, plastic lumber, flower pots, and trash cans.

Compost Material

Some biodegradable materials from natural sources can be composted. Composting involves mixing the biomaterial with soil. The bacteria in the soil then aerobically decompose the organic matter, and the resulting material is referred to

Table 14.2 Different Types of Recyclable Plastic and Their Codes

Recycle Code	Type of Plastic	Percent of Containers and Packaging with This Type of Plastic (%)
1 PETE	Polyethylene terephthalate	15–20
2 HDPE	High-density polyethylene	25–35
3 V	Vinyl (polyvinyl chloride)	<5
4 LDPE	Low-density polyethylene	25–35
5 PP	Polypropylene	10–15
6 PS	Polystyrene	<5
7 OTHER	All other recyclable plastics	5–10

Source: USEPA, 2003 data tables, in *Municipal Solid Waste in the United States: Facts and Figures*, U.S. Environmental Protection Agency, Washington, DC, 2003, Table 7, http://www.epa.gov/epawaste/nonhaz/municipal/msw99.htm.

as **compost**. Compost can then be used as a soil conditioner and fertilizer. Common composting materials are grass clippings, ground-up brush and tree cuttings, leaves, and paper.

Reuse

Recycling is a wonderful energy- and resource-saving idea, but why is it necessary to take a glass bottle, melt it down, and make a new bottle out of it? For many years, the soft drink industry operated by filling bottles with their product, selling the product, collecting a deposit on the bottle, retrieving the bottle, refunding the deposit, washing and sterilizing the bottle, and finally refilling the bottle. This saves energy over recycling because it takes a lot less energy to sterilize a bottle than to melt it down. As local bottlers were gradually replaced with regional bottlers, the use of returnable bottles became economically less profitable. More of the cost of the product was tied up in transportation costs. It cost the company less to ship the bottles one way than to ship them round trip. For this reason, the companies switched to disposable containers that the consumer throws away. It may be less expensive for the bottler, but how about for society as a whole? There are some good examples of items in our society that are reused. Such items include certain auto parts that are reconditioned, stained glass windows, brass fittings, fine woodwork, and bricks. It is likely that thousands of other applications could be developed.

No Use

Probably the ultimate waste reduction approach is not to use an item in the first place. A very high percentage of the trash discarded by modern Western society is packaging. How necessary is the packaging? What purpose does it serve? Could we get along without it? In the general stores of the 19th century, most items for sale were placed on shelves or in open jars or barrels. Clearly, for food items we can argue that there is a need for packaging on sanitation grounds, but what about things such as nails? In a general store, one scoops out the nails, weighs them, and pays for them. Little or no packaging is used. If one goes to a discount store to buy nails, one finds them packaged in little packages with a cardboard back and a plastic bubble to hold them in. Which makes more environmental sense?

The modern fast-food restaurant probably best typifies our throwaway society. Built on the premise that disposable food containers are less expensive than the labor required to wash nondisposable plates, these establishments increase the trash burden of our culture. For a number of years there was a debate about the use of biodegradable packaging in fast-food restaurants. Should they use paper or styrofoam? Styrofoam does not biodegrade; therefore, over time it has been used less and less. But, the process of manufacturing paper causes pollution and involves the use of trees as raw materials. It is all a phony argument, though, because the throw-away approach is only economically advantageous if society as a whole pays for or subsidizes the waste disposal. Traditional restaurants had only garbage and food-shipment cartons to dispose of. Obviously, these establishments had much less impact on the wastestream. Is our culture willing to go back to such a way of doing business?

DISCUSSION QUESTIONS

1. How can dumps and landfills be a threat to the groundwater supply? What steps are taken in modern sanitary landfills to stop groundwater contamination?
2. Why is methane gas production a problem for landfills? In modern sanitary landfills, what is usually done to combat this problem?
3. If one was to bury a phone book in a landfill and the landfill was not covered, what is the likelihood of finding a readable book after 50 years in the following cities?
   a. Phoenix, Arizona
   b. New Orleans, Louisiana
   c. El Paso, Texas
   d. Seattle, Washington
4. Poverty seems to affect the amount of domestic waste production per capita. What would be the difference in waste production between the poor in a highly industrialized society such as the United States and waste production in a very poor country such as Ethiopia. Why?
5. What are some of the advantages and disadvantages of incineration of solid wastes? If you were in charge of handling the domestic solid wastes for a small country with a high population density, which method would you prefer and why? What would be your response for a large country with a low population density and why?
6. Why is it that paper is degraded every time it is recycled? Is there a point at which paper cannot be recycled? What types of paper products are not recyclable?
7. Because the raw materials for glass are plentiful, why should glass be recycled? Is this ultimately an issue related to sound environmental policy or a matter of aesthetics?
8. Explain why the recycling of aluminum has such a great benefit and is so successful.
9. What is the difference between thermosetting plastics and thermoplastics? Why is one of these recyclable and the other is not?
10. Discuss some of the problems associated with plastics recycling. Explain the significance of the recycling numbers.
11. Why is reuse better than recycling? Why is no use better than reuse?
12. For certain businesses—for example, fast-food restaurants—there is an economic advantage to using disposable materials as opposed to cleaning and reusing nondisposable items. Should a method be developed so that these businesses pay more of the true costs of disposing of the solid waste they produce? Would it be possible to track the solid wastes that customers dispose of elsewhere? Should such businesses be taxed to cover the perceived cost?

BIBLIOGRAPHY

Anon., Danger lurks under Smokey Mountain, *The Manila Times*, August 2, 2006.

Anon., Manila squatters buried by garbage; more than 50 killed when giant trash heap collapses and crushes shantytown, *The Washington Post*, July 11, 2000, p. A16.

Arms, K., *Environmental Science*, Saunders, New York, 1990.

ASU, *Where Does It Go From Here? Recycling at ASU*, Arizona State University, Phoenix, http://property.asu.edu/recycle/how.html.

Celdran, A.M. and Sahakian, M., *Smokey Mountain Remediation & Development Program (SMRDP)*, Asian Development Bank, Manila, Philippines, 2007 (www.povertyenvironment.net/files/37715-REG-TACR.pdf).

Cunningham, W.P. and Saigo, B.W., *Environmental Science: A Global Concern*, Wm. C. Brown, Dubuque, IA, 1990.

Hill, J.W., *Chemistry for Changing Times*, 6th ed., Macmillan, New York, 1992, p. 334.

Joesten, M.D., Netterville, J.T., and Wood, J.L., *World of Chemistry: Essentials*, Saunders, New York, 1993, p. 209.

Kreith, F. and Tchobanoglous, G., *Handbook of Solid Waste Management*, McGraw-Hill, New York, 2002.

Lindstrom, P., Methane emissions, in *Emissions of Greenhouse Gases in the United States 2003*, Energy Information Administration, U.S. Department of Energy, Washington, DC, 2004 (ftp://ftp.eia.doe.gov/pub/oiaf/1605/cdrom/pdf/ggrpt/057303.pdf).

Mazzoni, M., Aluminum can recycling rate highest in a decade, *Earth911.com*, September 1, 2011, http://earth911.com/news/2011/09/01/aluminum-can-recycling-rate-highest-in-a-decade/.

Melosi, M.V., *Garbage in the Cities: Refuse, Reform, and the Environment*, University of Pittsburgh Press, Pittsburgh, PA, 2005.

Nshimirimana, J., Attitudes and Behaviour of Low-Income Households towards the Management of Domestic Solid Waste in Tafelsig, Mitchell's Plain, PhD thesis, University of the Western Cape, Capetown, South Africa, 2004.

Tchobanoglous, G. and Kreith, F., Eds., *Handbook of Solid Waste Management*, 2nd ed., McGraw-Hill, New York, 2002.

USEPA, 2003 data tables, in *Municipal Solid Waste in the United States: Facts and Figures*, U.S. Environmental Protection Agency, Washington, DC, 2003, Table 7, http://www.epa.gov/epawaste/nonhaz/municipal/msw99.htm.

USEPA, *Municipal Solid Waste*, U.S. Environmental Protection Agency, Washington, DC, 2012, http://www.epa.gov/epawaste/nonhaz/municipal/index.htm.

World Bank, *Urban Poverty and Slum Upgrading: Environment: Poverty and Urban Environmental Conditions*, The World Bank, Washington, DC, 2011, http://web.worldbank.org/WBSITE/EXTERNAL/TOPICS/EXTURBANDEVELOPMENT/EXTURBANPOVERTY/0,,contentMDK:20234025~menuPK:7173807~pagePK:148956~piPK:216618~theSitePK:341325~isCURL:Y,00.html.

CHAPTER 15

Hazardous Wastes

In the late 1800s, William T. Love started work on a canal to take water around Niagara Falls to provide for river traffic and hydroelectric power; however, he ran into financial problems shortly after the work was begun, and all that was ever completed was a section about a mile long. In the years that followed Love's attempt at development, others built electric power plants near the falls. The inexpensive power provided by these electric power plants attracted many chemical companies. As the chemical companies grew, many of them needed places to dump waste chemicals. In the early years of the 20th century, there were very few controls on the dumping of materials, and various sites around the Niagara Falls area were used. Love's unfinished canal was a particularly attractive dumpsite. It was a trench about a mile long, about 50 ft (15 m) wide, and from 10 to 40 ft (3 to 12 m) deep. In 1942, people of the City of Niagara Falls began to use the old canal as a dump. Following the city's lead, the Hooker Chemical Company also began dumping a variety of chemicals into the canal. Many of the materials were chemical residues of the various pesticides and other chemicals manufactured by the company. Hooker's dumping needs were such that in 1946 they purchased the canal and began large-scale dumping into the canal. They eventually filled it up with 20,000 tons of chemical wastes. In the early 1950s, the site was sealed by a cap of clay soil and abandoned.

The urban area began to expand. The canal, which had once been an abandoned ditch far outside the town, was soon surrounded by more and more urban development. The Board of Education needed locations on which to build new schools and in 1953 bought Love Canal in exchange for $1 cash and the grant of a liability release to Hooker Chemical. The construction of a new school on the property attracted housing development to that area. Homes and schools were built near the old dump site, and the school playground was situated directly above the old canal.

As early as the 1950s there were reports of problems among children playing on the site. The complaints included respiratory tract irritation and eye irritation. There was even an occasional report of chemical burns from material seeping up through the ground. By the 1970s, it was clear that there was a significant problem. The situation was aggravated by several years of excessive precipitation; the rainwater collected in depressions in the ground and produced pools of oily water. Dogs roaming around that area developed skin diseases that caused their hair to fall out in clumps. The reports of chemical burns in children playing in the area continued, and occasionally chemical drums would drift to the surface.

The first outside objective evidence of a toxic chemical problem came from an engineering firm that was employed to look into the nature of seepage found in basements in the area. The groundwater was found to contain a variety of toxic chemicals, including some that were suspected carcinogens. Parents in the area became concerned and circulated a petition to close the school. Those circulating the petition soon discovered that other families in the area were also experiencing health problems. These problems included birth defects, asthma, bronchitis, chronic infections, miscarriages, and stillbirths. Many authorities dismissed much of this information as "housewife research," but in the end the weight of the evidence could not be denied. On August 2, 1978, the State of New York ordered 240 families to evacuate the area; however, this action did not resolve the matter. There were still many occupied homes in locations near the evacuation area that were at some risk. In May of 1980, President Carter declared the area a federal disaster area, and 711 families were moved out. After years of legal maneuvering, a $250 million settlement was reached in 1988 between the Love Canal Homeowners Association and the owner of Hooker Chemical, Occidental Petroleum. After 13 years and considerable cleanup expense, the Love Canal Revitalization Agency renamed the area Black Creek Village and began selling houses in the area. By 1998, 232 out of 239 renovated homes available had been sold. These were the houses that remained after the houses in the immediate area of the dump were torn down, and they were made available for sale at 20% less than market value. Debate continues as to whether this area is safe for habitation, but most of the new residents seem satisfied with the situation. On September 30, 2004, the U.S. Environmental Protection Agency (EPA) removed **Love Canal** from the National Priorities List (NPL) of serious hazardous waste pollution sites under the Comprehensive Environmental Response, Compensation, and Liability Act (CERCLA).

Love Canal was the hazardous waste event that catapulted the American government into action. It was not the first hazardous waste problem, nor will it be the last. Although Love Canal was a large abandoned waste site, it was not by any measure the largest, but Love Canal had the publicity and the exposure necessary to make the American public ask questions about chemical wastes and demand answers. Before we can study the disposal of hazardous wastes, though, we need to look at just what a hazardous waste is.

WHAT ARE HAZARDOUS WASTES?

As much as most of us have a feel for the term **hazardous waste**, writing an exact definition is much more difficult. Generally, the term implies a substance that somehow compromises our health or safety. The **Resource Conservation and Recovery Act** (**RCRA**) of 1976 defines hazardous wastes as

> A solid waste, or combination of solid wastes, which because of its quantity, concentration, or physical, chemical, or infectious characteristics may—(a) cause, or significantly contribute to an increase in mortality or an increase in serious irreversible, or incapacitating reversible illness; or (b) pose a substantial present or potential hazard to human health or the environment when improperly treated, stored, transported, or disposed of, or otherwise managed.

Under RCRA, solid wastes are not limited to items we usually think of as solid. Solid wastes are partly defined as

> *any garbage, refuse, sludge* from a waste treatment plant, water supply treatment plant or air pollution control facility *and other discarded material, including solid, liquid, semisolid, or contained gaseous materials* resulting from industrial, commercial, mining and agricultural activities and from community activities ...

Although the complete definition is not given, the statement does make it clear that some liquid and gaseous materials are included; thus, this definition includes liquids and gases contained in drums or cylinders.

Generally, **hazardous materials** exhibit one or more of the following characteristics: *ignitability*, *corrosivity*, *reactivity*, or *toxicity*. Ignitability refers to the ability of a material to cause a fire during routine handling or to accelerate a fire once started. Corrosivity refers to the ability of a material to react with metal. Reactivity refers to any kind of dangerous chemical interactions not covered under ignitability and corrosivity. Toxicity refers to substances that if released into the environment could pose a health threat to humans or other life forms.

WHERE DO HAZARDOUS WASTES COME FROM?

A major source of hazardous wastes is the production and distribution of pesticides. These chemicals by their very design are toxic to certain species. Because it is rare to find a toxic chemical that is absolutely species specific, most of these pesticides are somewhat toxic to other species. In addition, many of the chemicals used in the production of these pest-eliminating chemicals are themselves toxic. Any company manufacturing pesticides will generate toxic byproducts, excess unsold pesticide products, and other related toxic chemicals. (One of the businesses that Hooker Chemical was involved in was pesticide production.)

Many other industries have the capability to produce hazardous wastes. As one might expect, most of these are chemical companies. These companies are involved in very diverse activities. They produce wood products, rubber products, textiles, batteries, leather, plastics, paints, ink, and petroleum products. All of these are products that we want or need and pay good money to obtain. These companies do not produce hazardous wastes just for the heck of it, but rather the wastes are byproducts of the products in which our society shows an interest. Some companies are responsible and some companies are irresponsible with regard to the disposal of these wastes, but the production of the wastes is a result of the demands of our consumer society.

Some of the hazardous chemicals are used to manufacture products that do not themselves qualify as hazardous materials, but many other hazardous substances find their way into the marketplace. They are used for many purposes, from paint thinner to insecticides. Obviously, many of these are purchased for household use. Most people would be surprised to find out how many hazardous chemicals can be found in their cupboards, closets, and utility rooms. These hazardous household chemicals can have significant environmental impact.

In the United States, hazardous household chemicals are specifically exempted from RCRA and other federal hazardous waste statutes; therefore, there is no federal prohibition against improper disposal of these wastes. Improper disposal is common because many citizens are unaware of the hazardous nature of many of their discarded products. The variety and commonness of hazardous household materials are truly amazing, as seen in Table 15.1.

The seriousness of the problem resulting from improperly disposing of hazardous wastes is an issue of some debate. Some individuals argue that the amounts of wastes from households are so small as to have negligible impact on municipal landfills, but others note that it is not just the wastes from an individual citizen but rather the total wastes from all the citizens of an area that should be the concern. All of the hazardous household products, collectively dumped, can amount to a large amount of toxic materials, which when leached out can cause serious problems.

Even if individuals become aware of the dangers of some household chemicals, there are often very few options for disposal. Many communities have not set up hazardous household waste disposal programs, and it is often difficult to properly dispose of some chemicals. There are, however, many things that individuals can

Table 15.1 Hazardous Household Substances

Paint Products	**Auto, Boat, and Equipment Maintenance**
Solvent-based paints	Batteries
Solvents and thinners	Motor oil (new or used)
Paint removers and strippers	Additives
	Gasoline
Cleaning Agents	Flushes
Bleach	Some auto repair materials
Degreasers and spot removers	Lubricating oil
Toilet, drain, and septic tank cleaners	Air-conditioning refrigerants
Kitchen cleaners and disinfectants	
Polishes and waxes	**Hobby and Recreation**
Deck, patio, and chimney cleaners	Photo chemicals
Metal cleaners and polishes	Pool chemicals
	Glues, cements, and adhesives
Pesticides	Inks and dyes
Insecticides	Glazes
Fungicides	Chemistry sets
Rodenticides	Bottled gas
Herbicides	White gas
Mollusicides	Charcoal fluid
Wood preservatives	
Moss retardants	**Miscellaneous**
	Asbestos-containing material
	Some medicines

do. Some require a local hazardous waste program, but many do not. The following suggestions are taken from copies of various flyers found in the book *Alternatives to Landfilling Household Toxics* (Purin et al., 1987):

1. Buy only what one needs.
2. Read and follow instructions exactly.
3. Keep unused products in their original containers and store in a safe place away from children, pets, or sources of ignition.
4. Use the least toxic product available. Many nontoxic products are safe and effective and can be easily disposed of. Look for the "nontoxic" label on the product. Use nonchemical housekeeping, gardening, and hobby techniques to reduce the exposure to and need for disposal of hazardous materials.
5. Recycle used hazardous substances whenever possible. Used motor oil, brake fluid, and transmission fluid can be taken to a gas station that recycles (call first to check if they accept oil) or to a local oil-recycling center if one is located nearby. Let paint particles settle in paint thinner, then decant the thinner for reuse.
6. Donate paints, household cleaners, or other usable and safely packaged products to a local charity or service organization.
7. Use up, rather than throw out, hazardous substances that have not been banned.
8. Never mix different products, as explosive or poisonous chemical reactions may occur.
9. Do not pour toxics down the drain. They can disrupt sewage treatment plants and household septic systems.
10. Do not dump toxics onto the ground. Do not pour wastes or rinse containers in the street. This can pollute the soil and water. Only clean water should go down storm drains.
11. Do not throw toxics in the trash. They can hurt sanitation workers and contaminate landfills and surrounding areas.
12. If one does not know what to do with a hazardous material, call the environmental services department of the local government.

HISTORICAL AND TRADITIONAL APPROACHES TO HAZARDOUS WASTE DISPOSAL

The production of hazardous wastes is largely a consequence of the Industrial Revolution. Till the middle of the 20th century, the attitude toward hazardous wastes was about the same as that of the solid wastes in general: out of sight, out of mind. Waste chemicals were dumped in about any out-of-the-way place that seemed convenient. One of the worst hazardous waste sites in the United States is a 60-acre site near Woburn, Massachusetts. Chemicals have been dumped there since 1853. Later on glue factories and tanneries were built nearby which added their contributions to this chemical dump site. One can find heavy elements such as arsenic, chromium, and lead in this waste site.

Uncontrolled and inappropriately managed hazardous waste sites are a legacy from our past. The EPA has listed approximately 1300 hazardous waste sites in the United States. The actual number is probably higher, and although recent data are not readily available the General Accounting Office (GAO) estimated the number at 28,000 in 1989. Some of these sites are huge, such as the 40-square-mile Tar Creek drainage basin in northeastern Oklahoma containing zinc and lead deposits from mining operations. Most sites, of course, are much smaller. Some of these old and inappropriate disposal sites are remote areas where chemical drums were dumped and abandoned, whereas others are landfills where the hazardous wastes were covered with soil but little else was done.

Another approach to disposing of hazardous wastes is to discharge liquid wastes directly into a pond or lagoon. One of the most notorious examples of this practice is the Stringfellow Acid Pits near Riverside, California. Between 1956 and 1972, over 200 companies and public agencies dumped wastes into the pit. Over this period of time, about 35 million gal (130 million L) were pumped into this shallow 22-acre pit. About one-fifth of the deposits were made by a company which was the nation's largest manufacturer of dichlorodiphenyltrichloroethane (DDT). The cost for cleanup of this site alone may exceed $300 million.

Following the Love Canal disaster, there was significant public pressure to do something to clean up hazardous waste sites. In 1980, the U.S. Congress passed the **Comprehensive Environmental Response, Compensation, and Liability Act (CERCLA)**. This act came to be known as the **Superfund Act**. The Act, among other things, provided for the creation of a fund for the cleanup of dangerous hazardous waste sites. Because the cleanup of the larger sites is a very expensive proposition, such a fund would have to be very large. In the original act, the plan was to raise $1.6 billion over 5 years, but in the 1986 reauthorization it was necessary to seek $9 billion over 5 years, with $1.25 billion coming from general revenue and the remaining raised mostly through special taxes on the chemical industry. The Superfund was not intended to pay for the entire cleanup of these sites, as those who created the sites also have some liability. The government has the right to collect damages from (1) past and present owners or operators of the waste site, (2) transporters of wastes to the site, and (3) those who sent the wastes to the site (generators).

Even with the huge amounts of money raised through the Superfund Act, the EPA did not have enough money to clean up all of the most serious hazardous waste sites at once; therefore, the EPA has developed a Superfund Comprehensive Accomplishment Plan (SCAP) to determine the sites targeted for cleanup in a given fiscal year. The sites to be cleaned up are chosen from a list prepared by the EPA known as the **National Priorities List** (NPL). The funds available for cleaning up hazardous waste sites have continued to decline. The special tax on the chemical industry expired in 1995, so the only funds available for cleanup have come from funds appropriated by the Congress. In 2004, the GAO estimated that the spending power of the Superfund had declined by 34% over the previous decade when adjusted for inflation. With funds severely limited, the eventual cleanup of all identified sites may take a very long time.

Abandoned waste sites strike fear into our hearts, particularly if they are located close to where we live or work, although it is not clear how dangerous many of these sites are to the health of the general public. Dr. Bruce Ames of the University of California, Berkeley, stated in a video series quoted in the book *The World of Chemistry: Essentials* that, "People have been very worried about toxic waste dumps but, in fact, the evidence that they're really causing any harm is really minimal; there's not very much evidence. And the levels of chemicals are very tiny, so we don't really know whether there's no hazard or a little bit of hazard" (Joesten et al., 1991). The inclusion of this quote is not so much to suggest that hazardous waste sites are not a problem as it is to suggest that the real risk from these waste sites and the perceived risk may not be the same thing. In prioritizing the cleanup of waste sites, we should target the sites that are actually the worst rather than those that are *perceived* to be the worst.

Land Spreading

One old way of disposing of liquid hazardous wastes was to spread them out over a large area of land. Unfortunately, it is often very difficult to identify where this has been done, and as a result cleanup is often unlikely. If the site is identified, cleanup would involve the removal of acres and acres of topsoil. It is not entirely clear just how dangerous this practice is, though. If the material can be broken down in the soil by natural processes, it may not pose a high risk; however, if the toxic chemical persists in the soil long enough to move into the groundwater or if it evaporates into the air, then there may be significant possibilities of water or air pollution.

Illegal Disposal

With the passage of the RCRA of 1976, the improper disposal of hazardous material became a federal crime in the United States. Up to that time, various states had statutes regarding toxic waste disposal that were enforced to varying degrees. Today, the regulatory environment is stricter. Most companies try to comply, but it is expensive. At times when competition is intense and profit margins are small, it can be very tempting for some companies to dispose of their waste illegally. They can either carry out the illegal act themselves or contract with an unscrupulous disposer to do it for them. Sometimes, though, the disposal company is the culprit. These companies might charge an unsuspecting chemical company high fees for legal disposal but then dispose of the material illegally and cheaply, pocketing a huge profit.

Many methods are used by illegal disposers to get rid of this unwanted material. It can be dumped in a landfill as municipal wastes if the material looks like municipal wastes. If the toxic chemicals are liquids, the material can be mixed with a large amount of municipal trash and then disposed of in an ordinary sanitary landfill. Another approach is "midnight dumping" into lakes, rivers, storm sewers, or along deserted roads using a tanker truck. One driver commented, "About 60 miles is all it takes to get rid of a load, and the only way I can get caught is if the windshield wipers or the tires of the car behind me start melting."

Many times illegal disposers simply store the drums of chemicals in large deserted warehouses. This process would not necessarily be unsafe if it was done properly with the appropriate controls, but that costs money and that is exactly what illegal operators are trying to avoid. So, they pile up drums of all kinds of chemicals, and some of these chemicals can react with each other. If some of these drums are leaking or develop leaks, then fires or explosions can result from the mixing of chemicals. The potential for disaster is very real. One such disaster took place in Elizabeth, New Jersey, in 1980. For years, the Chemical Control Corporation plant had a reputation as an environmental nightmare waiting to happen. The site contained tens of thousands of barrels of a large variety of chemicals, including explosives and toxic materials. Shortly before this disaster, the New Jersey Department of Environmental Protection had begun to act and had removed about 10,000 barrels, but on April 20 the remaining barrels (estimated at about 24,000) exploded. A huge fireball went hundreds of feet high into the air, hurling 55-gallon drums high up into the night sky. Many of these drums became bombs and exploded in the air. Thirty people were injured in the explosion, and the fire raged for 10 hours before it was finally put out. The fire produced a plume of toxic gases that threatened Elizabeth and Staten Islands. Officials closed schools in these communities, and residents were urged to stay home with their windows closed. Many of the chemicals that did not burn flowed into the Elizabeth River, creating additional water pollution. An interesting note on this episode is that the Chemical Control Corporation was apparently under the control of John Albert, who was reported to have connections with organized crime. Some writers have suggested that organized crime is deeply involved in illegal hazardous waste dumping, but there is little hard evidence.

CURRENT PRACTICES IN HAZARDOUS WASTE MANAGEMENT

In the United States, legislation controlling the disposal of hazardous wastes has tightened considerably. Many chemical companies are concerned that regulations may affect their competitiveness against foreign companies operating where regulations are not as strict. Although the Superfund legislation is certainly important, the centerpiece of federal hazardous waste legislation has been the RCRA of 1976, as well as the 1984 and the 1986 amendments to that act. Two of the key thrusts of the RCRA as amended are the encouragement of waste reduction and prohibitions on the land disposal of hazardous waste. The 1984 amendments to RCRA make it clear that land disposal of hazardous waste is the least desirable method of disposal, and that if hazardous substances are allowed to be disposed of on land some type of pretreatment will be required. What options then are available to the hazardous waste generator? The first and best option is to reduce or eliminate the waste production. Failing that, the generator can consider waste treatment, incineration, permanent storage, exportation, deep well injection, or landfilling.

Waste Reduction or Elimination

The process of disposing of chemical wastes has become so expensive that there is considerable economic advantage for a company to eliminate or reduce the amount of any waste that they can. Such an effort begins with research into the chemical processes used in chemical manufacture. Few chemical processes are absolutely perfect. Most processes produce a small amount of something other than the desired product. If the reaction is carried out on a very large scale, the small amount of something else may be a large amount of material. The business of identifying alternative chemical processes for the manufacture of various products that produce less hazardous waste has become very important in the chemical industry. The field of chemistry that attempts to find environmentally less-dangerous chemical manufacturing procedures is called **green chemistry**.

With reasonable efforts, companies can make improvements to reduce the production of hazardous wastes. The 3M Company of St. Paul, Minnesota, between the years of 1975 and 1985, eliminated 140,000 tons of solid and hazardous wastes, as well as 1 billion gallons of wastewater and 80,000 tons of air pollution. At their Bishop, Texas, plant, the BHC Company, a subsidiary of the chemical giant BASF, was able to reduce a wasteful six-step process for making the analgesic ibuprofen to a much more efficient three-step process that produces a much smaller quantity of chemical waste products to be disposed of.

Another approach to waste elimination is to realize that one company's waste may be another company's starting material. Waste exchange programs are being developed between companies to identify wastes that can be used by other companies. This approach is common in Europe.

Waste Treatment

Two factors make hazardous wastes dangerous: the inherently hazardous nature of the substance and the ability of the substance to move away from its place of disposal. The goal of hazardous waste treatment is to make disposal of these wastes safer.

Physical Treatment

Any process that would trap a hazardous material in a strong, solid, nonreactive environment would render the substance essentially nondangerous and allow it to be handled as a nonhazardous solid waste. Some of the materials that could be used for trapping hazardous substances include ceramics, glass, or cement; for example, one can pour a toxic substance into a vat of molten glass, mix it up, and then allow the glass to solidify. The toxic material is trapped in the glass in a harmless form. The glass can then be disposed of as ordinary solid waste.

Chemical Treatment

Chemical treatment involves finding a chemical reaction that will convert a dangerous substance into an innocuous material that is no longer classified as hazardous. Many times some of the simple classes of reactions that we have learned may be used; for example, an acid can be neutralized with a base, or a base can be neutralized with an acid. Many toxic substances can be converted into nontoxic substances by oxidation or reduction. Combustion will convert most organic materials into carbon dioxide and water. This is the process of *incineration*, which is discussed separately.

Biological Treatment (Bioremediation)

We are all aware that many things biodegrade in nature in the presence of water and natural bacteria; for example, wood rots, paper deteriorates, and dead creatures decompose. Not all hazardous wastes can be attacked by natural organisms, but some can. Biological treatment, or *bioremediation*, is gaining in popularity in the United States because of the increasingly strict regulations of hazardous waste disposal by the EPA. Most of the available alternatives to bioremediation are very expensive, severely restricted, or suspect for one reason or another. An additional advantage of bioremediation can be found in the products obtained from the process. Most organic molecules are broken down into carbon dioxide and water, which are ordinary, completely safe molecules. Bioremediation of hazardous material can be carried out generally in one of four ways: land treatment, offsite bioreactors, onsite bioreactors, or direct onsite treatment. Land treatment, which was mentioned earlier in this chapter, depends on the natural microorganisms present in soil and must be carried out on a restricted site where the contaminants cannot reach the water supply. This approach is severely restricted by the EPA, as it comes under the heading of land disposal, but this does not mean that it is not used at all; rather, land spreading will probably remain a minor player in the waste disposal business.

Bioreactors are highly controlled tanks or lagoons containing bacteria and nutrients. Mixing and aeration ensure that maximum consumption of wastes occurs. The best bioreactors are permanent facilities built to exact specifications. One common disadvantage of all types of permanent hazardous waste facilities is that the wastes must be transported to the site, and the transport of hazardous wastes is a restricted and expensive proposition. Onsite bioreactors, however, eliminate hazardous waste transport, and, although the treatment is not as efficient, it is effective.

Another even more direct approach is to add microorganisms directly to the wastes. This works well in the case of liquid wastes dumped directly into a pit or lagoon. By adding the bacteria directly to the contaminated water, nothing has to be moved. Key to such a process is identifying and having available bacteria that will in fact break down the chemicals present. In some cases, bacteria that are naturally present in the reactor will suffice. In other cases, specific organisms must be added that will work for the materials at hand. These can be other local microbes or organisms shipped in from other locations. In situations where no microorganisms are available for decomposing the hazardous waste of interest, then one solution would be to use genetically engineered microorganisms designed to consume the waste in question. Because it is difficult to predict the effect of genetically engineered organisms on the surrounding ecosystems, the use of bioengineered organisms is tightly controlled by the EPA in the United States.

Incineration

If there is a most effective method of choice for hazardous waste treatment and disposal, it would be incineration. The appeal is the same as for the incineration of municipal solid wastes. The material is principally converted into harmless carbon dioxide and water. The process of hazardous waste incineration is similar to municipal incineration except that hazardous waste incinerators are under much tighter regulation than municipal incinerators. For more details on the process of incineration, refer to Chapter 14.

Hazardous waste incinerators have the same drawbacks as municipal incinerators. First, not everything burns and the ash in this type of incinerator may be very toxic. Of particular note are substances that contain heavy metals. Because these heavy metals are inherently toxic, incineration may change the form of the heavy metal compounds but the toxicity remains; therefore, with incineration, the hazardous waste is reduced in volume but not eliminated. The hazardous waste incinerator ash still has to be disposed of in some manner. Furthermore, the problem of dioxin production in the incinerator continues to be an area of concern. Dioxins may be more of a problem for a hazardous waste incinerator because dioxins and furans are often among the wastes incinerated. Dioxin emissions must be monitored very carefully.

It is very difficult to find a location for hazardous waste incinerators (or hazardous waste anything for that matter). People are fearful because of the nature of the material the incinerator handles. These incineration facilities represent a particularly difficult not-in-my-backyard (NIMBY) problem.

Exportation

A U.S. businessman with several thousand gallons of hazardous waste to dispose of who has found it very expensive to dispose of them legally in the United States might consider removing it from this country altogether. Interestingly, it is relatively easy to do. All one needs to do is find another country that is willing to take it and get permission in writing to ship it there. It is also necessary to notify the EPA of the intent to ship the material and to file an annual report.

The serious problem here is that many developing countries will accept the wastes in exchange for hard currency, but they do not have the facilities to handle such wastes properly. An Italian businessman arranged to store several thousand barrels of waste in Nigeria, but Italy was later forced to take the wastes back when 19 people died after consuming rice contaminated when it was stored near the chemicals.

If a waste generator cannot find a country that will accept the hazardous material, it is often possible to ship the material to a foreign country anyway. The illegal shipment of hazardous wastes is relatively easy, as most customs officials have little training in identifying them. There has been some effort to cut down on the international trade in hazardous wastes. In 1989, representatives of 116 nations met in Basel, Switzerland, and drafted a treaty banning the international shipment of hazardous wastes except when the receiving country grants permission in writing. The treaty, titled "The Basel Convention on the Control of Transboundary Movements of Hazardous Wastes and Their Disposal," has been signed by 170 countries. The treaty went into force on May 5, 1992. The United States, Haiti, and Afghanistan have signed the treaty but so far have not ratified it and, therefore, have not officially accepted the terms of the treaty.

Deep Well Injection

Injection wells (discussed in Chapters 7 and 13) involve the controversial process of injecting wastes into very deep underground aquifers. The process is becoming increasingly popular as other processes become more expensive or difficult to use. Deep well injection is simple, inexpensive, and out of sight. Currently, 51 facilities are injecting hazardous waste into 163 wells. This process is tightly regulated under the Underground Injection Control (UIC) program, which sets the standards for any hazardous waste injections in the United States.

Landfilling

The burying of hazardous wastes requires a **secure landfill**. The term “secure” simply means that the landfill is built and operated in such a way that all of the waste deposited in it will remain secure at that location and will not contaminate surrounding soil or groundwater. At first glance, a comparison of the diagram of a secure landfill in Figure 15.1 with the diagram of a sanitary landfill in Figure 14.3 reveals very few differences. Both require a plastic liner to isolate the landfill and a leachate removal and treatment system, but in a sanitary landfill some biodegradation is expected to occur, whereas in hazardous waste landfills the wastes are simply permanently buried at the site. A sanitary landfill contains compacted wastes, whereas in a hazardous waste landfill one finds steel drums of liquid and other containers of wastes. These drums and containers must be carefully placed so they are not located near something with which their contents might react. In a sanitary landfill, one might find methane probes to remove methane from the biodegradation process, but with little biodegradation occurring, methane probes are generally not needed in a hazardous waste landfill.

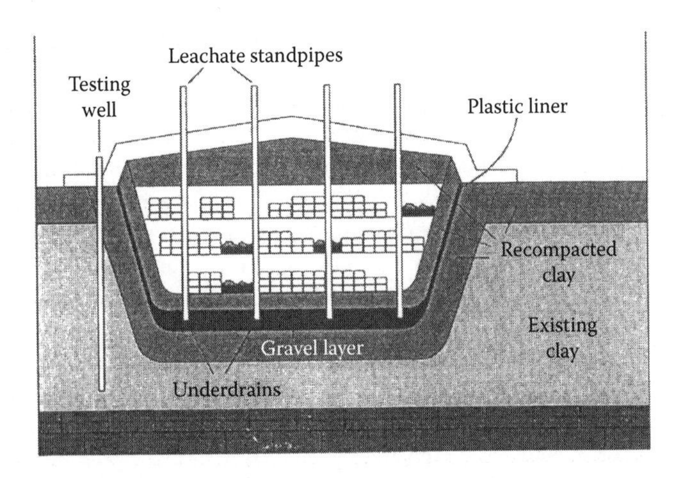

Figure 15.1 Diagram of a secure hazardous waste landfill. (From Cunningham, W.P. and Saigo, B.W., *Environmental Science, A Global Concern*, William C. Brown, Dubuque, IA, 1990. With permission of The McGraw-Hill Companies.)

Because of the nature of the contents of a hazardous waste landfill it is essential that none of the material placed at the site escapes. As a result, the requirements for hazardous waste liners are much tighter. The plastic liner must be situated between two layers of recompacted clay to provide maximum assurance that no leakage will occur. Testing wells are then drilled to allow a check of the water outside the landfill to see that no hazardous material is escaping.

Under the 1984 amendments to the RCRA, the use of landfills for hazardous wastes is strongly discouraged. Because most of the material will not degrade over any moderate period of time, all of the problems still exist, just buried; therefore, eternal vigilance is required.

Permanent Storage

If in fact landfill amounts to a kind of a permanent storage of chemicals underground, then why not admit that fact and keep control of the process? The material in a landfill is still with us but we have lost control of it. We do not know exactly where it is or what state it is in. Furthermore, if we discovered some use for it or a better way to treat it, we could not get at it easily. Permanent retrievable storage is a method of handling hazardous wastes where the wastes are stored in a secure building, salt mine, or cavern in such a way that every item stored there can be located. If necessary, a particular item can be retrieved. Such systems are very expensive, but so are most of the other options. The storage facility requires constant monitoring to detect any leaks or other issues that might develop. A security staff is necessary to keep unauthorized persons away from the wastes. Only time will tell if such storage facilities gain wider acceptance. Permanent storage has been tried in Germany and Sweden, among other countries.

SPECIAL CONSIDERATIONS FOR RADIOACTIVE WASTES

Radioactive wastes were addressed earlier in Chapter 7 in the discussion on nuclear energy. Although the disposal of spent nuclear fuel rods is a major problem, these rods are not the only radioactive wastes with which we could come into contact. Other types of radioactive wastes include waste materials from the manufacture of nuclear weapons and wastes from the use of radioisotopes in medical procedures.

Before World War II, little use was made of radioisotopes outside of scientific research, and there was little need to dispose of large amounts of radioactive material. This situation changed drastically with the production of nuclear weapons during and after World War II. In fact, because of the Cold War that followed, the United States produced nuclear weapons nearly continuously until 1989. In the race to make and develop nuclear weapons, the waste produced in the process was not always handled well. In addition, after World War II the amount and types of nuclear wastes increased dramatically due not only to the production of weapons but also to the advent of nuclear power and other uses of radioisotopes.

Beyond nuclear weapons and nuclear power, other applications of radioisotopes include their use as tracers in chemical and biological processes, nuclear imaging in medicine, radioisotopic dating, as sources of radiation for cancer therapy, and as sources of γ radiation for radiography, to kill pathogens, and to sterilize insect pests. All of these uses of radioactive material produce wastes in various amounts and with various intensities of radiation.

Classification of Radioactive Wastes

Generally, nuclear wastes are divided into high-level waste (**HLW**) and low-level waste (**LLW**). HLW is the simplest category to define. Basically, these are spent fuel rods, materials related to those rods, and any material that is similar enough to require long-term isolated storage. LLW includes all other nuclear wastes, an extremely diverse collection of materials. For the most part, these wastes have fairly low levels of radioactivity and require very little special handling; however, there is considerable variability, and LLW is usually divided into four subcategories: A, B, C, and greater-than-C, with A being the most benign, and C and greater-than-C being the most dangerous.

Current Waste Management Practices for Radioactive Wastes

Disposal of radioactive waste differs significantly from other hazardous waste dispoal. Radioisotopes are not affected by incineration, bioremediation, or chemical treatment as most other hazardous wastes are. About the only practical option for radioactive waste is burial. Radioisotopes will continue to degrade at a rate that is determined by the nature of each isotope. The only option for such waste is to isolate the material unless it is so benign as to be of no danger.

Probably no other area of environmental management better exemplifies the notion of "if we ignore the problem then maybe it will go away" than the disposal of radioactive waste. The disposal of both HLW and LLW suffers from severe political problems. We seem to want to benefit from the use and production of radioisotopes but not deal with the aftermath. We have already discussed the controversy surrounding the HLW site at Yucca Mountain, Nevada, in Chapter 7, and HLW (primarily nuclear power plant spent fuel rods) continues to pile up at various facilities around the country.

The situation for LLW is not much better. There are only three LLW disposal sites in the United States: Barnwell, South Carolina; Richland, Washington; and Clive, Utah. Currently, the sites at Richland and Barnwell only take wastes from certain states, and the site at Clive only accepts class A LLW. Generators of class B and C wastes in many states find themselves in the situation of having to store their wastes onsite.

Currently, one class of LLW falls through the cracks. These wastes are the greater-than-class-C type, which, although classified as LLW, are considered unsuitable for shallow land burial, which is typically done at most LLW disposal sites. Currently, the greater-than-class-C wastes are caught up in political issues similar to HLW.

Nuclear waste produced by the government and their contractors during the production and development of various weapons systems is completely exempted from all of these regulations. In the early years, urgent concern for the speedy manufacture of such weapons meant that little attention was paid to the proper disposal and storage of the resulting radioactive wastes. Many of the defense plants that dealt with radioactive material ultimately became major environmental concerns because of the possible leakage of dangerous radioactive material into the area around the plants. The Hanford site near Richland, Washington, is one of the most notorious, but several other sites have been or are being cleaned up. As these sites are cleaned up and the radioactive material removed, much of the waste is taken to Carlsbad, New Mexico. Wastes generated by the defense sector are disposed of in Carlsbad, independent of civilian rules and regulations. The disposal rooms are cut into a 2000-ft (600-m) thick layer of salt and are 2150 ft (650 m) below the surface.

DISCUSSION QUESTIONS

1. Indicate which of the following might be considered hazardous waste:
   a. Drums from a pesticide manufacturing plant
   b. Drums of broken glass
   c. Paint thinner
   d. Laundry detergent
   e. Coffee grounds
   f. Antifreeze
2. List at least four hazardous products that you might find in your home, and for each indicate how you might dispose of it.
3. You have a large drum of mercury-containing fungicide to dispose of. Suggest how you would dispose of the material and why.
4. Why is waste reduction or elimination considered the best approach to hazardous waste management?
5. Some methods of hazardous waste disposal and treatment really do get rid of the hazardous waste, whereas others more or less effectively simply move the waste to a "safe" place. Divide disposal and treatment methods into these two types and then comment on the effectiveness of each.
6. How does a secure hazardous waste landfill differ from a modern sanitary landfill? Why the differences?
7. In what ways are radioactive wastes fundamentally different from other hazardous wastes? With what type of nonradioactive hazardous wastes do radioactive wastes have the most in common?
8. What is the difference between high- and low-level waste when discussing radioactive wastes?
9. Your job is to establish a low-level radioactive disposal site near a rural community. How would you go about convincing the local residents that the disposal site would be safe and a positive thing for their community?

BIBLIOGRAPHY

Anon., GAO warns about Hanford cleanup, *Chemical & Engineering News*, 84(37), 25, 2006.

Anon., DOE resubmits nuclear waste bill, *Chemical & Engineering News*, 85(11), 28, 2007.

Anon., DOE to assess new radioactivity class, *Chemical & Engineering News*, 85(31), 44, 2007.

Anon., *Timeline—The Nuclear Waste Policy Dilemma*, YuccaMountain.org, Nuclear Waste Office, Eureka County, Nevada, 2013, http://www.yuccamountain.org/time.htm.

Applegate, J.S. and Laitos, J., *Environmental Law: RCRA, CERCLA, and the Management of Hazardous Waste*, Foundation Press, New York, 2006.

ATSDR, *Tar Creek Superfund Site—Ottawa County, OK*, Agency for Toxic Substances & Disease Registry, U.S. Department of Health and Human Services, Washington, DC, 2010, http://www.atsdr.cdc.gov/sites/tarcreek/.

Baker, R. and Simpson, F.S., State of California Reaches Historic Settlement for Stringfellow Hazardous Waste Site [news release], Department of Toxic Substances Control, California Environmental Protection Agency, Sacramento, January 8, 1999.

Cann, M.C. and Connelly, M.E., *Real World Cases in Green Chemistry*, American Chemical Society, Washington, DC, 2000.

Cooney, C.M., Illuminating the future for bioremediation, *Environmental Science & Technology*, 34(7), 162A, 2000.

Cunningham, W.P. and Saigo, B.W., *Environmental Science: A Global Concern*, Wm. C. Brown, Dubuque, IA, 1990, p. 488.

DePalma, A., Love canal declared clean, ending toxic horror, *The New York Times*, March 18, 2004, p. A1.

Eilperin, J., Lack of money slows cleanup of hundreds of Superfund sites; federal toxic waste program's budget is stagnant, *The Washington Post*, November 25, 2004, p. A1.

Engelhaupt, E., Happy birthday, Love Canal, *Environmental Science & Technology*, 42(22), 8179–8186, 2008.

FEA, *Background Paper on Permanent Storage in Salt Mines*, Federal Environment Agency, Berlin, Germany, 2004 (http://www.basel.int/techmatters/popguid_may2004_ge_an1.pdf).

GAO, *Hazardous Waste Sites: State Cleanup Status and Its Implications for Federal Policy*, GAO/RCED-89-164, U.S. General Accounting Office, Washington, DC, 1989.

Hogue, C., Toxics inventory, *Chemical & Engineering News*, 83(20), 9, 2005.

Hogue, C., EPA to remove dioxin-contaminated Hooker Chemical landfill from list of nation's most hazardous waste sites, *Chemical & Engineering News*, 90(35), 26, 2012.

Joesten, M.D., Netterville, J.T., and Wood, J.L., *The World of Chemistry: Essentials*, Saunders, New York, 1991, p. 261.

Johnson, J., Reprocessing key to nuclear plan, *Chemical & Engineering News*, 85(25), 48–54, 2007.

Kemsley, J., Bacterium snacks on vinyl chloride, *Chemical & Engineering News*, 81(27), 8, 2003.

Melfort, W.S., *Nuclear Waste Disposal: Current Issues and Proposals*, Nova Science, New York, 2003.

OEM, *Environmental Management: History*, Office of Environmental Management, U.S. Department of Energy, Washington, DC, 2013, http://www.em.doe.gov/pages/History.aspx.

Pichtel, J., *Waste Management Practices: Municipal, Hazardous, and Industrial*, Taylor & Francis, Boca Raton, FL, 2005.

Purin, G., Orttung, J., Van Stockum, S., and Page, J., *Alternatives to Landfilling Household Toxics*, Golden Empire Health Planning Center, Sacramento, CA, 1987, pp. A176–A178.
Richardson, R., *Home Hints and Tips: The New Guide to Natural, Safe and Healthy Living*, DK Publishing, New York, 2003.
Ritter, S.K., Green chemistry, *Chemical & Engineering News*, 79(29), 27–34, 2001.
Tuchman, G., Despite toxic history, residents return to Love Canal, *CNN.com*, August 7, 1998, http://www.cnn.com/US/9808/07/love.canal/.
USDOE, *Why WIPP?*, WIPP Information Center, U.S. Department of Energy, Washington, DC, 2007 (http://www.wipp.energy.gov/fctshts/Why_WIPP.pdf).
USEPA, *Love Canal, New York*, EPA Region 2 Report, NYD000606947, U.S. Environmental Protection Agency, Washington, DC, 2007.
USEPA, *Superfund: Basic Information*, U.S. Environmental Protection Agency, Washington, DC, 2012, http://www.epa.gov/superfund/about.htm.

Glossary

Abyssal zone: The deepest part of the open ocean that is in total darkness all of the time.

Acid deposition: All of the processes by which acidic materials are removed from the atmosphere.

Acid fog: Fog containing water droplets with a pH of less than 5.0.

Acid rain: Rain with a pH of less than 5.0.

Acid: Any substance which, when dissolved in water, produces hydronium ions (H_3O^+).

Activated sludge process: Process that uses microorganisms in sewage sludge to break down organic materials in sewage effluent while air is mixed with the sewage and sludge materials.

Active solar heating: The process by which solar energy is captured in specially designed collectors that concentrate the solar energy and store it as heat for space heating and possibly water heating.

Acute toxicity: Toxicity that manifests its effects in a relatively short period of time (i.e., hours or days).

Aerobes: Organisms that use oxygen in producing energy from biochemical energy molecules. The products of this process are principally carbon dioxide and water.

Agriculture: The cultivation of crops and raising of livestock on a relatively large scale using larger tools for assistance.

Air toxics: Also known as hazardous air pollutants, or HAPs, a group of air pollutants as defined by the Federal Clean Air Act that are usually an issue only in a small number of areas of the country and may have to be controlled in those specific locations.

Albedo: The percentage of solar energy reflected from the surface of a planet or a moon or some part of the surface of a planet or a moon.

Alchemy: The pursuit of methods to convert common metals into gold, which led to many of the experimental techniques of today.

Alkali: A synonym for base, which is any substance which, when dissolved in water, produces hydroxide ions (OH^-).

Alkane: Compound containing only carbon, hydrogen, and single bonds.

Alkene: Compound containing carbon, hydrogen, and at least one double bond.

Alkyne: Compound containing carbon, hydrogen, and at least one triple bond.

Allotrope: A form of element that differs from other forms of the same element in its molecular arrangement.

Alloy: A homogeneous mixture of a metal with another element or elements.

Alpha (α) particle: A particle emitted from many radioactive nuclei that corresponds to the nucleus of a helium atom. These particles have very little penetrating power but can do great damage over a short distance.

Alveoli: The microscopic air sacks in the lungs where atmospheric gases are exchanged with the bloodstream.

Anaerobes: Organisms that do not use oxygen in producing energy from biochemical energy molecules. The products of this process include principally methane and related compounds.

Annex I countries: A list of the countries in the United Nations Framework Convention on Climate Change that are considered developed and major emitters of greenhouse gases. These are the nations specifically regulated under the Kyoto Protocol.

Antagonism: A decrease in the total toxicity of two toxins in the presence of each other.

Anthracite coal: The oldest form of coal with the highest energy content and the highest carbon content. This type produces the least amount of pollutants when burned. Often referred to as *hard coal.*

Anticholinesterase poisons: Poisons that act by inhibiting the enzyme acetylcholinesterase, the effect of which blocks the destruction of the neurotransmitter acetylcholine.

Applied science: The application of science to some practical purpose.

Aquifer: Soil and water positioned above an impermeable layer of rock or clay.

Aristotle (384–322 BCE): An influential Greek philosopher who had a powerful influence on the development of scientific theory. He was a student of Plato.

Aromatic compound: Any compound that is similar or related to benzene. These compounds have unusual stability.

Arrhenius, Svante (1859–1927): An influential Swedish chemist and Nobel Laureate from the late 19th and early 20th centuries who developed one of the theories of acids and bases as well as other important concepts and ideas.

Asbestos: A complex set of silicon- and oxygen-containing minerals that has extremely good high-temperature insulating properties.

Asbestosis: A benign chronic degenerative lung condition related to the inhalation of fine asbestos particles in the air.

Atmosphere: The air above the surface of the Earth.

Atom: A small nearly indestructible particle of which all matter is composed.

Atomic Energy Commission: The post-World War II commission formed to direct the production of more and better nuclear weapons; also oversaw the development of nuclear power. Later became the Nuclear Regulatory Commission.

Atomic mass unit: A unit of mass defined in such a way that a common hydrogen atom has a mass of one. The exact definition is 1/12th the mass of one atom of the most abundant isotope of carbon.

Atomic theory: A theory that asserts that all matter is made up of nearly indestructible particles known as atoms.

Bacon, Sir Francis (1561–1626): A proponent of experimental planning and record-keeping that evolved into the scientific method; one of the founders of natural philosophy.

Base: Any substance which, when dissolved in water, produces hydroxide ions (OH^-).

Bathyal zone: The section of the open ocean that is in a condition of constant twilight during the daytime because of its depth.

Becquerel, A.H. (1852–1908): The French scientist who discovered radioactivity. He won the Nobel Prize with the Curies in 1903.

Bessemer, Henry (1813–1898): Developer of the Bessemer process for the large-scale conversion of cast iron to steel.

Bessemer process: The first major process developed for the conversion of cast iron to steel on a large scale.

Beta (β) particle: A particle emitted from many radioactive nuclei that corresponds to the electron. These particles have moderate penetrating power.

Bioaccumulation: The accumulation of toxins in the fat of an organism that results from the organism coming into contact with water containing the toxin.

Biochemical oxygen demand (BOD): The oxygen demand in a sample of water due to the breakdown of various organic materials by various aerobic microorganisms.

Biodiesel: A form of diesel fuel generated from biologic sources, most commonly fats and oils.

Biomagnification: The increase in the concentration of fat-soluble toxins as one moves to higher levels of the food web.

Biomass energy: Energy produced from currently existing biological materials.

Biosphere: All of the biological matter on Earth, both living and dead.

Bisphenol A (BPA): A common ingredient in various plastics; it is a well-known environmental estrogen.

Bituminous coal: Second oldest form of coal with the second highest energy content and the second highest carbon content. It is the highest fuel content coal available in large quantities. This type of coal is often referred to as *soft coal*.

Black lung disease: A chronic obstructive lung disease caused by inhaling coal dust.

BOD: *See* Biochemical oxygen demand.

Boyle, Robert (1627–1691): Natural philosopher who contributed significantly to the early foundations of science, including physics and chemistry.

Breeder reactor: A nuclear reactor that uses fast neutrons and the more common uranium-238 to produce fissionable fuel in the form of plutonium.

British anti-Lewisite (BAL): A compound produced by the British during World War I to combat the poisonous effects of the gas Lewisite; it is currently used as an antidote for mercury poisoning.

Cap and trade: A program designed to limit the emission of a pollutant by issuing allowances for the amount of a pollutant individual emitters can release. These allowances can be used immediately, saved for future use, or sold to other emitters. The total number of allowances issued each year is capped at a specific number.

Carbon cycle: A series of chemical reactions that allow the continual reuse of carbon. The major pathway involves carbon dioxide in the atmosphere and photosynthesis in plants.

Carcinogens: Compounds with a demonstrated ability to cause cancer.

Carnot, Sadi (1796–1832): French engineer and physicist who developed a theoretical theory of the heat engine which became the foundation of the field of thermodynamics.

Carrier: Facilitates the movement of substances in the environment.

Carson, Rachel (1907–1964): A scientist and writer whose book *Silent Spring* had a significant influence on promoting the environmental movement, particularly with regard to the unnecessary use of synthetic chemicals.

Cast iron: Iron produced by heating iron ore with charcoal or coke. Usually contains 2 to 4% carbon.

Catalyst: A substance that speeds up a reaction without being consumed in the reaction.

CERCLA: *See* Comprehensive Environmental Response, Compensation, and Liability Act.

Chemistry: A study of matter and energy and their relationships with regard to changes in matter.

Chernobyl: Site of the 1986 nuclear disaster in the Ukraine.

Chloracne: A nonfatal disfiguring disease, usually affecting the face, caused by certain chlorinated chemicals, such as dioxins and PCSs.

Chlorofluorocarbons (CFCs): Compounds containing only carbon, fluorine, and chlorine. The smaller of these compounds have found extensive use in, for example, the refrigeration industry. These compounds are often associated with the depletion of ozone in the stratosphere.

Chronic toxicity: Toxicity that manifests it effects over a relatively long period of time (i.e., years or decades).

Chrysotile: A form of asbestos that can be drawn out in fibers.

Cilia: Fine hairlike fibers in the nasal passages and the upper respiratory tract that filter out most particulates inhaled from the air.

City: Settlement containing a relatively large number of people and governed by some sort of administrative structure.

Classical air pollutants: The constituents of air that are clearly unhealthy and that have been traditionally considered as contributing to air pollution.

Clausius, Rudolph (1822–1888): German mathematician and physicist who is credited with formulating the second law of thermodynamics.

Clean Air Act: The federal law originally enacted in 1970 and amended in 1977 and 1990 that is designed to regulate the release of pollutants into the air.

Clean zone: The area in a stream or other body of water where the biochemical oxygen demand and dissolved oxygen levels are at or near safe and acceptable levels.

Climate: The pattern of weather conditions over a region or regions for a long period of time.

Coal: A solid fossil fuel formed from ancient plant life and made up principally of carbon.

Coastal zone: The relatively shallow parts of the ocean that are defined as beginning at the high tide mark and extending to the edge of the continental shelf. This zone contains about 90% of all ocean life.

Coke: Relatively pure carbon-containing material produced by heating coal to drive out many of the impurities.

Combustion: The rapid combination of a material with oxygen accompanied by the evolution of considerable amounts of heat and gases.

Compact fluorescent bulbs: Light bulbs made from fluorescent tubes with a compact shape that can be used in place of typical incandescent bulbs.

Compartment: An arbitrarily defined part of the environment used to discuss environmental processes.

Compost: Organic waste materials that have been collected and allowed to decompose for use as natural fertilizer.

Compound: Matter made up of molecules containing the atoms of more than one element.

Comprehensive Environmental Response, Compensation, and Liability Act (CERCLA): Passed in 1980, the Act regulates and establishes liability for the cleanup of hazardous waste sites. The Act also provided for the development of a fund contributed to by the government and the chemical industry to cover the cost of cleanups. The fund came to be known as the Superfund and the act as the Superfund Act.

Conference of Parties (COP): Governing body of the Framework Convention on Climate Change.

Confined aquifers: Aquifers with little or no exposure to the surface where water can infiltrate.

Control rods: The rods in a nuclear reactor used to slow down or stop the reactor. The farther these rods are inserted into the core of the reactor the lower the energy output.

Coolant: Fluid used to remove heat from the core of a nuclear reactor.

COP: *See* Conference of Parties.

Core: The part of a nuclear reactor containing the fuel rods and related materials necessary for operation of the reactor.

Coriolis effect: Changes in apparent wind direction due to rotation of the Earth.

Coriolis, Gaspard (1792–1843): French scientist who studied force in a rotating reference frame and developed theories regarding what is now known as the Coriolis effect.

Covalent bonding: Bonding between atoms that involves the sharing of electrons.

Criteria air pollutants: Air pollutants regulated in all areas of the country under the Clean Air Act.

Criteria document: A document produced for an air pollutant by the Environmental Protection Agency before that pollutant can be controlled under the Clean Air Act. The document outlines the scientific criteria under which the pollutant is to be regulated.

Critical mass: A quantity of material with the mass, density, and shape required to support nuclear fission.

Crust: The solid layer that makes up the outermost layer of the Earth.

Cultural eutrophication: The addition of large amounts of nutrients to a body of water by human activity.

Curie, Marie (1867–1934): A chemist and physicist who did early and significant research on the existence and nature of radioactivity. She won the Nobel Prize along with her husband and A.H. Becquerel in 1903 and won a second Nobel Prize in 1911.

Curie, Pierre (1859–1906): Influential physicist who, along with his wife, Marie Curie, and A.H. Becquerel, made early and significant discoveries about radioactivity. He won the Nobel Prize along with his wife and A.H. Becquerel in 1903.

Cyclone: An area of rotating winds around a region of low pressure. Also, the name used for hurricane-like storms found in the Indian Ocean.

Dalton, John (1766–1844): English scientist credited with the initial formulation of modern atomic theory.

DDT: *p*-Dichlorodiphenyltrichloroethane is a well-known organic insecticide that was discovered in 1939 and used extensively after that time. Its use has now been banned in most parts of the world.

Decay heat: Heat produced solely from the radioactive decay of fission products found in used nuclear fuel rods.

Decomposition zone: The area just downstream from a major waste discharge where the biochemical oxygen demand is very high and the dissolved oxygen levels are declining rapidly.

Democritos (ca. 460–370 BCE): A Greek philosopher who was the first to formulate the theory that matter is made up of indivisible atoms. Democritos credited the idea to his teacher, Leucippos.

Dermal absorption: Direct absorption of chemicals into the bloodstream through the skin.

Desalination: The process of removing salt from water to produce freshwater.

Dioxins: In environmental toxicology, generally refers to a variety of polychlorinated benzodixoins that are formed as common byproducts in reactions involving chlorinated hydrocarbons.

Dissolved oxygen (DO): The concentration of oxygen dissolved in any sample of water.

Distillation: The process of boiling a liquid and condensing the vapors to yield a liquid without the nonvolatile impurities. Can be used to remove salt from water.

DO: *See* Dissolved oxygen.

Doldrums: Areas near the equator where the air tends to be calm, or if there are winds they are very light and variable.

Domestic waste: Waste generated by households and commercial businesses.

Ecotoxicology: The study of the effects of toxins on ecosystems.

ED_{50}: The dose of a toxin at which 50% of the test animals exhibit a particular effect.

Electric power: Power developed from the electrical potential of an electric current, usually an alternating current.

Electron: One of the particles of which an atom is composed. This particle has a charge of –1 and is very light, with a mass of only 1/1837 atomic mass unit.

Element: Matter composed of only one type of atom.

Element cycles: A series of chemical reactions that allow the continual reuse of an element.

Emission allowances: The amount of a pollutant allowed to be emitted by an individual producer on an annual basis under a cap and trade program.

Endocrine hormones: Hormones within an organism that are produced by endocrine glands.

Endocrine system toxins: Toxins that have a biological effect by mimicking the activity of endocrine hormones.

Energy: The ability to do work or raise the temperature of a sample of matter.

Energy conservation: The intentional use of strategies to produce the same work or heating using less energy than conventional approaches.

Entropy: A measure of disorder or randomness of a system.

Environmental estrogens: Chemicals found in the environment that can mimic the activity of natural estrogens.

Environmental health toxicology: The study of the effects of environmental toxins on human health.

Environmental Protection Agency (EPA): A federal agency charged with regulation and enforcement of the United States' environmental protection laws.

Environmental tobacco smoke (ETS): Smoke dispersed from individuals smoking tobacco.

Environmental toxicology: The study of the effects of toxins on the environment.

Enzymes: Protein-based biological catalysts.

EPA: *See* Environmental Protection Agency.

Ethanol: The specific alcohol commonly used as a beverage and as a fuel source.

ETS: *See* Environmental tobacco smoke.

Euphotic zone: The top level of water in the open ocean that receives plenty of light and contains most of the life in the open ocean.

Eutrophic: Lakes and streams with a large amount of nutrients and an abundance of life.

Eutrophication: The increase in the nutrient level in a lake or stream over time.

Fat-soluble toxins: A class of toxins having low water solubility and which are often found dissolved in fatty tissue.

FCCC: *See* Framework Convention on Climate Change.

Fibrosis: The sealing off of lung tissue by scar tissue as a result of irritation by foreign particles.

First law of thermodynamics: Energy can be neither created nor destroyed; it can only be converted from one form to another. Also known as the *law of the conservation of energy.*

Fixed-film biological process: Process that uses microorganisms attached to a surface to break down organic materials in sewage effluent being passed over the surface.

Fly ash: Lightweight ash that can be carried up the chimney with the flue gases; typically a part of smoke.

Fossil energy: Energy derived from using ancient biological materials as the fuel source.

Fossil fuels: Fuels derived from ancient biological materials.

Fracking: The release of trapped gases or liquids (e.g., natural gas or oil) from rock structures by the injection of fluids under high pressure. Also, known as *hydraulic fracturing.*

Framework Convention on Climate Change (FCCC): Organization created to develop international methods to mitigate and control climate change.

Fuel rods: Rods in a nuclear reactor that contain the nuclear fuel, usually uranium-235; they are normally 15 to 20 feet long and about the diameter of a pencil.

Fukushima Daiichi: Nuclear power plant that was the site of a disastrous nuclear accident following the March 2011 earthquake and tsunami on the northeast coast of Japan's major island.

Fungicides: Substances used to control the growth of various types of fungi.

Gamma (γ) radiation: High-energy electromagnetic radiation emitted from radioactive nuclei that has very high penetrating power and significant effects over large distances.

Gastrointestinal absorption: Absorption of substances into the bloodstream via the gastrointestinal tract.

General metabolic toxins: Toxins with multiple modes of action, one of which is interference with some essential biochemical process.

Generators: Devices that convert mechanical energy into electricity.

Geosphere: Nonbiological spheres of the environment, including the lithosphere, hydrosphere, and atmosphere.

Geothermal energy: Energy produced by utilizing the steam found near the surface of the Earth.

Global warming: Gradual, general warming of the Earth's atmosphere believed to be caused by the increased emissions of certain gases.

Gray water: Water that has been used for various types of washing and is then reused to water gardens or flush toilets, etc.

Green chemistry: An approach to industrial chemistry that designs processes that will produce less hazardous waste either by producing nonhazardous byproducts or by minimizing the amount of hazardous waste that is produced.

Greenhouse effect: The insulating effect of certain gases in the air which can lead to higher temperatures in the lower atmosphere. These gases are believed to act in a manner analogous to the glass in a greenhouse.

Grit removal: A treatment process where the rate of flow of incoming sewage is slowed sufficiently to allow moderately large particles to settle out.

Groundwater: Water found underground mixed with the soil. This water is generally in sufficient quantities to saturate the soil in which it is found.

Haber, Fritz (1868–1934): German chemist who developed a process for converting atmospheric nitrogen into ammonia, which can then be used in fertilizers or explosives.

Haber process: A process for converting atmospheric nitrogen into ammonia.

Hadley cell: A global air circulation pattern where air tends to rise in regions near the equator and flow north in the upper atmosphere, followed by sinking near the poles and flowing back toward the equator near the surface.

Hadley, George (1685–1768): English lawyer and meteorologist credited with development of the first theory of global air circulation.

Half-life: The time it takes for one-half of something to disappear in a natural process; often applied to the radioactive decay of nuclei.

Hall, Charles Martin (1863–1914): An American chemist who was one of two people to independently develop a large-scale practical method to obtain aluminum metal from its ore by electrolysis. He founded Aluminum Company of America (ALCOA).

Hall–Héroult process: A large-scale practical method to obtain aluminum metal from its ore by electrolysis. Developed independently by Charles Martin Hall and Paul Héroult in 1886.

HAPs: *See* Hazardous air pollutants.

Hazardous air pollutants (HAPs): A group of air pollutants as defined by the federal Clean Air Act that usually are an issue only in a small number of areas of the country where they may have to be controlled.

Hazardous chemicals: Substances that are flammable, explosive, corrosive, allergenic, or toxic.

Hazardous materials: Materials that are ignitable, corrosive, reactive, and/or toxic.

Hazardous waste: As defined by the RCRA, a solid waste, or combination of solid wastes, which because of its quantity, concentration, or physical, chemical, or infectious characteristics may (1) cause or significantly contribute to an increase in mortality or an increase in serious irreversible or incapacitating reversible illness; or (2) pose a substantial present or potential hazard to human health or the environment when improperly treated, stored, transported, or disposed of, or otherwise managed.

Heat energy: The energy that flows from one body to another when there is a temperature difference between the two.

Heavy oil: An unusually viscous oil found in considerable quantities in Venezuela. Contains significant amounts of sulfur and heavy metals.

Herbicides: Substances that kill plants.

Héroult, Paul (1863–1914): A French entrepreneur who was one of two people to independently develop a large-scale practical method to obtain aluminum metal from its ore by electrolysis.

Heterogeneous mixture: A mixture in which the substances in the mixture are not evenly distributed throughout the mixture.

HLW: High-level wastes, nuclear wastes consisting of spent fuel rods, materials related to those rods, and any material that is similar enough to require long-term isolated storage.

Homogeneous mixture: A mixture in which the substances in the mixture are evenly distributed throughout the mixture.

Horizontal drilling: Drilling in which the direction is turned from the normal vertical drilling to drilling in a horizontal direction.

Horse latitudes: Areas near the 30°N and 30°S latitudes where the air tends to be calm, or if there are winds they are very light and variable.

Horticulture: The cultivation of mixed crops and raising of livestock on a relatively small scale using only a few tools for assistance.

Hubbert, Marion King (1903–1989): American geologist who developed a model describing peak U.S. oil production and predicted the occurrence of peak production with some accuracy.

Hubbert's Peak: Peak U.S. oil production predicted by Marion Hubbert in 1956.

Hunter–gatherers: Groups of individuals who survive by moving from place to place and either hunting game or gathering plant materials for food.

Hurricane: The name given to tropical storms driven by rising warm moist air and sustained wind speeds in excess of 74 miles per hour (119 kilometers per hour). This name is used in the Atlantic Ocean and the eastern Pacific Ocean.

Hybrid electric vehicle: A vehicle powered by both a gasoline engine and an electric motor. The electric motor is powered by batteries that are charged by the motion of the car and gasoline engine in such a manner as to maximize fuel efficiency.

Hydraulic fracturing: The release of trapped gases or liquids (e.g., natural gas or oil) from rock structures by the injection of fluids under high pressure.

Hydrocarbon: A compound containing only carbon and hydrogen.

Hydroelectric power: Electric power generated using water, usually held behind a dam, to turn electric generating turbines.

Hydrogen cycle: A series of chemical reactions that allow the continual reuse of hydrogen. The major pathway involves water and photosynthesis in plants.

Hydrogen fuel cell: An electrochemical cell that generates electric current by combining hydrogen and oxygen to produce water.

Hydrosphere: All of the Earth's water, including liquid, frozen, and vapor forms. This sphere includes among other things oceans, lakes, rivers, and glaciers.

Hypothesis: An educated guess regarding the solution to a problem that can be verified or disproved by experiment.

Incandescent light bulb: Bulbs that produce light by using an electric current to heat a resistant filament to a high temperature.

Incineration: Reducing the volume of waste by combusting the waste materials.

Industrial Revolution: An era of rapid changes in how goods are produced and how society is organized to produce those goods; it occurred between 1760 and 1950.

Industrial smog: Smog that is generally associated with coal burning; it usually develops during an air inversion that traps air near the ground.

Injection well: A well into which something is injected. The material can be waste that is injected into a deep aquifer or a substance used to force oil or natural gas to the surface.

Insecticides: Substances that kill insects.

Intergovernmental Panel on Climate Change (IPCC): An international group of atmospheric scientists and other experts dedicated to analysis of the current state of the science related to climate change. The organization was created by the United Nations Environment Programme and the World Meteorological Organization.

Intertropical convergence zone: A region that wraps around the Earth near the equator where air flows in and is warmed and rises.

Inversion layer: A layer of air near the ground where temperature increases with increasing altitude rather than decreasing with altitude, which is the normal expectation.

Ion: A chemical species containing a net charge.

Ionic bonding: Bonding between oppositely charged ions that is created by the transfer of electrons from an atom, atoms, or group of atoms to another atom, atoms, or group of atoms.

Ionic substances: Substances made up of a collection of positive and negative ions.

IPCC: *See* Intergovernmental Panel on Climate Change.

Irrigation: The diversion of freshwater from its source, such as a river, lake, or groundwater, to provide the water required to grow crops.

Isomers: Compounds with the same number of atoms of elements as another compound or compounds but which differ in structural arrangement and properties.

Isotope: A form of an element that differs from the other forms of the same element in the atomic weight of the atoms.

Jet streams: Relatively high-velocity winds that flow at the highest altitudes of the troposphere. These winds flow completely around the globe in a west to east direction in the temperate zones, generally between 30 and 60° latitude.

Kinetic energy: Energy that matter has because of its motion.

Kyoto protocol: A treaty developed in 1997 at COP3 in Kyoto, Japan, designed to limit the emission of gases that lead to global warming.

Lapse rate: The rate of decrease in the temperature of air with increasing altitude.

Lavoisier, Antoine (1743–1794): French scientist who is considered by many to be the father of modern chemistry. He investigated the process of combustion and developed satisfactory theories to explain the process. He is often credited with developing the law of conservation of matter.

Law of conservation of matter: Matter can be neither created nor destroyed; it can only be converted from one form to another.

Law of unintended consequences: The observation that actions taken to solve one environmental problem often have other effects which in many cases have a negative impact.

LD_{50}: The dose of a toxin at which 50% of the test animals die.

Leachate: Water that has passed through solid waste materials. This water often contains objectionable and perhaps dangerous materials picked up from the waste materials.

Leucippos (first half of 5th century BCE): Greek philosopher who is credited as being the first to suggest the theory that matter is made up of indivisible atoms. The theory was written down and elaborated upon by his student Democritos.

Light-emitting diode bulbs: Bulb that uses solid-state technology to produce light and produces much less heat compared to the standard incandescent bulb.

Lignite: The youngest and lowest energy content form of coal.

Liquefied natural gas: Natural gas that has been converted to liquid form by cooling it to temperatures below its boiling point. This conversion is usually done for storage and shipment purposes.

Lithosphere: The crust of the Earth, which includes soil, rocks, and all other materials on the surface.

LLW (low-level waste): Nuclear wastes that are *not* those wastes consisting of spent fuel rods, materials related to those rods, and any material that is similar enough to require long-term isolated storage.

LNG: *See* Liquefied natural gas.

Love Canal: A site near Niagara Falls, New York, that was a major dump site for hazardous chemicals. A school and a housing development were built in the area. In the late 1970s, it became the center of a major hazardous waste incident.

Lumens: A unit of measure of total light production from a luminous source.

Mainstream smoke: Smoke that is drawn through a cigarette, cigar, etc. and through the lungs of the person smoking before being released back into the air.

Mantle: The viscous layer in the structure of the Earth between the core and the crust. It is made up of rocky material that is partly solid and partly a viscous liquid.

Mendeleev, Dmitri Ivanovitch (1834–1907): A Russian chemist who was one of the first to state the periodic law. From this law he developed a periodic table that evolved into the modern periodic table.

Mesosphere: A layer of the atmosphere with a negative temperature gradient located roughly between 30 and 50 miles (50 and 80 km) in altitude.

Mesothelioma: A rare cancer of the lining of the lungs usually associated with inhalation of asbestos particles.

Mesotrophic: Lakes and streams with intermediate amounts of nutrients.

Metabolic toxins, general: Toxins with multiple modes of action, one of which is interference with essential biochemical processes.

Methane: The simplest hydrocarbon, with one carbon and four hydrogen; it is a major component of natural gas and is a powerful greenhouse gas, 21 times more potent than carbon dioxide.

Milankovitch cycles: Climate cycles that are predicted based on the tilt of rotation of the Earth on its axis and the path of the Earth around the sun. These cycles help us to understand the occurrence of ice ages.

Milankovitch, Milutin (1879–1958): Serbian mathematician who developed the predictive climate cycles that bear his name.

Mixture: Two or more substances mixed together.

Moderator: The material of the core of most nuclear reactors that slows down neutrons so nuclear fission can occur. In a vast majority of U.S. reactors the moderator is water.

Molarity: A measure of the concentration of a substance in solution. It is the number of moles of a substance per liter of solution.

Mole: The number of atomic size particles found in 12 grams of the most abundant isotope of carbon, which corresponds to 602,200,000,000,000,000,000,000 particles (6.022×10^{23} particles).

Molecule: An atom or any combination of atoms that is capable of existing for some amount of time in the presence of other molecules.

Montreal Protocol: The first international treaty on the protection of stratospheric ozone; it was drawn up in Montreal in 1987.

Mutagens: Substances that damage or alter the genetic information in cells.

NAAQS: *See* National Ambient Air Quality Standards.

National Ambient Air Quality Standards (NAAQS): The limits for criteria pollutants that serve as the basis for their control under the federal Clean Air Act.

National Priorities List: A list of priorities for the cleanup of hazardous waste sites prepared by the EPA under the Superfund Act (CERCLA).

Natural gas: A gaseous fossil fuel pumped from the ground that is predominately made up of methane.

Natural philosophy: A branch of philosophy devoted to the study of the natural universe; it is considered by many to be the forerunner of modern science.

Natural science: The pursuit of knowledge about the natural world by observation and testing. Most commonly includes biology, chemistry, and physics, although sometimes psychology is included.

Negative temperature gradient: A temperature gradient in which the temperature decreases with altitude.

Neurons: Nerve cells that are typically very long and thin, essentially microscopic fibers.

Neurotoxins: Toxins whose principal site of action is the nervous system.

Neurotransmitters: Chemicals that transmit nerve impulses across synapses between neurons.

Neutralization reaction: A reaction between an acid and a base that leads to a neutral result (i.e., a final product that is neither acidic nor basic).

Neutron: One of the particles of which an atom is composed. This particle has no charge and has a mass of roughly one atomic mass unit.

NIMBY (not in my backyard): Refers to the tendency to oppose the placing of sites that might be objectionable close to an individual's home.

Nitrogen cycle: A series of chemical reactions that allow the continual reuse of nitrogen. The major pathway involves nitrogen in the atmosphere and various processes of nitrogen fixation.

Nitrogen fixation: The process of converting gaseous nitrogen in the atmosphere into nitrogen compounds.

Nitrous oxide: A small biologically active molecule with properties similar to carbon monoxide but usually present in the atmosphere in much lower levels.

No-observed-effect level (NOEL): The highest dose of a toxin at which no effect is observed. Usually associated with chronic toxins.

Nonpoint source: Water pollution that cannot be pinpointed to a specific location but instead is introduced by general runoff over a large area.

Nonpolar substances: Substances made up of molecules with homogeneous distribution of electrical charge.

NRC: *See* Nuclear Regulatory Commission.

Nuclear energy: Energy produced by either splitting large nuclei or fusing small nuclei.

Nuclear fission: The process of splitting nuclei into smaller nuclei of similar sizes. The process is usually accompanied by the release of large amounts of energy.

Nuclear fusion: The process of merging nuclei together to form larger nuclei. The process is usually accompanied by the release of large amounts of energy.

Nuclear reactor: A reactor used to produce electric power in a controlled fashion from nuclear fission reactions.

Nuclear Regulatory Commission (NRC): A post-World War II commission formed to oversee the development and production of more and better nuclear weapons and nuclear power. Originally known as the Atomic Energy Commission.

Nucleus: In chemistry and physics, refers to the very dense center of an atom which contains most of the atom's mass.

Octet rule: The theory that atoms react in such a way as to gain eight valence electrons.

Offsets: The amount by which a facility may exceed an annual permitted limit of a hazardous air pollutant if another facility remains below its limit by the same amount. These offsets may not be bought or sold.

Oil: Common term for the liquid fossil fuel obtained from underground reserves by drilling wells and pumping the liquid out from the ground.

Oil shale: A type of shale rock that is impregnated with oil. It can be mined and the oil extracted from the rock.

Oligotrophic: Refers to lakes and streams with few nutrients and a relatively low abundance of life.

Open sea: The deep ocean outside the continental shelves.

Open-hearth process: A large-scale process for the conversion of cast iron to steel developed by William Siemens. It was the major process for the production of steel from the mid-19th century to the mid-20th century.

Organic chemistry: Branch of chemistry dealing with the compounds of carbon.

Organochlorine pesticides: Chlorinated aromatic hydrocarbons used to control various types of organisms.

Outgassing: The gradual release into the air over time of volatile materials from various types of manufactured materials such as plywood, chipboard, particleboard, etc.

Overburden: The ground that must be removed to reach coal or another resource in strip mining.

Oxidation: The chemical combination of a substance with oxygen, the chemical removal of hydrogen from a substance, or the loss of electrons from a chemical species.

Oxidation number: The number of electrons an atom generally loses when forming an ionic bond; if the atom gains electrons, this is shown as a negative number. These numbers are the same as the charge on the ion that is formed.

Oxygen cycle: A series of chemical reactions that allow the continual reuse of oxygen. The major pathway involves oxygen in the atmosphere and photosynthesis in plants.

Oxygen sag: The decline in dissolved oxygen levels in a decomposition zone due to a major waste discharge.

Ozone layer: Region in the stratosphere that contains relatively high levels of ozone.

Passive collection of solar energy: The process by which solar energy is captured directly within a structure without any energy storage or circulating devices.

Passive solar heating: The process by which solar energy is captured directly within a structure without any energy storage or circulating devices.

PCBs: *See* Polychlorinated biphenyls.

***p*-Dichlorodiphenyltrichloroethane (DDT):** A well-known organic insecticide that was discovered in 1939 and used extensively after that time. Its use has now been banned in most parts of the world.

Periodic law: A law based on the idea that when elements are arranged in the order of their atomic numbers many properties will recur on a periodic basis.

Perkin, William Henry (1838–1907): English chemist who developed the first chemical process for the production of purple dye.

Permafrost: Layer of soil, sediment, and/or rock that is permanently frozen.

Permit: Document stating the amount of a hazardous air pollutant that a facility may legally release on an annual basis.

Persistent: Refers to compounds that break down at an extremely slow rate in the environment.

Pesticides: Substances that kill or otherwise control unwanted organisms.

Petroleum: Liquid fossil fuel obtained from underground reserves by drilling wells and pumping the liquid out from the ground.

Phosphorus cycle: A series of chemical reactions that allow the continual reuse of phosphorus. The major pathways have no atmospheric component.

Photochemical smog: Smog that is generally associated with automobile exhaust gases, warm weather, and sunshine. Usually develops during an air inversion that traps air near the ground.

Photosynthesis: Process carried out in green plants, which use sunlight to convert water and carbon dioxide from the air into carbohydrates and gaseous oxygen.

Point source: Water pollution where the source can be pinpointed to a specific location.

Polar substances: Substances made up of molecules with uneven distribution of electrical charge.

Polychlorinated biphenyls (PCBs): A group of related synthetic chemicals developed for use in electrical transformers, plastics, and paints and which have been found to produce adverse environmental impacts.

Polymer: Very large molecules made up of repeating units related to smaller molecules from which these molecules are produced.

Positive temperature gradient: A temperature gradient in which temperature increases with altitude.

Potential energy: Energy that is stored and can be released to do work or produce heat.

Precautionary principle: Any potentially harmful substance should not be used until it is proven safe.

Prevention of Significant Deterioration (PSD): Clean Air Act requirement that the Environmental Protection Agency must regulate air emissions in an area in such a way that already clean air is not allowed to deteriorate, even if the level of a given pollutant is well below its NAAQS.

Primary air pollutants: Pollutants that are released directly into the air.

Primary energy source: An energy source whose energy cannot be clearly traced to another source.

Primary NAAQS: Air pollution limits for criteria pollutants based on human health effects.

Primary sedimentation: The process in sewage treatment that occurs before any biological or chemical processes are begun; the flow rate is slowed sufficiently such that almost all of the undissolved materials will either float to the surface or sink to the bottom for removal.

Primary treatment: The first step in sewage treatment that depends on physical separation of solids from liquids.

Proton: One of the particles of which an atom is composed. This particle has a charge of +1 and has a mass of roughly one atomic mass unit.

PSD: *See* Prevention of Significant Deterioration.

Public trust: Policy handed down by the California Supreme Court stating that government has some obligation to maintain the natural state of public lands.

Pure substance: Matter containing only one type of molecule.

Quantitative structure–activity relationship (QSAR): A method of estimating the toxicity of a molecule by assigning toxicity factors to various parts of the molecule.

Radioactivity: The property of certain unstable nuclei to eject particles or emit electromagnetic radiation.

Rain shadow: An area on the leeward side of mountains where little rain falls because normal air flow over the mountain causes the rain to be deposited on the windward side of the mountain.

RCRA: *See* Resource Conservation and Recovery Act.

Receptors: Sites on cells that interact with extracellular substances to produce an effect in the cell, such as transmitting a nerve impulse, changing metabolism, changing the ability of substances to enter the cell, etc.

Recharge zone: The part of an aquifer that is exposed to the surface such that water can infiltrate into the aquifer.

Recovery zone: The area downstream from a major waste discharge just beyond where most of the organic material will have been decomposed and where the biochemical oxygen demand is declining and dissolved oxygen levels are increasing.

Recycling: The processing of discarded materials into new products that can serve the same or different purposes.

Reduction: Chemical removal of oxygen from a substance, chemical combination of a substance with hydrogen, or a gain of electrons by a chemical species.

Reserve: The amount of any resource that has yet to be removed from the ground.

Residence time: The average amount of time that a chemical species remains in a given environmental compartment.

Resource Conservation and Recovery Act (RCRA): Federal act that regulates the disposal of hazardous wastes in the United States.

Respiration: Process carried out in living organisms that converts biomaterials and gaseous oxygen into water, carbon dioxide, and energy in the form of heat and/or work.

Respiratory inhalation: Refers to toxins entering the bloodstream through inhalation into the lungs.

Respiratory toxins: Toxins that normally produce their effects after being inhaled.

Reuse: The processing of discarded products back into their original form for use without major change in form or appearance.

Reverse osmosis: The process of forcing saltwater through a semipermeable membrane such that salt can be removed from the water.

Rutherford, Ernest (1871–1937): A scientist who was instrumental in early understanding of the structure of the atom. He was born in New Zealand but did most of his work in England and Canada.

Salt: An ionic compound where the negative ion is something other than the hydroxide ion (OH^-).

Saltwater intrusion: Intrusion of saltwater into a freshwater aquifer such that freshwater wells are contaminated with saltwater.

Sanitary landfill: Disposal site for solid waste where the refuse is covered over each night to keep rats, birds, insects, and other pests out of the material.

Science: The pursuit of knowledge by observation and testing.

Scientific method: Allows conclusions to be drawn about the universe by carrying out experiments and making observations.

Screening: The use of a bar screen to remove very large objects from a sewage stream.

Second law of thermodynamics: The entropy (disorder) of the universe moves toward a maximum. Alternatively, for any process using energy to do work some part of that energy will be converted to, or remain as, heat energy, which will be less useful for doing work than the original energy.

Secondary air pollutants: Pollutants that appear in the air due to chemical reactions of primary pollutants with other compounds in the air.

Secondary energy source: An energy source that gets its energy from another clearly identifiable energy source.

Secondary NAAQS: Air pollution limits based on environmental and property effects such as damage to structures, plants, and animals or reductions in visibility.

Secondary sedimentation: The process in sewage treatment that follows biological treatment; it is designed to remove dissolved organic materials by slowing the flow rate sufficiently so that almost all of the undissolved materials will either float to the surface or sink to the bottom for removal.

Secondary treatment: The second step in sewage treatment; it uses biological methods to break down organic materials dissolved in the effluent.

Secure landfill: Landfill designed for the disposal of hazardous waste. These landfills are designed in such a way that all materials in the landfill will remain secure at that location and will not contaminate surrounding soil or groundwater.

Semipermeable: Refers to membranes that allow some substances to pass through but not others.

Septic tank: An underground tank used by individual households to collect raw sewage and allow the solids to settle. Liquid is removed near the top of the tank.

Septic zone: The area downstream from a major waste discharge where the biochemical oxygen demand is still very high and the dissolved oxygen levels are very low and have yet to recover.

Sidestream smoke: Smoke that is not drawn through a cigarette, cigar, etc. but instead is released directly into the air at the point of combustion.

Siemens, William (1823–1883): Developer of the open-hearth process for the large-scale production of steel.

Smith, R. Angus (1817–1884): Author of the book *Air and Rain*, published in 1872, in which he coined the term *acid rain*.

Smog: Term coined by Dr. H.A. Des Voeux to apply to the air pollution of London around the turn of the 20th century. It is a contraction of the words *smoke* and *fog*. This term is now used to describe nearly all air pollution.

Smoke: A mixture of finely divided particles emitted from the combustion of carbon-based materials. The mixture usually includes soot and fly ash.

Soil moisture: Moisture found in soil that is not at saturation.

Solar energy: Energy derived by capturing energy produced by the sun.

Soot: Finely divided graphite particles usually emitted from the combustion of carbon-based materials.

Steam engine: A mechanical device that uses steam to convert heat energy to mechanical energy.

Steel: An alloy of iron and 0.5 to 1.5% carbon.

Stratosphere: A layer of the atmosphere with a positive temperature gradient that begins somewhere between 5.5 and 10 miles (9 and 16 km) and ends at about 30 miles (50 km) in altitude. It contains the ozone layer.

Strip mining: A method of mining that involves removing the ground above the desired resource and then digging it out from above.

Subbituminous coal: Coal that is intermediate in age between lignite and bituminous coal; it has an energy content less than bituminous and greater than lignite.

Subsidence: Occurs when land sinks due to the removal of underground liquids such as water or oil.

Sulfur cycle: A series of chemical reactions that allow the continual reuse of sulfur.

Superfund Act: *See* Comprehensive Environmental Response, Compensation, and Liability Act.

Sustainable: Refers to processes that can be carried out over and over again for an indefinite period of time.

Synapse: The narrow gap between nerve cells over which a nerve impulse must be transmitted.

Synergism: An increase in the total toxicity of two toxins when they are in the presence of each other.

Tar sand: Sands that are mixed with large amounts of oily material that can be extracted and converted to crude oil.

Technology: Has been defined as the sum total of processes by which humans modify the materials of nature to better satisfy their needs and wants.*

Temperature gradient: Change in air temperature as altitude changes.

Teratogens: Substances that induce birth defects by causing direct damage to the developing embryo.

Tertiary treatment: Sewage treatment other than the standard physical separation and biological treatment. Usually involves chemical methods to destroy pathogens or remove specific substances.

Theory: A hypothesis that has been tested and verified enough times that the scientific community feels comfortable with its veracity.

Thermodynamics: The study of heat transfers and related forms of energy.

Thermoplastics: Plastics that can be melted and reshaped for new purposes.

Thermosetting plastics: Plastics that cannot be reshaped by heating.

Thermosphere: A layer of the atmosphere with a positive temperature gradient; it begins at roughly 50 miles (80 km) and extends until the atmosphere ceases to exist for all practical purposes. This is the last layer of the atmosphere. This layer contains the ionosphere.

Three Mile Island: Site of the 1979 nuclear accident in Pennsylvania that nearly led to a meltdown. Some radioactive steam was released. It was the worst nuclear accident on U.S. soil.

Threshold dose: A dose below which a toxin shows no observable effect.

Tidal energy: Energy produced by using the movement of the tides to turn a water wheel or turbine.

Tile field: Pipes into which effluent from a septic tank is delivered. These pipes are perforated and laid out underground. The effluent percolates through the soil for purification.

Tornado: Extremely intense windstorm that takes the form of a rotating column of air extending from a thunderstorm.

Toxicology: The study of toxins.

Toxin: Chemical that is dangerous to organisms at fairly low concentrations or in small amounts.

Trade winds: Prevailing easterly winds that occur between the equator and 30°N or 30°S latitude.

Troposphere: The lowest level of the atmosphere; it has a negative temperature gradient and extends up to between 5.5 and 10 miles (9 and 16 km) in altitude. This is where active weather phenomena occur.

Tunnel mining: A method of mining that involves digging tunnels underground to reach the desired material to remove it.

* Hill, J.W., *Chemistry for Changing Times*, 6th ed., Macmillan, New York, 1992, p. A37.

Typhoon: The name given to tropical storms driven by rising warm moist air and sustained wind speeds in excess of 74 miles per hour (119 kilometers per hour). This name is used in the western Pacific Ocean.

Ultraviolet (UV) radiation: Electromagnetic radiation closely related to visible light but containing more power per packet of electromagnetic energy.

Unconfined aquifer: Aquifer with a large area exposed to the surface such that water can infiltrate it.

Valence: The combining power of an atom in a chemical reaction; most commonly used in covalent bonding.

Valence electrons: The outer shell electrons, which are commonly involved in chemical reactions.

VOCs: *See* Volatile organic compounds.

Volatile organic compounds (VOCs): Organic materials that are likely to produce vapors when exposed to the air.

Volta, Alessandro (1745–1827): Italian physicist who developed the first useful electrochemical cell, the forerunner of the modern battery.

Water cycle: The cycle that describes the ways in which water moves into and out of the atmosphere and how it moves on the surface.

Water pollution: Any chemical or biological change in water that adversely affects the aquatic life normally found in the water or that makes the water unfit for a desired use.

Water power: The use of water usually held behind a dam to produce power by turning a waterwheel or a turbine.

Water-soluble toxins: A class of toxins with fairly high water solubility.

Water table: Dividing line between the zone of aeration and the zone of saturation.

Watt, James (1736–1819): Credited with making considerable improvements to the steam engine in 1769 that revolutionized industrial energy production in England at that time.

Weather: The physical state of the troposphere at any given time and place.

Wetlands: Land that is generally covered by a shallow layer of saltwater or freshwater for most or all of the year.

Wilkinson, John (1728–1808): English entrepreneur who was instrumental in advocating the use of iron.

Wind farm: A site with a collection of windmills used to produce electricity.

Wind power: The use of wind to turn windmills to produce either mechanical energy or electricity.

Work: The result of using force to move matter over some distance.

Wrought iron: Relatively pure iron that is produced from cast iron by pounding out the impurities from the iron at high temperature.

Yucca Mountain: Proposed site for a federal nuclear waste repository until a decision was made in 2012 to abandon the site and begin the search anew.

Zone of aeration: A layer of soil that contains moisture and air and is, therefore, not saturated with water.

Zone of saturation: A layer of soil that contains moisture and no air and is, therefore, saturated with water.

Index

A

D

E

J

K

L

M

N

O

Q

R

S

T

U

V

W

Y

Z